CONTEXT-AWARE COMPUTING AND SELF-MANAGING SYSTEMS

Chapman & Hall/CRC
Studies in Informatics Series

SERIES EDITOR

G. Q. Zhang

Case Western Reserve University
Department of EECS
Cleveland, Ohio, U.S.A

PUBLISHED TITLES

Stochastic Relations: Foundations for Markov Transition Systems
Ernst-Erich Doberkat

Conceptual Structures in Practice
Pascal Hitzler and Henrik Schärfe

Context-Aware Computing and Self-Managing Systems
Waltenegus Dargie

Chapman & Hall/CRC
Studies in Informatics Series

CONTEXT-AWARE COMPUTING AND SELF-MANAGING SYSTEMS

Edited by

Waltenegus Dargie

CRC Press
Taylor & Francis Group
Boca Raton London New York

CRC Press is an imprint of the
Taylor & Francis Group, an **informa** business

A CHAPMAN & HALL BOOK

CRC Press
Taylor & Francis Group
6000 Broken Sound Parkway NW, Suite 300
Boca Raton, FL 33487-2742

First issued in paperback 2019

ISBN-13: 978-1-4200-7771-1 (hbk)
ISBN-13: 978-0-367-38584-2 (pbk)

Library of Congress Cataloging-in-Publication Data

Context-aware computing and self-managing systems / [edited by] Waltenegus Dargie. -- 1st ed.
    p. cm. -- (Context-aware computing and self-managing systems)
    Includes bibliographical references and index.
    ISBN 978-1-4200-7771-1 (alk. paper)
    1. Autonomic computing. I. Dargie, Waltenegus.

QA76.9.A97C65 2009
004--dc22   2008038056

Visit the Taylor & Francis Web site at
http://www.taylorandfrancis.com

and the CRC Press Web site at
http://www.crcpress.com

To Pheben, with love. Welcome to the world.

Preface

This book brings two research issues together: context-aware computing and the self-managing aspect of autonomous computing. Context-aware computing is an extensively researched area, while self-managing systems are emerging. The goal of this book is to investigate the various roles context-aware computing can play to develop self-managing systems, where a self-management system can be a device, a middleware, an application, or a network.

The first chapter of the book identifies aspects that are common to both context-aware computing and autonomous computing. It offers a basic definition of context-awareness and provides several examples — more focus is given to the acquisition, presentation and management of context information. It presents as well basic aspects of self-managing systems and offers a few examples of self-managing systems.

The remaining part of the book is divided into context-awareness and self-management. The context-awareness subpart demonstrates how a context can be employed to make systems smart; how a context can be captured and represented; and how dynamic binding of context sources can be possible. The self-management subpart of the book demonstrates the need for "implicit-knowledge" to develop fault-tolerant and self-protective systems. It also presents a higher-level vision of future large-scale networks.

Several researchers have participated in editing this book. I would like to acknowledge the contributions of Prof. Noriaki Kuwahara (Kyoto Institute of Technology), Prof. Ren Ohmura (Keio University), Prof. Markus Endler (Catholic University of Rio de Janeiro), Prof. Antonio Alfredo F. Loureiro (The Federal University of Minas Gerais), Prof. Mieso Denko (University of Guelph), and Dr. Daniel Schuster (Technical University of Dresden) for reviewing some of the chapters. Of course, there were also a plethora of reviewers whose names I have not mentioned here, but who have reviewed each chapter of the book and provided critical feedbacks.

I would like to acknowledge the contribution of my former post graduate student, Rami Mochaourab, who worked tirelessly with LaTeX to provide the book the shape it now has. He was always available, always willing to try new ideas, and always on time. Without his support, the book would never be finished on time.

Dr. Waltenegus Dargie
Technical University of Dresden
Germany

Contributors

Abe, Akinori
ATR Knowledge Science
Laboratories
Kyoto, Japan

Beltran, Victoria
Wireless Networks Group
Department of Telematics
Technical University of Catalonia
Barcelona, Spain

Breitman, Karin
Departamento de Informática
Pontifícia Universidade Católica
(PUC-RJ)
Rio de Janeiro, Brazil

Briot, Jean-Pierre
Departamento de Informática
Pontifícia Universidade Católica
(PUC-RJ)
Rio de Janeiro, Brazil

Bouvry, Pascal
University of Luxemburg
Luxemburg

Casademont, Jordi
Wireless Networks Group
Department of Telematics
Technical University of Catalonia
Barcelona, Spain

Catalan, Marisa
Wireless Networks Group
Department of Telematics
Technical University of Catalonia
Barcelona, Spain

Charif, Yasmine
Laboratoire d'Informatique de Paris
Université Paris
Paris, France

Dargie, Waltenegus
Chair of Computer Networks
Faculty of Computer Science
Technical University of Dresden
Dresden, Germany

Davoli, Franco
Department of Communications,
Computer, and Systems Science
University of Genoa
Genoa Italy

Ding, Jianguo
University of Luxembourg
Luxembourg

El Fallah Seghrouchni, Amal
Laboratoire d'Informatique de Paris
Université Paris
Paris, France

Endler, Markus
Departamento de Informática
Pontifícia Universidade Católica
(PUC-RJ)
Rio de Janeiro, Brazil

Guan, Haibing
School of Information Security
Engineering
Shanghai Jiao Tong University
Shanghai, P. R. China

Hadjiantonis, Antonis M
Centre for Communication Systems
Research
Department of Electronic
Engineering,
University of Surrey
Guildford, UK

Hartel, Pieter
University of Twente
Enschede, The Netherlands

Klauser, Bruno
Cisco Europe
Glattzentrum, Switzerland

Kogure, Kiyoshi
ATR Knowledge Science
Laboratories
Kyoto, Japan

Krämer, Bernd J.
Fern Universität in Hagen
Hagen, Germany

Kuwahara, Noriaki
Kyoto Institute of Technology
Kyoto, Japan

Liang, Alei
Software School
Shanghai Jiao Tong University
Shanghai, P. R. China

Liu, Lei
Sun Microsystems, Inc.
Menlo Park, CA, USA

Luis Ferrer, Jose
Wireless Networks Group
Department of Telematics
Technical University of Catalonia
Barcelona, Spain

Mazuel, Laurent
Laboratoire d'Informatique de Paris
Université Paris
Paris, France

Naya, Futoshi
ATR Knowledge Science
Laboratories
Kyoto, Japan

Ohboshi, Naoki
Kinki University
Osaka, Japan

Ozaku, Hiromi Itoh
ATR Knowledge Science
Laboratories
Kyoto, Japan

Paradells, Josep
Wireless Networks Group
Department of Telematics
Technical University of Catalonia
Barcelona, Spain

Pavlou, George
Centre for Communication Systems
Research
Department of Electronic
Engineering
University of Surrey
Guildford, UK

Sabouret, Nicolas
Laboratoire d'Informatique de Paris,
Université Paris
Paris, France

Sanchez-Loro, Xavier
Wireless Networks Group
Department of Telematics
Technical University of Catalonia
Barcelona, Spain

Scholten, Hans
University of Twente
Enschede, The Netherlands

Sundramoorthy, Vasughi
Lancaster University
Lancaster, UK

Viterbo, José
Departamento de Informática
Pontifícia Universidade Católica
(PUC-RJ)
Rio de Janeiro, Brazil

Wolter, Ralf
Cisco Systems
Duesseldorf, Germany

Contents

List of Tables

List of Figures

Chapter 1

Context and Self-Management

Waltenegus Dargie

Technical University of Dresden, 01062, Dresden, Germany

Abstract

This chapter provides an introduction to Context-Aware Computing and Self-Managing Systems. It begins by explaining why self-management is desirable in complex systems and by describing self-management aspects (self-configuration, self-optimization, self-healing and self-protection). For all these features, a self-managing system's needs to have a perpetual awareness of what is taking place both within itself and without. It is this duly awareness of one's state and surrounding that leads to self-adaptation. As a result, the chapter tries to demonstrate the scope and usefulness of context-aware computing in developing self-managing systems.

1.1 Introduction

Computing systems are becoming very complex, highly heterogeneous and distributed. At the same time, the users of these systems are usually mobile and demand greater flexibility and efficiency in terms of response time, resource utilization, robustness, etc., to achieve critical business goals. The implication is that operating and maintaining computing systems is becoming an increasingly expensive business. In fact, Fox and Patterson claim that annual outlays for maintenance, repair and operations far exceed total hardware and software costs, for both individuals and corporations [1].

This high cost of ownership of computing systems has resulted in a number

1

of industry initiatives to reduce the burden of operations and management by making computing systems - at least gradually - self-managing. A few examples are IBM's Autonomic Computing, HP's Adaptive Infrastructure and Microsoft's Dynamic System Initiatives [2].

Self-management derives its basic principles from the autonomous nervous system, which governs our heart rate and body temperature, *thus freeing our conscious brain from the burden of dealing with these and many other low-level, yet vital, functions* [3]. This essential principle, if transferred well, enables computing systems, whether acting individually or collectively, to receive higher-level objectives from their operators (users) but manage to maintain and adjust their operation in the face of changing components, workloads, demands and external conditions as well as imminent hardware and software failures.

According to Kephart and Chess [3] and Tesauro et al. [4], a self-managing system contains an autonomic manager software and a (hardware or software) managed element. The managed element is what is being made self-managing and provides a sensing and actuating interface. Through the sensing interface, an array of sensors measure vital internal as well as external (environmental) phenomena which may potentially influence the system's short and long term performance. The actuating interface provides a way for the autonomic manager to modify the behavior of the managed element. The autonomic manager itself contains components for monitoring and analyzing sensor data and for planning and executing management policies. Common to all of these components is knowledge of the computing environment and service-level agreement as well as other related facts.

The monitoring component inside the autonomic manager is responsible for reducing the amount of raw sensor data by applying filtering and correlation operations on the data. The analysis component gets refined data from the monitoring component in order to identify emerging or foreseeable problems or potential causes of adaptation. The planning component accommodates workflows that specify a partial order of actions which should be carried out in accordance with the results of the analysis component. And finally, the execution component controls the execution of such workflows and provides coordination if there are multiple concurrent workflows.

1.2   Aspects of Self-Management

A system is said to be self-managing if it exhibits one or more of the following characteristics: self-configuration, self-optimization, self-healing, and self-protecting.

Self-configuration refers to the capability of a system to dynamically adjust one or more parameters to accommodate expected or unexpected change within itself or in the operating environment. The change may be due to departure, arrival or failure of a component; a change in the business policy of the user; or environmental, social or political constraints. A self-configuration capability of a system enables it to keep on functioning in the presence of continually changing and unforeseen obstacles.

Self-optimization refers to the ability of a system to tune its parameters so that it can function most efficiently. Efficiency can be measured in terms of cost, quality of service, throughput, etc. A self-optimizing system improves its performance by finding, verifying and applying the latest software updates.

Self-healing refers to the ability of a system to detect, localize, diagnose, and repair problems resulting from bugs or failures in software and hardware.

Finally, self-protection refers to the ability of a system to defend itself as a whole against large-scale, correlated problems arising from malicious attacks or cascading failures that remain uncorrected by self-healing measures. This also includes the anticipation of potential dangers and the carrying out of predictive measures to avoid or mitigate premeditative attacks.

1.3 Examples of Self-Managing Systems

So far, the motivation for self-managing systems was discussed conceptually. In the following two subsections, we will present examples of self-managing systems. In the first subsection, we will present a self-optimizing system which autonomously regulates the transmission power and data rate of distributed and independent wireless local area network access points in densely deployed metropolitan environments. In the second subsection, we will present a self-recovering satellite-receiver which localizes problems and dynamically reinitializes only those components which are the direct causes of the problem instead of considering component-specific problems as global phenomena.

1.3.1 Self-Managing Chaotic Networks

Akella et al. [5] propose self-managing algorithms to manage what they call *spontaneous* or *chaotic* wireless networks. As opposed to carefully planned and deployed wireless networks, chaotic networks are typically deployed spontaneously by individuals or independent organizations that set up one or a small number of APs. This type of unplanned, uncoordinated and unmanaged deployment results in highly variable densities of wireless nodes and APs and causes considerable interference and inefficient utilization of valuable resources such as spectrum and energy. Akella et al. report that in some metropolitan cities in the US as much as 8000 APs are deployed randomly in close proximity, each AP having more than 80 interfering APs. Moreover, users of these networks use default (factory-set) configurations for key parameters such as the transmission channel and transmission power.

The researchers propose two algorithms to autonomously manage transmission power levels and data rates. In combination with a careful channel allocation technique, the power control system attempted to minimize the interference between neighboring APs by reducing transmission power on individual APs. The power management algorithm reduces transmission power as long as the link between an AP and its client could maintain the maximum possible speed. An experiment result shows that the power control system improves throughput from 0.15 Mbps to 3.5 Mbps.

1.3.2   Recovery-Oriented Computing

Fox and Patterson report that operator error was a leading cause of problems of Internet systems [1]. They remarked that traditional efforts to boost the dependability of software and hardware have for the most part overlooked the possibility of human mistakes. Motivated by these observations, they propose four recommendations for developing self-recovering computing systems: Accordingly, developers should assume operator errors as inevitable problems and should therefore design systems that recover quickly. Second, operators should be provided with tools to localize the sources of faults in multicomponent systems. Third, the systems should provide support of an *undo* function so that operators can correct their mistakes. Forth, the systems should accommodate the injection of test errors to evaluate and predict system behavior.

To demonstrate the usefulness of this guideline, the researchers implemented a number of self-recovering systems. As an example, they built a satellite receiver in a *traditional* fashion, by employing inexpensive ground receivers assembled from commonly available PCs, low-cost ham radios and home developed software to capture incoming satellite data. Not surprisingly, whenever the system experienced failures, the operators had to restart it either preemptively (because the system was behaving strangely), or reactively (because it had crashed or seized up). The researchers reported that without a human operator to reactivate the equipment manually, the satellite signal could be lost, and with it, all the data for that orbit. Later, the researchers acquired domain knowledge about the most frequent causes of failure and modified each receiving-station software module so that only a subpart of the system's components should be reinitialized in the event of imminent failure. They made them succeed to automate the recovery process for a range of recurring problems. Furthermore, they improve the average restoration time from 10 minutes to 2 minutes.

1.4   Context-Aware Computing

An essential aspect of a self-managing system is the employment of sensors to unobtrusively measure and report relevant system properties. As far as dealing with sensor data is concerned, there are four main challenges [6]:

- The data can be incomplete, representing only a partial view of the state of the system or the operating environment;

- The data can be imprecise due to the limitation of the employed sensing elements. Different sensors have different resolutions, accuracies and sensing ranges. Besides, the performance of the sensors can be influenced by external factors such as surrounding noise or temperature;

- The data may not directly represent the desired aspect to be observed or measured. For example, end-to-end response time during a business transaction over the Internet is difficult and expensive to measure. Thus, surrogate metrics such as CPU queue length can be used to infer it [2]; and,

- There can be multiple measurement sources that produce both interval and event data, and the intervals may not be synchronized, for example, 10 seconds vs. 1 minute vs. 1 hour [2].

A significant body of work exists on sensor data fusion, filtering, interpolation and correlation. In this book, we will be investigating the role of context-aware computing in dealing with sensed data, in particular, and developing self-managing systems, in general. Some of the aspects of context-awareness we will be investigating more closely include declarative specification of sensors; dynamic binding of data sources; modeling and representing sensed data; and data analysis and reasoning.

1.4.1   Context-Awareness

The initial motivation for context-aware computing was the reduction of the explicit information a user needs during an interaction with a computer. The premises for this are the mobility and activity of users in ubiquitous computing environment. For example, a mobile user can interact with a computer while driving, talking to other people, holding a lecture or attending to a child [7].

Earlier approaches in the HCI community attempted to address this issue by (1) presenting to a user multiple modalities of interaction (gesture, voice, graphic, tactile, etc.) to make interaction intuitive; and (2) increasing the vocabularies of each modality to make interaction rich. In both cases, however, the interaction model requires explicit input from the user because the computer is entirely unaware of and quite unable to utilize background information which can be vital to the understanding of the user's intentions or wishes.

The complementary approach is the use of implicit information which can be vital to the understanding of explicit inputs from a user [8]. The information may related to the user directly or to the physical surrounding wherein the interaction is unfolding. This information can be obtained from a variety of sources, including sensors and software services monitoring the status of a device, an application, a computing platform, a network or a part of the physical world.

The idea of using implicit information is taken from the way human beings communicate with each other, and how they exploit implicitly available information to increase their communication bandwidth. For example, when people attend a meeting, their eyes communicate to convey agreements or disagreements to what is said or unsaid; voices are whispered to exchange impromptu opinions; facial expressions reveal to the other participants fatigue, boredom or disinterest. More importantly, speeches may not be grammatically correct or complete. Previous as well as unfolding incidents enable the participants to capture what cannot be expressed verbally. Speakers shift from one language to another and use words with multiple meanings, and still the other participants can follow.

Flexibility and adaptation is possible because the social and conceptual setting (i.e., the context) encompassing human-human interaction is effortlessly recognized by all participants. As a result, within the perceived context, many activities unfold, some of which are unpremeditated, yet consistent with the context, while other activities express the freedom associated with the recognition of the context - for example, using incomplete or incorrect statements, or using words with multiple meanings. Still other activities reflect the participants' adjustment of behavior in compliance with the context of the setting - for example, participants whispering to exchange impromptu ideas.

Dey calls systems that use implicit information to provide useful services in a proactive manner context-aware [9] and categorizes context-awareness in one of the following aspects:

1. The presentation of information and services to a user;

2. The automatic execution of a service; and

3. The tagging of context to information for later retrieval.

In the first category, a context-aware system employs implicitly available information (for example, the location of a mobile user) to provide a service that is associated with the user's current activity. For example, a user whose present activity is printing a document will be presented with the list of nearby available printers; and a tourist will be provided with a description of a point of interest that matches his preference and location.

In the second category, a service is dynamically executed in accordance with his present context (activity or location). A typical example is the dynamic adjustment of the physical and ambient setting of a car (seat and mirror position; cabin temperature; and preferred radio station, etc.) according to a user's identity. In this case the context of interest is the identity of the user and it can be represented by an RFID.

In the third category, specific information is associated with activity, identity, location and time contexts, among others. The information will be displayed to the user when his context matches the ones with which the information is associated. For example, a user can request an email application to list all the emails he sent while attending a particular meeting. For this to happen, the email application must associate outgoing emails with the current setting (activity) of the user.

In all of the three categories the acquisition of a context of interest remains the same. Physical or software sensors are used to capture certain phenomenon. The sensor data are processed to extract meaningful information, and this information will be used as an implicit input for the system to carry out a task with minimum user's involvement. This aspect fits well with the vision of autonomous computing or self-managing systems.

In the following subsection, we provide some examples of context-aware systems to demonstrate how a context of interest is captured by employing physical sensors.

1.4.2   Surrounding Context

Eronen et al.[10] classify auditory scenes into predefined classes by employing two classification mechanisms: 1-NN classifier and Mel-frequency cepstral coefficients with Gaussian mixture models. The aim is to recognize a physical environment by using audio information only. The audio scene comprises several everyday outside and inside environments, such as streets, restaurants, offices, homes, cars, etc.

The features to be extracted for the purpose of classification are: zero-crossing rate (ZCR) and short-time average energy in time domain; band-energy ratio, spectral centroid, bandwidth, spectral roll-off and spectral flux in frequency domain; and linear prediction and cepstral features such as linear prediction coefficients (LPC), cepstral coefficient and Mel-frequency cepstral coefficients (MFCC). The classification systems classify 17 out of 26 indoor and outdoor scenes with an accuracy of 68.4% with analysis duration of 30 seconds. Each classified scene has at least five samples from different recording sessions before a classification process started. The classification performance is evaluated using leave-one-out cross-validation, where a classifier is trained with all instances except the one that is left out for the classification.

Korpip et al. [11] propose a multi-layered context-processing framework to carry out a similar work. The bottom layer is occupied by an array of sensors enclosed in a small sensor. The sensor board is attached to a shoulder strap of a backpack containing a laptop. When collecting scenario data, a user carries the backpack. A cordless mouse controls the measurement system to mark the scenario phase. Nine channels are used to obtain data pertaining to a physical environment: three of the channels for a three-axis accelerometer, two for light intensity and one for temperature, humidity, skin conductivity and audio.

The other layers in the context processing hierarchy include a feature extraction layer incorporating a variety of audio signal processing algorithms. A naive Bayesian classifier reasons about a higher-level context by classifying the features extracted by the DSP algorithms. A total of 47 quantized audio features, including harmonicity ratio, spectral centroid, spectral spread, spectral flatness and fundamental frequency are used to describe seven audio related contexts: speech, rock music, classical music, car, elevator, running tap water or other sounds.

The framework provides support for quantifying the uncertainty associated with a recognition process. An additional merit of the framework includes training the model with data to recognize new contextual states.

1.4.3   Activity on a Street

Moenne-Loccoz et al. [12] propose architecture for modeling the temporal evolution of visual features characterizing a human behavior, and to infer their occurrences. The architecture consists of a vision module, an interpretation module, and a knowledge base. The vision module performs segmentation and classification on a video stream input. It tracks individuals or groups of individuals. The interpretation module recognizes a set of behaviours such as the fighting of individuals or vandalism. To ease the interpretation task, three entities are introduced to the knowledge base:

- State: it refers to the property of a mobile object. Examples are: seating/standing, still/walking/running.

- Event: it characterizes a change of state. Examples are: to sit down/to stand up, to stop/to begin running - to sit down, for instance, is the change of the state standing into the state seating.

- Scenario: it is a combination of states, events, and/or sub-scenarios. Examples are: running towards a train, following someone.

Additionally, the knowledge base defines a detailed description of the scene environment. This knowledge is used by the interpretation module. Knowledge of a scene environment includes the nature and position of still environment such as walls, benches and doors. The expert knowledge defines a complex scene in terms of simple scenes. For example, running towards a train is described by a combination of the chain of events: running, train present and trajectory is towards the train.

The prior knowledge along with the representation of the scene presented by the vision module is supplied to a Bayesian network inside the interpretation module to recognize hierarchically all the occurrences of states, events, and scenarios, which signify human activities.

Over 600 frames were used to train the network to recognized violent behaviors such as people fighting or show some pronounced agitation. 80% of the frames contain anticipated behaviour (violence) while the remaining 20% were spurious (no violence).

The architecture provides support for the dynamic definition of higher-level contexts; the input contexts are, unfortunately, limited to video features. Since a Bayesian network is employed for a recognition purpose, the uncertainty associated with a recognition task can be quantified. Belief revision is not treated in the architecture.

1.4.4 User's Attention in a Meeting

Wu [13] extends the functionalities of Dey's Context Toolkit [14] to support context fusion. Even though the Aggregator proposed by Dey gathers all relevant contexts of a particular entity, it does not actually process these contexts to achieve a meaningful understanding of the situation of the entity. This assignment is left to the applications themselves. Dey's argument for this is that an aggregation task is specific to each application. While this holds true, due to physical limitations of sensing elements and other external factors, propositions made by context sources (sensors) may lack the appropriate precision or abstraction.

Wu applies Dempster-Schafer's theory of evidence to deal with uncertainty associated with context sensing. In his implementation, an Aggregator receives video and audio features from a camera and a set of microphone widgets to determine the likelihood of a participant's focus of attention in a meeting.

The application scenario comprises a small round table in a small meeting room, where a few people sitting around the table participate in a discussion. An omni-directional camera at the center of the table captures the activities of the participants, while a microphone in front of each participant measures relative sound strength. A skin-color based face detector recognizes the face location, from which a participant's head pose is estimated using neural network algorithms. A Gaussian model is assumed to describe the head pan angle distribution. The head pose estimated from this process is the basis for estimating a participant's focus of attention.

Meanwhile, the relative sound strength from each microphone is used to determine the speaker at any given time. Hence, the audio widget takes signal strength from all microphones as input to determine who is speaking at a given moment and who has been speaking a short while before. An essential assumption to infer a participant's focus of attention is that non-speakers focus their attention on the present speaker.

Accordingly, a participant's focus of attention is estimated independently by two different sensing modalities. The Dempster-Schafer theory of evidence is used to combine the beliefs of the two sources in order to arrive at a reliable proposition.

The strength of this approach lies in its ability to improve the quality of a context obtained from various sources. It also quantifies the uncertainty associated with context aggregation.

1.4.5   Activity Context from Multiple Sensors

Mntyjärvi et al. [15] proposes a four-layered framework for recognizing a user's activity. At the lowest level there are context information sources. These sources deliver sampled raw measurements which map to physical properties. The middle layers are occupied by the context measurement and context atoms extraction unites - the raw sensor data are sampled and pre-processed in these two layers. In the case of sensor measurements, signal values are calibrated and rescaled. Pre-processed signals are used as inputs to various feature extraction algorithms in time and frequency domains, producing features to describe context information. For example, the root mean square (RMS) value of an audio signal describes the loudness of a surrounding. The first task in context extraction is to abstract raw sensor signals and compress information by using different signal processing and feature extraction algorithms. The features to be extracted are chosen according to how well they describe some parts of the real world context. Extracted features are called context atoms since they contain the smallest amount of context information. The upper layer is occupied by the context information fusion unit, which manipulates the context atoms to produce higher-level contexts that represents a real-world event.

An implementation of the context information unit employs k-means clustering and minimum-variance segmentation algorithms. Sensor data are logged from a self-contained device that encloses an array of sensors comprising three accelerometer sensors, illumination sensors, humidity sensors, thermometers, skin conductivity sensors and a microphone.

The higher-level contexts recognized include various user's activities such as running, walking and climbing a flight of stairs; contexts related to a mobile device includes whether it is being held in a hand or being placed on a table; and so on.

1.4.6   IBadge

The iBadge [16] wearable system monitors the social and individual activities of children in a kindergarten. It incorporates sensing, processing, communication and actuating units. The sensing unit includes a magnetic sensor, a dual-axis accelerometer, a temperature sensor, a humidity sensor, a pressure sensor and a light sensor. It also includes a ultrasound transceiver and a RF transceiver for position and distance estimations. The processing unit includes speech and sensor data processing. A server side application assists a teacher by receiving and processing location, orientation, ambient and audio contexts from the iBadge to determine the social and learning status of a child. The location and orientation contexts are used to determine whether a child is isolated or associates with other children while the audio context is used to determine whether a child is sociable or aggressive.

1.4.7 Mediacup

Mediacup [17] is an ordinary coffee mug in which a programmable hardware for sensing, processing and communicating context is embedded. The hardware is a circular board designed to fit into the base of the cup and incorporates a processor subsystem, a sensing (accelerometer and temperature sensors) subsystem and a wireless transceiver. The mug continuously monitors its state by aggregating data from the two sensors, producing a higher-level context and communicating the result to a remote application via a wireless link. Heuristic-based rules are employed to reason about movement related contexts. The various propositions include whether a cup is stationary or not; whether someone is drinking from it or playing with it; or whether it is being carried around. The higher-level contexts related to temperature include whether a cup is freshly filled with coffee or whether a coffee is cooling off.

1.5 Context-Aware, Self-Managing Systems

So far, context-aware computing and self-managing systems are emerging independently. The application domains for which they are studied are different as well. The two approaches have several features in common. For example, both approaches aim at reducing human involvement: while context-aware computing aims at reducing the amount of explicit input a user should provide to computing systems, autonomous computing (self-management) aims at reducing the operational and maintenance cost of a system.

If one takes the conceptual framework of Kephart and Chess [3] as a reference framework of self-managing system, the sensing, actuating and analysis component are typical components of context-aware computing. Much work has been done by the research community of context-aware computing to support context acquisition, context modeling, context representation, context reasoning and context management. Researchers of autonomous computing can benefit a great deal by considering the usefulness of this work to develop self-managing systems.

1.6   Organization of the Book

The aim of this book is twofold:

1. To enable researchers of context-aware computing to identify potential applications in the area of autonomous computing; and

2. To support researchers of autonomous computing in defining, modeling and capturing dynamic aspects of self-managing systems.

The Merriam-Webster's English Dictionary defines a system as *a regularly interacting or interdependent group of items forming a unified whole.* We adopt this definition to define a *system* in the context of this book. Therefore, we use the word system to refer to a composition of software and hardware components which work together as a whole to accomplish a specific task on behalf of a user. In this regard a system can be a device, an application, a middleware, or a network.

The book is organized as follows: Chapter 2 provides a context-aware applications that assists nurses in hospitals to efficiently and safely administer medicaments. Chapter 3 provides a detail account of service discovery approaches and their place in self-managing systems. Chapter 4 discusses how heterogeneous context sources and their content can be managed in a distributed manner. It discusses also the usefulness of ontology as a context representation and exchanging tool. Chapter 5 discusses content negotiation in web environments.

The remaining 4 chapters focus on self-managing networks. Chapter 6 presents a higher-level vision for future self-managing networks. Chapter 7 presents in detailed the use of context-aware computing and policy-based approach to build self-managing networks. The remaining 2 chapters, namely, chapter 8 and 9, focus on two aspects of self-managing networks, i.e., self-protection and self-healing.

References

[1] A. Fox and D. Patterson. Self-repairing computers. *Scientific America,* 228(6), 2003.

[2] Y. Diao. Self-managing systems: A control theory foundation. In *The 12th IEEE International Conference and Workshops on Engineering of Computer-Based Systems,* 2005.

[3] J.O. Kephart and D.M. Chess. The vision of autonomic computing. *Computer*, 36(1):41–50, 2003.

[4] G. Tesauro. Reinforcement learning in autonomic computing. *IEEE Internet Computing*, 11(1), 2007.

[5] A. Akella, G. Judd, S. Seshan, and P. Steenkiste. Self-management in chaotic wireless deployment. In *The 11th Annual International Conference on Mobile Computing and Networking*, 2005.

[6] W. Dargie and T. Springer. Integrating facts and beliefs to model and reason about context. In *In Proceedings of the 7th IFIP International Conference on Distributed Applications and Interoperable Systems*, 2007.

[7] W. Dargie and T. Hamann. A distributed architecture for reasoning about a higher-level context. In *In Proceedings of the 2nd IEEE International Conference on Wireless and Mobile Computing, Networking and Communications (WiMob 2006)*, 2006.

[8] W. Dargie. *Architecture for computing context in mobile devices*. PhD thesis, Technical University of Dresden, 2006.

[9] A.K. Dey. Understanding and using context. *Personal Ubiquitous Comput.*, 5(1), 2001.

[10] A.J. Eronen, V.T. Peltonen, J.T. Tuomi, A.P. Klapuri, S. Fagerlund, T. Sorsa, G. Lorho, and J. Huopaniemi. Audio-based context recognition. *IEEE Transaction on Audio, Speech, and Language Processing*, 14(1), 2006.

[11] P. Korpipää, M. Koskinen, J. Peltola, S.-M. Mäkelä, and T. Seppänen. Bayesian approach to sensor-based context awareness. *Personal Ubiquitous Computing*, 7(2):113–124, 2003.

[12] N. Moenne-Loccoz, F. Bremond, and M. Thonnat. Recurrent bayesian network for the recognition of human behaviours from video. In *In Proceedings of the 3rd International Conference on Computer Vision Systems*, 2003.

[13] H. Wu. Sensor data fusion for context-aware computing using Dempster-Schafer theory, 2003.

[14] A. Dey. *Providing architectural support for building context-aware applications*. PhD thesis, Georgia Institute of Technology, 2000.

[15] J. Mäntyjärvi, J. Himberg, and P. Huuskonen. Collaborative context recognition for handheld devices. In *In Proceedings of the 1st IEEE international Conference on Pervasive Computing and Communications*, 2003.

[16] A. Chen, R. Muntz, S. Yuen, I. Locher, S. Park, and M. Srivastava. A support infrastructure for the smart kindergarten. *IEEE Pervasive Computing*, 1(2):49–57, 2002.

[17] H.-W. Gellersen, A. Schmidt, and M. Beigl. Multi-sensor context-awareness in mobile devices and smart artifacts. *Mob. Netw. Appl*, 7(5):341–351, 2002.

Chapter 2

Verifying Nursing Activities Based on Workflow Model

Noriaki Kuwahara, Naoki Ohboshi, Hiromi Itoh Ozaku, Futoshi Naya, Akinori Abe, and Kiyoshi Kogure

ATR Knowledge Science Laboratories
2-2-2, Hikaridai, Seika-cho, Soraku-gun, Kyoto 619-0288 JAPAN

Abstract

In this paper, we propose an algorithm for observing nursing activities with a ubiquitous sensor network and detecting errors in nursing care by comparing observed data with the above model to provide nurses with context-aware support: namely, warnings about the possibility of errors. First, we discuss the usefulness of context-aware technology in a medical field where many temporally and spatially distributed personnel are precisely performing various medical practices. Next we introduce applications of context-aware technology in a medical domain. Then we show our research and development for observing and understanding nursing activities with a ubiquitous sensor network in a hospital and propose a robust error detection algorithm for nursing care procedures. This algorithm warns of mistakes or neglected tasks in nursing procedures using observed data from a ubiquitous sensor network. We assume two types of nursing care errors: "neglected tasks" and "out-of-sequence tasks." The algorithm focuses on these tasks in potentially fatal errors because such situations demand immediate attention. Its validity has been confirmed by the observed data for nursing care before and after cataract operations.

2.1   Introduction

Medical errors annually cause many deaths. In the United States, for example, it is estimated that more people die every year as a result of medical errors than from car accidents, breast cancer, or AIDS [5]. From 13,601 nursing care cases examined, the Japanese Ministry of Health, Labor and Welfare recently defined 10,564 as incidents and accidents. Many medical accidents are caused by neglect or carelessness when confirming safety during medical and nursing care [1], [2], [3], [4]. Such neglect or carelessness often happens when medical personnel are too busy. In Japan, for example, a nurse failed to notice a problem with some artificial respiration equipment because she was too busy dealing with other patients, even though hospital regulations require that such respiratory patients must be checked every fifteen minutes.

One approach to prevent such cases is to encourage other medical staff to help nurses who have become too busy, even if no explicit help message has been issued. However, this is difficult because medical staff in a hospital essentially work in a spatially and temporarily distributed manner. This difficulty can be reduced by awareness support technology that enables medical personnel in distant locations to share situations. Here, awareness, defined as "understanding the activities of other people who provide the context information affecting your activity," plays an important role when people are working cooperatively. Especially in medical environments where many people work in a spatially and temporarily distributed manner, understanding the context of other medical personnel is necessary to take proper action for patient care. Such awareness support can be realized using ubiquitous sensor network technology that captures and understands the activities of medical staffs.

Another approach to prevent medical accidents caused by neglect or carelessness is using a context-aware support system to remind nurses of important events that tend to be neglected or to warn them about the possibility of accidents or incidents. Such a system can also be realized with the above ubiquitous sensor network.

Ubiquitous computing technology continues to develop rapidly, and the fundamental infrastructures for ubiquitous sensor networks are being established for collecting an enormous amount of data of people's daily activities. Research is being conducted to recognize and understand human activities based on such collected data, for example, [6] and [7].

Under the above background, we are conducting research on understanding people's activities and their surrounding situations using data observed from a ubiquitous sensor network. We are focusing on medical and nursing care.

Our ubiquitous sensor network being developed for a hospital consists of wearable sensors attached to medical personnel and environmental sensors deployed in nurse stations and on hospital wards. Wearable sensors attached to medical personnel are connected by a body area network that is a part of a ubiquitous sensor network. In this body area network, wearable devices gather data from wearable sensors. Our goal is to identify the medical context of the personnel using the above observed data and to provide nurses with critical support in a just-in-time manner based on the identified contexts of medical staff, since medical and nursing environments are critical application domains for context-awareness technology. To achieve our goal, our system must understand the tasks performed by nurses along with nursing care contexts.

The above nursing care contexts, which are not so simple, have complicated structures. Therefore, we previously discussed the importance of modeling nursing activity based on the domain knowledge of nursing care to understand the context of such complicated activities [13]. We also presented an example of such a model for a procedure called "Nursing Care before/after Cataract Eye Operation" that we described with the OWL-S [14] framework.

In this paper, we propose an algorithm for observing nursing activities with a ubiquitous sensor network and detecting errors in nursing care by comparing observed data with the above model. The proposed algorithm focuses on "neglected" and "out-of-sequence tasks" that include potentially fatal errors that must be rectified immediately. It can also effectively handle situations where observation problems cause the system to generate false alarms.

In the rest of this paper, we first introduce research on context-aware technology and workflow management in medical and nursing domains as related works and clarify our goals with this background. Second, we explain the details of our proposed algorithm. Third, we show that it can detect the errors observed in the "Nursing Care before/after Cataract Eye Operation" procedure carried out in actual nursing care. Finally, we state our conclusion and discuss future works.

2.2   Related Works

From the viewpoint of the informatization of medical care procedures (in other words, workflow studies) for improving the quality of medical care, research efforts can be roughly categorized into the following three areas:

1. Standardization of medical and nursing care

2. Hospital information systems

3. Regional clinical information systems

The first area, the so-called "Clinical Path," provides standard medical and nursing care based on the patient's condition [18], [19], [20], and the other two focus on information system design involving interaction between medical staffs and hospital information systems (HIS) [21], [22], [23], [25], [26], [30]. Standard models of such interaction have also been discussed [32]. However, they only represent a small part of the daily activities of medical personnel. For example, when an intravenous drip injection is specified in the clinical path and ordered by a doctor, nurses must engage in many tasks and subtasks: receiving the order, preparing and mixing medicines, giving the drip injection to the patient, and monitoring the patient's condition after completing the drip injection. During this procedure, a nurse interacts not only with HIS but also with patients, medicines, and objects including bottles, lines, and syringes. In using a model to understand a nurse's activities and to confirm that she/he is correctly performing the entire drip injection procedure, the model must express the above interactions. We use the term "Workflow Model" in such a broad sense. One research goal is to provide a framework for developing such models based on the documents of medical and nursing care procedures and the observed data from ubiquitous sensor networks.

In general, the above HIS provides a workflow management function for monitoring the progress of medical and nursing orders and checking whether they are processed correctly. If errors are detected when the orders are processed, these systems usually require confirmation by a member of the medical staff. However, when the system monitors medical and nursing care procedures based on recognition results using observed data from sensor networks, the detected errors might cause false alarms. If the system required human confirmation of detected errors whenever they happen, medical personnel might be too overwhelmed to attend to their duties. Our next goal is to overcome this problem by proposing an algorithm that enables the system to determine whether a medical staff person should be alerted, based on the medical and nursing workflow model.

On the other hand, the following advanced research is already utilizing context-aware technology in medical fields:

- iHospital (information sharing using context-awareness) [15], [16]

- Instant messaging system on Context-aware Hospital Information System (CHIS) [24]

- Surgical Information Strategy Desk [17]

The following elemental technologies have also been developed for realizing context-aware environment in medical fields.

- Middleware for sensor networks (MILAN) [27]

- Wearable systems in nursing home care [28]

- RFID-applications to enable research in real-life environments [29]

- Wireless remote healthcare monitoring using body-area network [31]

- Bluetooth-based architecture for location-positioning services [34]

As application examples of context-aware technology in the medical field and such elemental technology, note "iHospital" and "MILAN," respectively. "iHospital," developed by the University of Aarhus in Denmark for the Horsense Hospital, is a medical surgery support system that enables medical staffs to share the situations of surgery rooms. Dual 44-inch displays, set in the headquarters, are called "Aware Media," where, for example, the schedule of medical surgeries, the progress of each procedure, and messages from medical personnel are displayed and shared. However, in the "iHospital" system, since medical staffs explicitly inputted and modified the situations presented on the above display, the shared information does not necessarily reflect the latest situations. On the other hand, we automatically observe and understand medical staff activities by analyzing observed data from the ubiquitous sensor network in the hospital.

MILAN is middleware that allows sensor network applications to specify their quality needs and adjusts the network characteristics to increase application lifetime while still meeting these quality needs. It can remove negative influences caused by differences among connected applications and infrastructure. It can also remove gaps between updated applications and old ones. MILAN itself does not function as a core system, but as the control center of a distributed system. In the framework of the E-Nightingale Project, we did not consider introducing any mechanisms or schemes that can effectively combine, control, and maintain applications and sensor networks. However, since the framework consists of various types of applications, some of which are frequently updated, as Heinzelman pointed out, in the future, such middleware as MILAN must be introduced to combine, control, and maintain independently designed and developed applications and infrastructures.

2.3   Overview of Research Goals

Using the just-in-time nursing advice system, the proposed algorithm performs its key function with one of the three main target systems of the E-Nightingale Project shown in Figure 2.1. We are now researching the following basic technologies:

1. Ubiquitous sensor network technology consisting of wearable and environmental sensors that constantly monitor nurses without impeding or interrupting their activities [33]

2. Data processing technologies for recognizing nursing activities using observed data from sensor networks [35], [44]

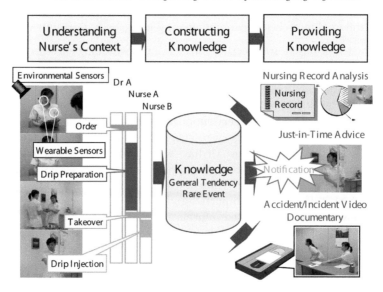

FIGURE 2.1: Overview of E-Nightingale Project's target systems

3. Knowledge processing technology for understanding the situation (context) of nursing activities and detecting errors by comparing the recognized results with the nursing workflow model [13]

4. User interface technology that alerts nurses to errors using the most suitable method based on nursing contexts [36]

Figure 2.2 shows the modular structure of our proposed system. The "Sensor Network Management Module" (SNMM) stores the raw data from sensors in the "Sensor Data Database," simultaneously interprets the observed data for the "Nursing Resource Management Module" (NRMM), and notifies its API of the interpreted results in RDF [37] format.

NRMM manages the status of objects including tools, medicines, and patients in nursing environments based on SNMM's notification of observed data. Such objects are expressed using OWL [38], instantiated in the "Nursing Resource Database" and called "Nursing Resources." Furthermore, we can describe the situation (context) of nursing activities using "Nursing Resources." For example, for a drip injection, a nurse handles bottles of drugs, lines, needles, and so on. SNMM notifies NRMM of the observed data as the status of each object, and the combination of the statuses of instances determines the nursing activity context. We use SWRL [39] to represent such contexts as rules. When one of the rules is applied, NRMM notifies the "Nursing Workflow Management Module" (NWMM) of the nursing context as a "Workflow Change Event."

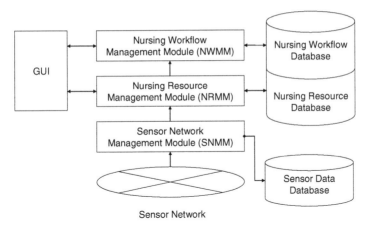

FIGURE 2.2: Modular structure of proposed systems

Figure 2.3 shows an example of the OWL descriptions of "Nursing Resources" and the SWRL descriptions of rules for generating "Workflow Change Events." These descriptions are all stored in the "Nursing Resource Database." NWMM verifies the nursing activities reported from NRMM as "Workflow Change Events" based on the nursing workflow model.

2.4 Case Study of Intravenous Medication Process Performed by Nurses

Our proposed system aims to reduce medical accidents at hospitals by utilizing ubiquitous computing and developing a system that can monitor every moment of nursing activities without impeding workflow to analyze their root causes. We must identify the following conditions to design such a system:

1. What types of errors frequently occur?

2. Why did previously cited errors occur?

3. Why were errors not detected?

Among medical errors, the largest number is medication administration errors and includes wrong dosages, wrong times, wrong patients, etc. Nurses are also prone to error because they directly administer medication to patients. This paper reports an analysis of a case study of the complete IV medication process performed by nurses and discusses possible solutions from a ubiquitous computing point of view for reducing risks in a hospital.

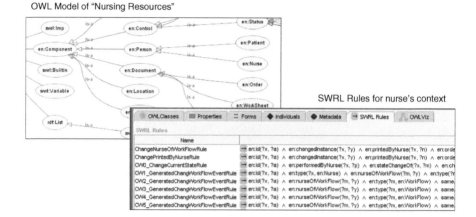

FIGURE 2.3: "Nursing Resource" and rule descriptions

2.4.1 Survey Method and Results

This survey analyzes the workflow of a scheduled IV medication to investigate possible errors in each step of the workflow and to clarify whether errors can be prevented with a ubiquitous sensor network.

We studied the scheduled intravenous [IV] medication process at a large hospital in Tokyo. In this hospital, an electronic patient record system has not been introduced, so at the beginning of the IV medication process, the nurse is only required to read the IV medication orders from the doctor's order form and copy them to different sheets (transcription) by hand. Because many kinds of orders can be found in the doctor order forms, not only IV medications written by the doctor, she cannot remove this order from the nurse station. Then the nurse prepares the meds for the IV based on the transcription and double-checks the transcription with the original orders to prevent transcription error. Next, the nurse mixes the meds with the main IV bottles, brings them to the patient bedside, and injects the patient. After the injection, the nurse periodically observes the patient's condition. We discussed what types of errors were most likely to happen with the nurses to determine when they frequently occurred in scheduled IV medication processes. Figure 2.4 shows the scheduled IV medication process and the most likely errors in each step of this process.

Through these interviews, we discovered that illegible orders might cause transcription errors in the copying order step, transcription errors and attaching incorrect labels might cause medicine error in the mixing of the IV drip step, and so on.

We also conducted questionnaire surveys on the surgical ward of this hospital from March 7th to 13th, 2004 to determine what caused nurses stress in

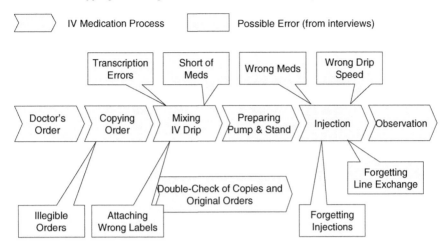

FIGURE 2.4: Scheduled IV drip process and possible errors in each step

their medical environments. Each nurse involved with any part of IV medications filled out such forms as when she performed the activities, what part of the IV medication process she performed, what errors happened, what interruptions occurred, what events she noticed, etc. Nurses worked three different shifts in this hospital, so 21 shift samples were expected, but only 13 shift (62%) samples were collected because some nurses were too busy to fill out the form due to emergency admissions and surgeries. The returned samples included 55 nurse awarenesses, and among these, 42 cases might cause mistakes. Table 2.1 shows the top five categories and the number of cases for each category.

Table 2.1 shows the types of cases listed in Figure 2.4, but interruptions are most frequently perceived by nurses. "Illegible orders from the doctor" means that nurses could not read the doctor orders, so they had to suspend their jobs to confirm the orders with other medical personnel. In other words, the nursing and other medical staffs experienced job interruptions.

Based on the results of our questionnaire surveys, except for possible errors found by interviews with nurses, there were many possibilities for disturbances or interruption of nurses. As shown below, interruptions during the IV medication process are highly dangerous and lead to accidents. Therefore, priority must be given to solutions using context-aware technology.

2.4.2 Possible Solutions from Ubiquitous Computing Point of View

As mentioned in the previous section, in both the cases of "Interruptions by nurse calls or patients" and "Illegible orders from doctors," nurses are re-

Table 2.1: Top five categories of collected cases of nurses' awareness

Categories of nurses' awareness	Number of cases / Total cases
Interruptions by nurse calls or patients	19 /42
Illegible orders from the doctor	9 / 42
Wrong IV drip speed	6 / 42
Wrong patients	3 / 42
Transcription errors	2 / 42

peatedly forced to stop what they are doing, and they are also interrupted by other medical personnel. Since interruptions are potential causes of adverse events [8], nurses strongly need protection from such events. When a nurse is performing a task that should not be interrupted because the risk of causing an adverse event is great, the system protects this nurse against interruptions from other medical staffs and patients including nurse calls because it is aware of the context. Such a system will make hospitals much safer, and utilizing the location information of each member of the medical staff in a hospital [9] is one solution. However, nurses should not be required to manually update their location information because the number of times they switch tasks exceeds 100 during their daily working hours. Without additional interaction, our system must automatically update nurse location information using ubiquitous computing. Otherwise, they will never use the system.

For preventing medical accidents, protecting nurses from interruptions when they are double-checking orders, mixing meds for IV medications, and injecting meds is crucial. The problem remains how to identify such nursing contexts of activities. While studying the workflow of the scheduled IV medication process, we found a strong correlation between their activities, their workplaces, and items they were handling. Table 2.4.2 shows the nurse's contexts and the features that characterize each context.

Therefore, a possible scenario for understanding the nursing context is shown below.

It is easy to deploy nurse location sensors connected to WLAN all around a hospital, so we expect to gather nurse location data. As for "Double-checking doctor's orders," there are two items (papers) shared between nurses. However, nurses only read them, and so detecting such interaction with the current technology is almost impossible. As for "Mixing meds of IV drips," nurses write on paper to check meds. Such interaction can be caught by digital pen and paper [10]. Also, if sensors are attached to the drawers of the medicine cabinet, it is easy to obtain the data that show the nursing behavior of selecting medicine. As for "Injecting meds to patients," if the RF-ID tag is attached to a tray of IV bottles and the reader is on the bedside table, we can determine situations where nurses bring the tray with IV bottles to bedsides.

Table 2.2: Nurse's contexts and their features

Context of the nurse	Place	Nurse's behaviors	Items interacted with nurse
Double-checking doctors' orders	Nurse station	One nurse reads doctor's order aloud and other nurse check the transcription of the order.	Original order form from the doctor Transcription of the order
Mixing meds of IV drips	IV drip mixing area	The nurse checks meds by striking through items of meds list on the transcription of the order. The nurse mixes meds by using injection syringes. The nurse puts IV bottles in the tray.	Transcription of the order Pen Meds IV bottles Tray
Injecting meds to the patient	Patient's bedside	The nurse set up the IV bottle, IV stand and lines. The nurse sticks a needle into the patient's arm. The nurse fixes the lines in order not to withdraw the needle. The nurse controls the drip speed.	Tray IV bottles IV stand Lines Pump

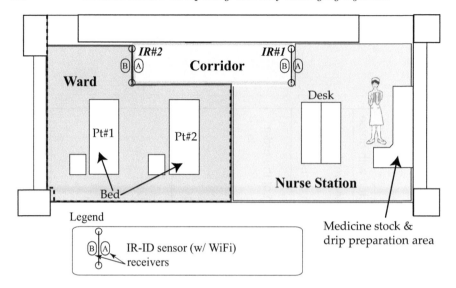

FIGURE 2.5: Experimental room layout

However, imagining a situation where RF-ID tags are attached to every object with which nurses interact in medical environments is rather difficult. We are planning to measure the movements of nurse's arms and bodies to infer the nursing task being performed.

We are planning to construct an experimental system for evaluating how accurately the system can understand the context of nurses and how much it reduces the interruptions of nurses using the above correlations between context, workplace, and items.

2.5   Prototype of Ubiquitous Sensor Network System

2.5.1   Experimental Room Description

We set up a laboratory room for capturing the location and activity data during the experiment. The room was divided into three regions: *Nurse Station*, *Corridor*, and *Ward* (Figure 2.5). Several medical apparatuses, e.g., a stethoscope, a blood-pressure gauge, syringes, tubes, and medicines, were placed on the shelf and in the medicine stock area in the Nurse Station. Nurses select such equipment before making their work rounds for such tasks as checking vital signs and later return them. They also prepare intravenous injections at the drip preparation area in the Nurse Station.

2.5.2 Location Tracking by IR-ID

We tracked the location of each nurse with an Infrared Identification (IR-ID) system in which the subject nurse wears an IR-ID transmitter that sends ID signals at 50 Hz. One IR-ID unit has two IR receivers that are installed on both sides of the upper parts of a door that separates the above regions. Corrected ID data at each IR-ID unit are transmitted to an IR data server by an IEEE 802.11b wireless network.

In our IR-ID system, an IR transmitter worn by the nurse on her head transmits 8-bit ID signals at 50 Hz. When a subject nurse passes through two IR-ID receivers installed on both sides of the upper part of a door, the IR-ID receiver unit uploads data records including timestamps, labels of the receiver (A/B), and IDs. The location labels of the subject are then estimated from the numbers of received IDs at each IR-ID receiver unit as follows. Let $f_{r,id}^u(t), u \in \{Unit_1, Unit_2, ..., Unit_N\}, r \in \{A, B\}$ be the function defined as:

$$f_{r,id}^u(t) = \begin{cases} 1, \text{ if } \text{ReceiverLabel}(t) = r \ \wedge \ \text{ID}(t) = id \\ 0, \text{ otherwise} \end{cases}$$

where ReceivedLabel(t) is a receiver label (A or B) at time t and ID(t) is a received ID at that time in the IR-ID records. Receiver ID frequency function $h_{r,id}^u(t)$ is calculated by counting the number of IDs at receivers A and B within a time window:

$$h_{r,id}^u(t) = \sum_{\tau=t-\frac{T_w}{2}}^{\tau=t+\frac{T_w}{2}} f_{r,id}^u(\tau),$$

where T_w is a window length set to one second. Figure 2.6 shows an example of $h_{r,id}^u(t)$ of receiver unit $u = Unit_1$ (IR#1) when a subject nurse came through the door from the corridor to the Nurse Station and then returned to the corridor (Figure 2.5). With the transmitter attached to the back of the nurse's head, receiver ID frequency function $h_{A,id}^u(t)$ at the Nurse Station (label A) increases after entering the Nurse Station and vice versa.

When a nurse moves among rooms, the location information of a nurse having an id ID is updated from the last receiver labels having the largest $h_{r,id}^u$ value among all IR-ID units; each is associated with a particular room.

2.5.3 Activity Data Collection with Bluetooth-Based Wireless Accelerometers

The activity data of a nurse were measured using *B-Packs*, a Bluetooth-based wearable sensor network platform. We used triaxial accelerometer data captured by *B-Packs* for classifying nursing activities. Four *B-Packs* were attached on both upper arms, the breast pocket, and the side of the waist of a subject nurse. These body points were empirically decided from the results of

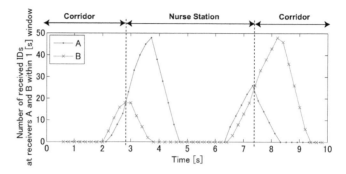

FIGURE 2.6: Location labeling from numbers of received IR-IDs at each receiver within a 1-second window in time line

past field investigations into the places where such wearable sensors wouldn't interfere with nurses while they cared for patients [11]. *B-Packs* were worn so that the Y-axis of each *B-Pack* extends vertically to the ground, and the Z-axis extends along the outer side of the wearer's body (Figure 2.7).

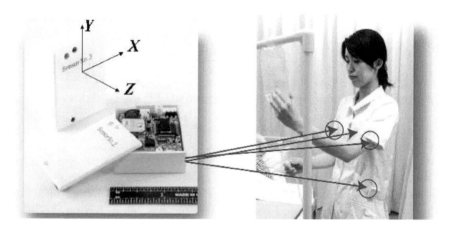

FIGURE 2.7: Subject wearing four Bluetooth-based triaxial accelerometers

The accelerometer data were sampled at 500 Hz with an average of 10 samples on each *B-Pack*, and then the averaged data of 50 Hz were uploaded and stored on a laptop PC in the experimental room by a Bluetooth wireless connection. The accelerometer data contain the timestamps of each sample measurement and the accelerometer readings of each axis.

2.5.4   Feature Extraction for Activity Recognition

We extracted the features of each nursing activity with a sliding window framework [12]. In a sliding window framework, features are calculated on N sample windows of acceleration data with M samples overlapping consecutive windows. From the previous results of setting these parameters, we chose $N = 256$ samples (which correspond to 5.12 seconds) and $M = N/2$ (50%) overlapping samples.

The following features are extracted for each accelerometer axis in a sliding window, a method that resembles [12].

- mean

- standard deviation

- energy

- frequency-domain entropy

- correlation

The DC component of the acceleration in each axis over the window is given by the mean value of the signal. Standard deviation is used for discriminating the range of possible acceleration values, which differ for different activities.

The energy feature is used for capturing the periodic nature in the data, and it's calculated as the sum of the squared discrete FFT component magnitudes of the signal. The sum is divided by window length W for normalization. If $s_1, s_2, s_3, ..., s_W$ are the FFT components excluding a DC component, energy feature $E = \sum_{i=1}^{W} |s_i|^2/W$. Using the mean and energy of acceleration has previously accurately recognized certain postures and activities [12].

Frequency-domain entropy is calculated as the normalized information entropy of the discrete FFT component magnitudes except for the DC component. This feature is used for discriminating activities with similar energy such as arm motions during drip preparation and injection.

The correlation feature is calculated between the two axes of each *B-Pack* and between all pairwise combinations of axes on different *B-Packs* to improve the recognition of activities involving multiple movements of multiple body parts. For example, the correlation between the x and y axis readings of the same accelerometer is calculated as the ratio of the covariance and the product of the standard deviations: $Corr = cov(x, y)/\sigma_x \sigma_y$.

2.6   Algorithm for Detecting Errors in Nursing Activities

In this section, we explain the "Nursing Workflow Model" framework and then show a model of the "Nursing Care before/after Cataract Operation" as an example. Next, we introduce an error detection algorithm in nursing activities based on hypothetical reasoning.

2.6.1   Nursing Workflow Model

We previously developed a framework for expressing a nursing care model based on OWL-S that was originally designed to describe combinations of web services [13]. OWL-S can describe complicated workflows by combining primitive services while considering execution order, concurrency, and repetition. We define "Job" classes to represent primitive tasks and their combinations in nursing care. Therefore, a "Job" class has features of both SimpleProcess and CompositeProcess in OWL-S. Furthermore, we use the following six classes to describe the control flow of nursing care tasks:

- **Any** allows Job classes in a bag to be executed in unspecified order but not concurrently.

- **Choice** calls for execution of a Job class from a given bag of Job classes.

- **Sequence** has a list of Job classes to be done in order.

- **Split** has a bag of Job classes to be executed concurrently.

- **While** tests for the condition, exits if it is false, does the operation if the condition is true, and then loops.

- **Until** does the operation, tests for the condition, exits if it is true, and otherwise loops.

- **If-Then-Else** is a control construct that has properties of If-condition, then, and else that involve different aspects of If-Then-Else. Its semantics is intended as "Test If-condition; if True do Then, if False do Else."

We can describe the "Nursing Care before/after Cataract Operation procedure" with the above classes. Figure 4 shows this example displayed on our proposed system's GUI. We developed this model in the following manner:

1. We extracted primitive tasks of nursing activities from hospital nursing manuals.

2. Since these primitive tasks must be done sequentially, we used the "Sequence" class to represent them.

3. On the other hand, since "Drip" and "Eye Drop" can be practiced concurrently, we used the "Split" class to describe such relationships between concurrent tasks.

4. Since "Eye Drop" is repeated until the pupils have sufficiently dilated, we used the "Until" class to describe such a task flow.

In Figure 2.8, "Eye Drop" is placed under the "Until" class. Then "Eye Drop" consists of the "Drops" of four different drugs in the following order: "Xylocaine," "Mydrin," "Nifuran," and "Neosynesin." "Eye Drop" becomes a composite "Job" that has a "Sequence" class containing these four primitive "Jobs."

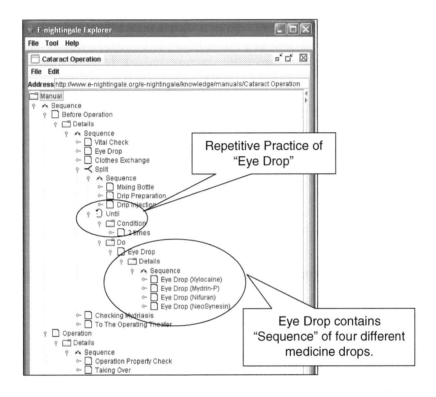

FIGURE 2.8: Nursing workflow model

2.6.2   Error Detection Algorithm

Our proposed algorithm compares nursing activities recognized using observed data from sensor networks with nursing workflow models and verifies whether nursing procedures are being done correctly. This algorithm enables the system to warn nurses who are on the verge of committing an inappropriate nursing procedure. However, due to the problems related to the observation of nursing activities, our system may generate false alarms. If nurses are forced to interact with the system to deactivate the alarm whenever an error is detected, they might be interrupted too often and become distracted from their duties. Therefore, we propose a robust error detection algorithm for verifying nursing activities that enables the system to infer the cause of the detected error and determine whether to immediately alert the nursing staff or to continue the verification process.

We previously proposed a method of risk management for nursing care based on hypothetical reasoning (abduction) [40] and extended it coupled with the concept of scenario [41] in a "Scenario Violation Model."

For a basic risk management method based on hypothetical reasoning, we proposed the following strategy (formalization) for risk determination. When we know all possible hypotheses and their ideal observations[1], we can detect malpractice beforehand because if someone selects an incorrect hypothesis set or fails to generate a necessary one, an ideal observation cannot be explained. When an ideal observation cannot be explained, an accident or incident will occur. With this mechanism, we can logically determine where accidents or incidents are likely to occur before they do. A simple logical framework to complete an activity is shown below (using the framework of Theorist [42]): If

$$F \cup h_1 \not\models ideal\_observation, \tag{2.1}$$

then find h_2 satisfying (2.2) and (2.3) from H.

$$F \cup h_2 \models ideal\_observation \tag{2.2}$$

$$F \cup h_2 \not\models \square. \tag{2.3}$$

$$h_1, h_2 \in H. \tag{2.4}$$

Here F is a set of facts that are always consistent, and h_1 and h_2 are hypotheses that are not always consistent with the set of facts and other hypothesis sets. Hypotheses are generated (selected) from hypothesis base H. $\square$ is an empty set. Therefore, the last formula means that F and h_2 are consistent.

If we can complete formula (2.2), the activity is successfully completed. On the other hand, if we cannot generate enough hypothesis sets to complete the formula (2.2), certain problems will disturb the completion of the activity. Thus, beforehand we can determine the possibility of risk by abduction. That is, when we cannot explain an *ideal_observation* with a current hypothesis set, an error might occur. If objective (*ideal_observation*) cannot be explained, it cannot be completed. This situation is caused by accidents or incidents. Thus, the possibility of risk is suggested.

"Scenario Violation Model" is an extension of an abductive risk determination model combined with the concept of a scenario. Ohsawa characterized a scenario as "a time series of events under a coherent context [43]." Accordingly, we can introduce a concept of time to abductive risk management. In

---

[1] If we use a workflow sheet for nurses or an electronic medical recording system, we can determine the ideal observations.

fact, all nursing activities or workflows involve the concept of time. In general, a workflow can be regarded as a set of scenarios. Therefore, the concept of time (scenario) must be introduced to a nursing risk management system. Based on a basic risk management method based on hypothetical reasoning, when no accident or incident occurs, the following formulas are obtained for the "Scenario Violation Model:"

$$F \cup \sum_{i(in\ chronological\ order)} O(i) \models O \tag{2.5}$$

$$F \cup s_i \models O(i) \tag{2.6}$$

,

where F is a set of facts (background knowledge), s_i is a scenario (a set of hypotheses considering time information), and $O(i)$ and O are observations (results of nursing activities). $O(i)$ can be regarded as a sub-observation of O.

As shown below, here we extend the "Scenario Violation Model" to the verification process of a nursing workflow after an error is detected. We regard F as a set of facts related to nurses and nursing workflows (e.g., patient information, doctor orders, and nursing care procedures) and a series of observed nursing activities. $h(i)$ can be regarded as a series of nursing activities that can explain $O(i)$ with F, where $O(i)$ is the i'th observation (result) of the nursing activities. It can be regarded as a series of nursing activities supplementing certain factors occurring between previous observation $O(i-1)$ and current observation $O(i)$.

Let $h(j)$ be a series of nursing activities that supplements certain factors occurring between previous observation $O(j)$ and current observation $O(j+1)$. Then let $\cup h(j)$ (from $j=0$ to k) be all $h(j)$ related to scenario violations $err(j)$ that are generated between $O(0)$ and $O(k)$. In the hypothetical reasoning framework, $\cup h(j)$ (from $j=0$ to k) can be logically generated based on the nursing workflow model. However, as shown above, since nursing workflows usually have branches as flow controls such as "If-Then-Else," "Choice," and "Split," more than two $\cup h(j)$ (from $j=0$ to k) can be generated.

Accordingly, to select the most appropriate one, we introduce hypothesis cost based on the influence of a mishap that might cause neglect or mistaken tasks on the hypothetical reasoning model.

When the cost of $h(j)$ for $O(j)$ of all candidates is lower than a certain threshold, the system selects the candidate with the minimum total cost. In this case, we assume that an error reflects data input failure. Therefore, we do not need to alert nurses to the possibility of an incident/accident, and the verification process of the observed data continues.

On the other hand, when hypothesis $h(j)$ is among the candidates whose cost is higher than the threshold, the system alerts nurses to the possibility of an incident/accident and requires them to confirm the hypothesis. Then the system feeds back the nurse's confirmation to choose the most appropriate

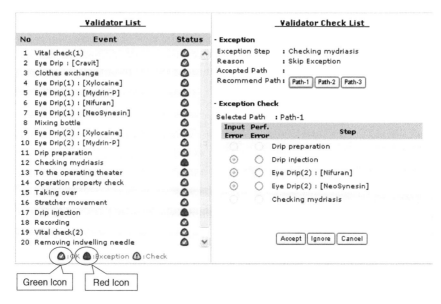

FIGURE 2.9: Confirming detected errors in observed nursing activities

candidate. Moreover, nurses can confirm a set of hypotheses and modify them to make their nursing records accurate.

Figure 2.9 shows an example of the GUI used for confirming the hypothesis set when errors are detected. Observed nursing activities are listed on the left side of the panel. Green icons indicate successful verification of the observed nursing activity, and red icons indicate that the observed activity has not yet been successfully verified. When selecting a nursing activity with a red icon, error information is displayed on the right side of the panel with hypothesis candidate paths. For the path selected by a nurse, the GUI confirms whether each task is an observation error (in other words, incorrectly practiced).

2.7   Testing Our Proposed Algorithm

First we explain the data correction method for recording the histories of nursing activities. Then we present test results that verify the observed nursing activities based on nursing workflow, focusing on "Nursing Care before/after Cataract Operation."

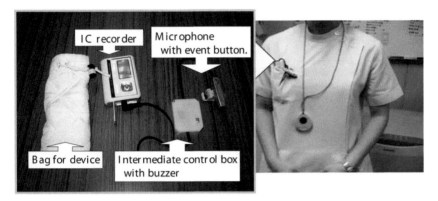

FIGURE 2.10: Event-driven voice recording set

2.7.1 Data Correction Method for Recording History of Nursing Activities

For job analysis, we have been intermittently collecting nursing activities with voice annotations at a Japanese hospital since 2004 [35]. Analysis results are used to increase business efficiency and to improve the quality of nursing care. We have already collected data from more than 400 nurses working in nine clinical divisions, including an ophthalmic and ear-nose-throat (ENT) clinic. Nurses work in three shifts and are assigned to the primary patients of each ward. All nurses were instructed how to use the recording set and the voice input format. They were also asked to include the patient's name and as many task details as possible in their recordings. Figure 2.10 shows our event-driven voice recording device that consists of an IC recorder, a microphone with an event button, and a control box with a buzzer between the microphone and recorder.

The event button is used for explicit voice annotation when nurses start or complete a task. When pushed, a buzzer sounds once to indicate that an annotation is being recorded. The buzzer is also programmed to beep every 10 minutes to prompt nurses to make such voice inputs. Simple signal processing extracts and classifies event-driven and periodic voice recordings as well as nurse call rings. Transcriptions of the segmented data are labeled and cross-checked based on the task classifications of the clinic-dependent workflow and Japanese Nursing Association Standards. Patient names were also extracted from their records.

2.7.2 Test Results

Among the observed data whose analysis has been completed, we sought data collected at an ophthalmic and ENT clinic that contained nursing ac-

Table 2.3: Test results verifying nursing activities based on nursing workflow model

	Rate of Neglected Task (%)	Rate of Disordered Task (%)	Assigned Nurses
Patient A	0	4	Nurse A (2 years)
Patient B	25	8	Nurse A (2 years)
Patient C	65	0	Nurse B (5 years) Nurse C (4 years)
Patient D	30	4	Nurse D (13 years) Nurse E (6 years)
Patient E	48	0	Nurse C (4 years)

tivities related to the "Nursing Care before/after Cataract Operation." We found data for five patients under these conditions, and Table 2.3 shows the verification results of these patient data. Our algorithm detected all errors contained in each set of the observed data and returned hypotheses to explain them. In Table 2.3, we categorized detected errors into two types: "neglected tasks" and "out-of-sequence tasks."

Based on the results, the nursing activities of every nurse contained errors. Most were "neglected tasks," two-thirds of which were unobserved. However, such frequent violations of protocol seem unlikely. Instead, these errors were probably caused by observation problems attributed to the data collection method. Nurses were often required to intentionally input voice annotations of tasks when they were very busy and may sometimes have forgotten to do so.

For example, as shown in Figure 2.11, the results obtained by analyzing the observed data of Nurse A show that she only performed premedication for the cataract operation of Patient A. On the other hand, she concurrently took care of many patients, including Patient B, after premedication for Patient A. Consequently, the observed data related to the nursing activities for Patient B contain many observation errors. The color indicates the type of individual task. Furthermore, all observed data contained some "out-of-sequence task" errors.

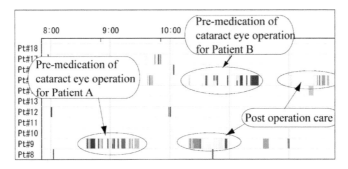

FIGURE 2.11: Nurse A's observed data

2.8   Conclusion and Future Works

We presented our proposed system for verifying nursing activities based on a "Nursing Workflow Model" for detecting errors. Using the observed data from actual nursing care, we tested our algorithm's ability to detect errors. We used five observed datasets related to "Nursing Care before/after Cataract Operation," and the algorithm successfully detected all errors contained in each dataset and returned hypotheses to explain them.

In this paper we only tested our algorithm using observed data generated through offline analysis of voice annotation data in actual nursing environments. However, in the near future we intend to achieve real-time verification of nursing activities. For this purpose, as shown in Section 2.5, we are now constructing a ubiquitous sensor network in a simulated nursing care environment that consists of environmental sensors identifying the locations of nurses, wearable sensors monitoring their actions, and RF-ID tags recording their handling of the objects to which the tags are attached. We are also establishing technology for recognizing nursing activities in real time using the observed data from this ubiquitous sensor network.

We also plan to deploy our system in an actual hospital nursing environment in the near future.

Acknowledgments

This research was supported by the National Institute of Information and Communications Technology.

References

[1] Reason J.: *Human Error*, Cambridge University Press (1992)

[2] Bogner M.S. eds: *Human error in medicine*, Hillsdale, NJ: Lawrence Erlbaum Associates (1994)

[3] Bates D. W., Cullen D. J., Laird N., et al.: Incidence of adverse drug events and potential adverse drug events, *JAMA, 274, pp. 29–34* (1995)

[4] Leape L. L., Bates D. W., Cullen D. J., et al.: Systems analysis of adverse drug events, *JAMA, 274, pp. 35–43* (1995)

[5] Kohn L. T., Corrigan J. M., and Donaldson M. S. eds.: *To Err Is Human: a Safer Health System*, The National Academy Press (2000)

[6] Patterson D., Liao L., Fox D., and Kautz H.: Inferring high level behavior from low level sensors, *Proc. of UBICOMP 2003* (2003)

[7] Duongo T. V., Bui H. H., Phung D. Q., and Venkatesh S.: Activity recognition and abnormality detection with the switching hidden semi-Markov models, *Proc. of CVPR=81f05* (2005)

[8] C.D. Chisholm, et al.: are emergency physicians interrupt-driven and "multitasking"?; *Acad of Emerg. Med. vol. 7, no. 11*, pp. 1239–1243 (2000)

[9] Favela, J., et al.: Integrating context-aware public displays into a mobile hospital information system; *Information Technology in Biomedicine, IEEE Trans. on Vol. 8, no. 3*, pp. 279–286 (2004)

[10] Anoto Pen, `http://www.anoto.com/`

[11] H. Noma, A. Ohmura, N. Kuwahara, and K. Kogure. Wearable sensors for auto-event-recording on medical nursing – user study of ergonomic design. In *Proc. the 8th International Symposium on Wearable Computers (ISWC'04)*, pages 8–15, October 2004.

[12] L. Bao and S. S. Intille. Activity recognition from user-annotated acceleration data. In *Proc. of PERVASIVE 2004*, pp. 1–17 (2004).

[13] Kuwahara N., Naya F., Ozaku H. I., and Kogure K.: Context-Awareness in a Real Working Environment: Model for Understanding Nursing Activities, *Proc. of 2nd International Workshop on Exploiting Context Histories in Smart Environments* (2006)

[14] OWL-S: Semantic Markup for Web Services: `http://www.w3.org/Submission/OWL-S/`

[15] J. E. Bardram and T. R. Hansen: The AWARE architecture: supporting context-mediated social awareness in mobile cooperation, *CSCW '04: Proceedings of the 2004 ACM conference on Computer supported cooperative work, New York, NY, USA, ACM Press, pp. 192–201* (2004)

[16] J. E. Bardram, T. R. Hansen, and M. Soegaard: AwareMedia: a shared interactive display supporting social, temporal, and spatial awareness in surgery, *CSCW '06: Proceedings of the 2006 20th anniversary conference on Computer supported cooperative work, New York, NY, USA, ACM Press, pp. 109–118* (2006)

[17] Iseki H., Muragaki Y., Nakamura R., Chernov M., and Takakura K.: Surgical Information Strategy Desk, *4th Symposium on Intelligent Media Integration for Social Information Infrastructure* (2006)

[18] Zander K.: Nursing Case Management: Resolving the DRG Paradox, *Nursing Clinics of North America, 23(3), pp. 503–520* (1988)

[19] Okada O., Ohboshi N., Kuroda T., Nagase K., and Yoshihara H.: Electronic clinical path system based on semistructured data model using personal digital assistant for onsite access, *J. of Med Syst, 29(4), pp. 379–389* (2005)

[20] Okada O, Ohboshi N., and Yoshihara H.: Clinical path modeling in XML for a web-based benchmark test system for medication, *J. of Med Syst, 29(5), pp. 539–553* (2005)

[21] Miller R. H. and West C. E.: The value of electronic health records in community health centers: policy implications, *Health Aff (Millwood), 26(1), pp. 206–214* (2007)

[22] Brandeis G. H., Hogan M., Murphy M., and Murray S.: Electronic health record implementation in community nursing homes, *J. of Am Med Dir Assoc, 8(1), pp. 31–34* (2007)

[23] Kimura M., Ohe K., Yoshihara H., Ando Y., Kawamata F., Hishiki T., Ohashi K., Sakusabe T., Tani S., and Akiyama M.: Patient information exchange guideline MERIT-9 using medical markup language MML, *Medinfo, 9 Pt 1, pp. 433–437* (1998)

[24] Munoz M. A., Rodriguez M., Favela J., Martinez-Garcia A. I., and Gonzalez V. M.: Context-Aware Mobile Communication in Hospitals, *IEEE Computer, September 2003, pp. 38–46*, (2003)

[25] Kimura M., Ohe K., Yoshihara H., Ando Y., Kawamata F., Tsuchiya F., Furukawa H., Horiguchi S., Sakusabe T., Tani S., and Akiyama M.: MERIT-9: a patient information exchange guideline using MML, HL7 and DICOM, *Int J. of Med Inform, 51(1), pp. 59–68* (1998)

[26] Yoshihara H.: Development of the electronic health record in Japan, *Int J. of Med Inform, 49(1), pp. 53–58* (1998)

[27] Heinzelman W., Murphy A., Carvalho H., and Perillo M.: Middleware to Support Sensor Network Applications, *IEEE Network Magazine, Special Issue, Jan* (2004)

[28] Drugge M., Hallberg J., Parnes P., and Synnes K.: Wearable Systems in Nursing Home Care: Prototyping Experience, *IEEE Pervasive Computing, pp. 86–91* (2006)

[29] Holzinger A., Schwaberger K., and Weitlaner M.: Ubiquitous Computing for Hospital Applications: RFID-Applications to Enable Research in Real-Life Environments, *29th Annual Int'l Computer Software and Applications Conference (COMPSAC'05), Vol. 2, pp. 19–20*, (2005)

[30] Takada A., Guo J., Tanaka K., Sato J., Suzuki M., Suenaga T., Kikuchi K., Araki K., and Yoshihara H.: Dolphin project—cooperative regional clinical system centered on clinical information center, *J of Med Syst, 29(4), pp. 391–400* (2005)

[31] Lubrin E., Lawrence E., and Navarro K. F.: Wireless Remote Healthcare Monitoring with Motes *Proc. of Int'l Conf. on Mobile Business (ICMB'05), pp. 235–241* (2005)

[32] Takada A.: *Research on the framework for standard clinical information system,* http://www.mhlw.go.jp/shingi/2004/03/s0319-6a.html (in Japanese)

[33] Noma H., Ohmura R., Naya F., and Kogure K.: Sensor Network Management for Understanding Everyday Activities, *Proc. of the Workshop on Knowledge Sharing for Everyday Life, pp. 17–24* (2006)

[34] Chan A. T. S., Leong H. V., Chan J., Hon A., Lau L., and Li L.: BluePoint: a bluetooth-based architecture for location-positioning services, *Proc. of the 2003 ACM symposium on Applied computing (SAC 2003), pp. 990–995* (2003)

[35] Ozaku H.I., Abe A., Sagara K., Kuwahara N., and Kogure K.: A Task Analysis of Nursing Activities Using Spoken Corpora. Advances in Natural Language Processing, *Research in Computer Science, Vol. 18, pp. 125–136* (2006)

[36] Miyamae M., Naya F., Noma H., Toriyama T., and Kogure K.: A Trial Design of an Information Display Method for Medical Nursing, *Adjunct Proc. of 18th International Conference on Ubiquitous Computing (UbiComp 2006) Poster Session* (DVD-ROM) (2006)

[37] Resource Description Framework (RDF), http://www.w3.org/RDF/

[38] OWL Web Ontology Language Guide, http://www.w3.org/TR/owl-guide/

[39] SWRL: A Semantic Web Rule Language Combining OWL and RuleML., `http://www.w3.org/Submission/SWRL/`

[40] Abe A. and Kogure K.: E-Nightingale: Crisis detection in nursing activities, in *Chance Discoveries in Real World Decision Making (Ohsawa Y. and Tsumoto S. Eds.), Data-based Interaction of Human intelligence and Artificial Intelligence Series: Studies in Computational Intelligence, Vol. 30, pp. 357–371* (2006)

[41] Abe A., Ozaku H. I., Kuwahara N., and Kogure K.: Scenario Violation in Nursing Activities—Nursing Risk Management from the viewpoint of Chance Discovery, *Soft Computing Journal, Vol. 11, No. 8, pp. 799–809* (2007)

[42] Poole D., Goebel R. and Aleliunas R.: Theorist: A Logical Reasoning System for Defaults and Diagnosis, The Knowledge Frontier: Essays in the Representation of Knowledge (Cercone N.J., McCalla G. Eds.), *pp. 331–352*, Springer-Verlag (1987)

[43] Ohsawa Y., Okazaki N., and Matsumura N.: A Scenario Development on Hepatics B and C., *Technical Report of JSAI, SIG-KBS-A301, pp. 177–182* (2003)

[44] Naya F., Ohumra R., Takayanagi F., Noma H., and Kogure K.: Workers' Routine Activity Recognition using Body Movements and Location Information, *Tenth IEEE International Symposium on Wearable Computers (ISWC2006), pp. 105–108* (2006)

Chapter 3

A Taxonomy of Service Discovery Systems

Vasughi Sundramoorthy

Lancaster University, Lancaster, Lancashire, LA1 4WA, UK

Pieter Hartel and Hans Scholten

University of Twente, Enschede, 7500AE, The Netherlands

Abstract

Service discovery is a platform for network entities, whether hardware or software, to spontaneously self-configure and self-heal in a volatile environment. An in-depth understanding of the fundamentals of service discovery is essential for system architects to effectively execute pervasive and autonomous context-aware applications. We offer this understanding by clarifying some fundamental concepts and issues in service discovery; the evolution of service discovery in the context of distributed systems, the basic architectures and their variance and the different functionalities that contribute towards satisfying the self-configuring and self-healing properties of service discovery. We then proceed to show how service discovery fits into a system, by characterizing operational aspects. Subsequently, we describe how existing state of the art performs service discovery, in relation to the operational aspects and functionalities, and identify areas for future work.

3.1   Introduction

Computer scientists can learn much from how the human body manages itself autonomously, and apply the same techniques to building distributed systems. This is how *autonomous computing*, an initiative of IBM [1], sees the future of computer systems. An autonomous system has *at least one* of the following four properties [2]:

1. Self-configuring. Systems that adapt automatically to dynamically changing environments. The systems can dynamically add ("on-the-fly") new hardware and software to the system infrastructure with no disruption of services.

2. Self-healing. Systems discover, diagnose and react to disruptions. The objective of self-healing is to minimize outages to keep applications available at all times.

3. Self-optimizing. Systems monitor and tune resources automatically. Self-optimization requires hardware and software systems to maximize resource utilization to meet end-user needs without human intervention. Resource allocation and workload management must allow dynamic redistribution of workloads to systems that have the necessary resources to meet workload requirements.

4. Self-protecting. Systems anticipate, detect, identify and protect themselves from attacks. Self-protecting systems must have the ability to define and manage access to computing resources, to protect against unauthorized access, to detect intrusions and report and prevent these activities as they occur.

Autonomy of distributed systems is also one of the fundamental characteristics of the more ambitious *pervasive (or ubiquitous) computing*. The vision of pervasive computing is elegantly articulated in Mark Weiser's acclaimed seminal paper published in 1991 by Scientific American [3]:

"The most profound technologies are those that disappear. They weave themselves into the fabric of everyday life until they are indistinguishable from it."

To realize the Weiser vision of pervasive computing, Satyanarayanan [4] poses system *invisibility* as an important research challenge. Invisibility in the context of pervasive systems means human intervention is so minimal, that technology disappears into the background of everyday life. Therefore, a pervasive system continuously meets user expectations and rarely presents unpleasant surprises.

Invisibility is achieved when the system applies autonomous behaviors: self-configuration, self-healing, self-optimization and self-protection. One of the most widely used techniques contributing to self-configuration and self-healing is *service discovery*. Service discovery is a fundamental step for intelligent applications, before they can collaborate to perform a certain function; entities are able to self-configure themselves by detecting other entities (hence, services), and are able to self-heal from failures, without the necessity for a professional system administrator.

One area of pervasive computing that is tightly coupled with service discovery is *context-aware computing*. According to Schilit et al. [5], context is "where you are, who you are with, and what resources are nearby," while Dey defines context "as any information that can be used to characterize the situation of an entity. An entity is a person, place, or object that is considered relevant to the interaction between a user and an application, including the user and applications themselves." [6]. Both definitions refer in some way towards consolidating information from different network entities to form the context for application execution. Discovering different network entities is a functionality offered by service discovery systems. Furthermore, many context-aware applications automatically execute their tasks for, or on behalf of the user; thus invisibility again becomes a critical requirement. Thus service discovery is an important medium for context-aware applications so that required services are discovered with minimum user intervention.

An example of a context-aware application that uses service discovery is an ambience controller for the home that uses time, temperature and location as inputs, to satisfy a particular condition: "in the *evening*, when the outdoor temperature is *hot* and *Alice is home*, set the mood of the home to *cool*." As a result of this condition, the ambience controller discovers a number of services (e.g., clock, location of Alice, temperature, lighting, audio, window) to receive inputs and to produce the output. Subsequently, when the time and temperature is appropriate, and Alice is home, the indoor temperature is reduced, lights are dimmed, windows are shut, curtains drawn and chamber music is played. One example of a context-aware home is the Aware Home project [7].

Another innovative context-aware application is a city tour guide system that enables PDAs or mobile phones to become aware of their location and orientation and present information about interesting sites based on the visitor's personal preferences found in her PDA or mobile phone. Besides the tourist attraction information, the visitor can also be offered other relevant information such as hotel booking, taxi information, and theater shows. Meanwhile the information from the PDA or mobile phone, such as location, is used by the different entities around the user to tailor the incoming information so that users are presented with the most convenient options. For such a system to be effective, the PDA or mobile phone should be able to discover nearby tourist spots, historical exhibits, hotel information, taxi companies and theaters. The PDA or mobile phone should also be discoverable by all these entities. To realize such a scenario, each entity would need to have a commonly understandable service interface that can be discovered and accessed, so the collaborating entities can present the most relevant information to the visitor. The Guide project conducted in Lancaster, UK implements some aspects of this application scenario [8].

To summarize, a service discovery system offers the handle on available network entities and the information they offer, to pervasive, autonomous and context-aware applications, by discovering required and available services, while providing a certain extent of invisibility through self-configuration and self-healing capabilities.

Contributions: This work diverges from existing surveys [9, 10, 11, 12] which categorize functional features of service discovery protocols, based on architectural and programming platform differences. Meanwhile the surveys done for mobile ad-hoc networks such as by Ververidis and Polyzos [13] center mainly on the different message propagation methods for multi-hop networks and cross-layer optimizations. Our work is more generic, where we aim to:

1. Simplify the fundamental concepts of service discovery systems

2. Specify the operational aspects that impact the design of a service discovery system, and the resulting design solutions.

3. Provide a taxonomy that first analyzes state of the art solutions with respect to the operational aspects, before comparing functional behaviors.

We hope that our work will contribute towards offering system architects and researchers in general an understanding of the fundamentals of service discovery, while presenting them with the advantages and limitations of existing state of the art so that they can decide on the system that suits their requirements the best.

This work focuses more on the *communication aspect of service discovery systems*, such as the type of architecture, topology, functional behaviors of the entities in the system and the operational aspects that impacts the design of the system. How services are described and accessed are issues outside the scope of this chapter. More information on service descriptions can be found by studying existing schemes such as WSDL [14], OWL [15], SSDL [16] and RDF [17]. Meanwhile, RPC [18] and service interface invocation as done in Java [19] are methods to access services once they are discovered. The description and access techniques can be integrated into the service discovery systems described in our work.

The chapter is organized as follows. In Section 3.2, we provide a new understanding of service discovery as a third generation name discovery system. In Section 3.3, we describe the service discovery system in terms of the participating entities, architecture and topology. We define the service discovery objectives and functions in Section 3.4. In Section 3.5, we analyze the operational aspects for a service discovery system. We proceed to summarize a selection of widely known service discovery systems in Section 3.6. In Section 3.7, we give a taxonomy of state of the art solutions to the operational aspects, and compare the functional implementations. Finally we conclude and identify areas for improvement.

3.2 Service Discovery: Third Generation Name Discovery

Service discovery enables invisibility in pervasive and autonomous systems, both of which results in the development of futuristic applications that can self-configure and self-heal without user intervention. However, fundamentally, service discovery is a distributed system that discovers and shares network entities. In fact, distributed systems that discover and share network entities have been around for more than two decades. We refer to this class of distributed systems as *name discovery systems*. In this section, we give an understanding on the evolution of service discovery.

In name discovery systems, every entity, whether hardware or software, is give a name. A *name* is a string of bits that is used to identify a variety of entities, such as computers, peripherals, applications, remote objects, files, etc. In the context of name discovery systems, a name is not the address of an entity (addresses are uninterpreted bit patterns such as Ethernet addresses [20]), but a name is the persistent identifier for the entity, or human understandable textual description [21]. Consistently named entities enable computers to communicate with one another via a distributed system, and share (access to) the entities [22]. To describe an entity, the name has a list of *attributes* associated with it. An attribute is a name-value pair that describes a property of the entity. For example, a device named "temperature sensor" may have an attribute "current temperature in Celsius" with a value of "27".

We classify *name services*, such as Grapevine [23], GNS [24] and DNS [25], developed in the 1980s as first generation name discovery systems. A *name server* stores a set of bindings between the name and the attribute list of an entity, and resolves queries for the entity. A query based on the name simply returns the list of attributes that describes the entity.

Directory services such as Profile [26], Univers [27], X.500 [28], LDAP [29] and CORBA's Interface Repository [30] developed in the 1990s are classified as second generation name discovery systems. Directory services provide a more powerful mechanism for querying entities. Directory services perform *attribute-based queries* that return the names associated with the attribute.

An example is a query that on the basis of a given telephone number returns the name of the associated employee.

The first two generations of name discovery systems enable the discovery and sharing of network entities. But they certainly cannot be categorized as "self-configuring" systems. As previously defined, a self-configuring system is a system that *adapts automatically to dynamically changing environment*, where new network entities are added to the system infrastructure with no disruptions [2]. The first two generations of name discovery systems are limited to a more stable environment, where connectivity is predominantly wired, nodes are mostly static, names and attributes rarely change and the systems are reliable. These systems *heavily depend on infrastructures* that consist of servers and directories that require configuration and maintenance by privileged system administrators. Therefore, the context of these two generations of name discovery systems is best suited for PC and server based environment.

Service discovery systems inherit the fundamental concepts of traditional name discovery systems, where entities are named according to a naming standard, and attribute-based queries return the names of the matching entities. A definition of a service that comes close to our understanding of *what is a service* is given by Coulouris and Dollimore [22] :

A service is a distinct part of a computer system that manages a collection of related entities and presents their functionality to users and applications.

We classify service discovery as a third generation name discovery system because service discovery satisfies the requirements of pervasive computing [3], by *enlarging the context and relaxing the limitations of traditional name discovery systems*. Unlike the stable, static and infrastructure based context of the first two name discovery systems, service discovery systems are best suited for an environment consisting of a variety of embedded devices (not just PCs and servers), with wired and wireless connectivity, static and mobile nodes, which undergoes frequent changes of resource availability and attribute values, and is vulnerable to failures. Succinctly put, service discovery provides **self-configuring** and **self-healing** capabilities for a **spontaneous network of devices**. A service discovery system allows entities to enter and leave the system automatically, with *minimum supervision and maintenance*. Service discovery systems also support access to services, where discovered service descriptions may contain executable programs, or URLs to the services.

Therefore, service discovery is a solution for naming and discovering resources in a versatile and spontaneous network of devices. Examples can be found in the relatively new areas of home entertainment networking and ambient intelligence.

In the rest of the chapter, we flesh out how service discovery systems implement self-configuring and self-healing capabilities. We will delve into the fundamental architectural considerations which results in different topologies and the functionalities that enable the automatic discovery of network entities.

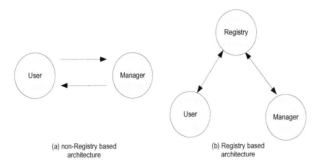

FIGURE 3.1: (a) The non-Registry architecture consists of Users and Managers that multicast queries and service advertisements. (b) The Registry architecture uses unicast for registering services and sending queries.

3.3 Service Discovery Architecture

A self-configuring system can dynamically add new network entities without any disruptions and with minimum (if at all) user intervention. In service discovery systems, network entities can discover each other at any point along the system lifetime. A service discovery architecture establishes how this is done.

Before we reveal the different service discovery architectures, let us first introduce the key elements in service discovery, along with their terminologies.

A service is specified by a *Service Description (SD)*, which typically describes the service in terms of: (1) device type (e.g. printer), (2) service type (e.g. color printing) and/or (3) attributes (e.g. location, paper size). There are three types of entities in a *service discovery system*; a *Manager* owns the SDs, a *User* has a set of requirements for the services it needs and a *Registry* caches available services so that Users can discover the services through *queries* to the Registry. A node can behave as a User, Manager, Registry or a combination of these roles.

There are several design considerations for service discovery architectures, which we will explore in detail in Section 3.5. Suffice to say for now, that most service discovery design revolves around how the User discovers a service in the system efficiently and effectively, so that the system is optimized for design considerations such as scalability, resource-constrained devices (memory, processing power, energy, etc), robustness, etc.

We simplify service discovery architectures according to how the User discovers a service: (1) Users can use a Registry (or a cluster of Registries) as the intermediate entity to discover services, therefore requiring a *Registry based architecture*. (2) Users can directly query Managers, and/or Managers can send multicast *advertisement* messages, therefore establishing a *non-Registry based architecture*. Figure 3.1 illustrates the two types of basic architectures. In the Registry-based architecture, Registries can be deployed by a system administrator, or automatically elected by the nodes in the system. Once the Registry is available, the rest of the nodes in the system will have to discover the Registry before services can be registered and queried. Unicast communication in the Registry-based architecture reduces network traffic, thus *increasing scalability*. In the non-Registry-based architecture, Users and Managers can perform multicast queries and service advertisements. Therefore, unlike the Registry-based architecture, the system is *not vulnerable to single point of failure* issues. However, since extensive multicast is used, network traffic increases, causing scalability issues. A combination of both these architectures can also be implemented, so that the system is more scalable and robust.

In the next section, we simplify the variances in both the Registry and non-Registry based architectures.

3.3.1   Logical Topologies (Overlays)

Service discovery entities habitually communicate with each other through *logical topologies* (also commonly known as *overlays*). A logical topology is typically used to optimize the way the system propagates and processes messages (such as queries and advertisements), thus optimizing one or more of the following: communication cost, energy efficiency, scalability, resource consumption, robustness, responsiveness and effectiveness (of the queries). We simplify existing logical topologies into four basic variances, based on query or advertisement message propagation: (1) *Meshed topology*, where messages are sent to all listening entities, (2) *Clustered topology*, where messages are sent to a cluster of listening entities, based on the type of service they provide, or the proximity to the source, (3) *Tree topology*, where messages are propagated along a hierarchy of Registries, (4) *Unconnected Registries topology*, where messages are sent to any discovered Registry. We describe how the non-Registry and Registry-based architectures use these topologies, and their advantage and disadvantage in the next few paragraphs.

3.3.2   Non-Registry Topologies

Non-registry architecture is inherently robust because it does not depend on a single point of reference, but it does result in high network traffic. To reduce network traffic, thus increasing scalability, communication can be restricted within a cluster (group) of entities. We identify two main topologies for the non-Registry architecture: (1) Meshed topology, as shown in Figure 3.2(a), where all entities receive each other's multicast queries and service advertisements. (2) Clustered topology, where entities form clusters based on some criteria (e.g. service type or location). Members of a cluster communicate only with each other; thus service advertisements and queries are limited within

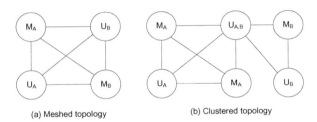

FIGURE 3.2: Logical non-Registry topologies. (a) In the meshed topology, Users, U, and Managers, M, can listen to each other's queries and service advertisements. (b) In the cluster-based topology, Users, U, and Managers, M, may form a logical cluster according to some criteria. A and B denotes two different clusters, where $U_{A,B}$ belongs to both clusters, and is able to discover services of both clusters.

the cluster. Figure 3.2(b) gives an example of the cluster-based topology.

The meshed topology improves the chances of discovering a service (queries are more effective), and the continued availability of the service, because it is not vulnerable to single point of failure issues [31]. However, it increases communication cost, and is less scalable because of the extensive use of multicast (queries and advertisements). On the other hand, the clustered topology makes the system more scalable, but increases the complexity of the system, because Managers and Users will have to establish clusters, and dynamically add and remove their membership. Furthermore, by limiting a query and advertisement to the members of a cluster, the effectiveness of service discovery is reduced. The scope of discovering a service can be widened by allowing the User to belong to a combination of clusters.

3.3.3 Registry-Based Topologies

The main benefit of the Registry architecture is scalability. However implementing Registries increases system complexity, as Registries may need to synchronize data among each other, while the network will have to be logically partitioned based on which Registry the entities associate themselves with. Furthermore, Registries will have to be automatically created, so that the system remains true to the principles of self-configuration. Meanwhile, to increase robustness, redundancy through multiple copies of the Registry is required.

In essence, the Registry architecture has four types of logical topologies: (1) An unconnected Registry topology, as shown in Figure 3.3(a). In this topology, Registries do not communicate with each other, but Managers and Users may associate themselves with multiple Registries. (2) A meshed Registry topology, as shown in Figure 3.3(c), where Registries communicate with each other as peers. A Registry forwards queries and replicas of its cache

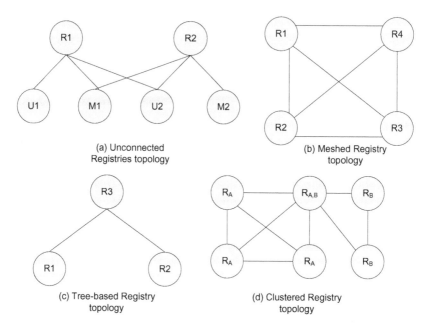

(a) Unconnected
Registries topology

(b) Meshed Registry
topology

(c) Tree-based Registry
topology

(d) Clustered Registry
topology

FIGURE 3.3: Logical Registry topologies. (a) In the unconnected Registry topology, Registries $R1$ and $R2$ do not communicate with each other, but User, $U1$, and Manager, $M2$, may register and discover services from both $R1$ and $R2$. (b) In the meshed Registry topology, Registries are peers to each other, and forward messages to all their peers. (c) In the tree-based Registry topology, Registries $R1$ and $R2$ are child Registries of $R3$. Child Registries may forward messages to parent Registries. (d) In the clustered Registry topology, Registries optimize the tree or mesh topology by limiting query processing to a select few Registries. A and B are two clusters, where Registries R_A and R_B can only communicate with the members of their own cluster. $R_{A,B}$ is able to communicate within both clusters.

to all its peers. (3) A tree-based Registry topology, where Registries form a parent-child relationship, based on some criteria (such as location or resource-constraints). A child Registry, such as shown in Figure 3.3(b), forwards queries to its parent when it does not find matching services within its own cache. The query may traverse all or parts of the hierarchy, depending on some query processing optimization criteria. (4) A clustered Registry topology, as shown in Figure 3.3(d), where Registries form clusters based on service type or location. This topology optimizes the tree-based and meshed topology, where query processing is done only by a select few Registries.

The unconnected Registry topology does not require Registries to synchronize registration data among each other. Therefore, adding a new Registry is not a complicated task. Redundant Registries also provide increased robustness against communication and Registry failures. However, when Users and Managers redundantly communicate with each discovered Registry, communication cost is increased, and scalability is reduced. The other three topologies allow the Manager and the User to communicate with a single Registry only, keeping the service discovery task simple on the side of the Managers and Users (especially suitable for resource constrained nodes). The meshed Registry topology allows all the Registries in the system to communicate with each other, so that the scope of service discovery is system-wide. However this topology is not practical for large systems that span long distances. The tree-based Registry and clustered Registry topology would be more suited for large systems. The tree-based Registry topology allows each Registry to store only a part of the available services, therefore allowing load balancing for systems with high density of nodes (such as by spawning child Registries to share the load). However, the system becomes more vulnerable to single points of failure issues when Registries fail. Registries will have to check continuously on the availability of the parent or child Registry. The clustered Registry topology is more robust against Registry failures, and has better query effectiveness because multiple Registries can respond to the query, from different areas of the system. However, Registries will have to be able to establish and dynamically add and remove memberships to the clusters, causing additional system overhead and complexity.

A service discovery system may implement one or a combination of several logical topologies. Therefore, it is necessary to define and prioritize the optimization parameters early in the system design stage, to satisfy the requirements of the system. We will discuss some of these design considerations in Section 3.5 and how the use of these topologies is part of the design solution.

h

3.4   Service Discovery Functions

In the previous section, we fleshed out the generic architectures and topologies (overlays) in service discovery systems, so that we can give the reader an overall picture of the service discovery system: the Manager, the Users, the Registry (if available) and the rationale for arranging these entities in a certain logical topology. In this section, we summarize the basic functionalities of these entities, and the relationship between them. However, we first explain why it is necessary to outline service discovery functionalities.

As mentioned in Section 3.1, context-aware computing is tightly coupled with service discovery because service discovery gives a handle on available and required context; context-aware applications consolidate information from different network entities to form the context for application execution. Therefore, to execute context-aware applications effectively, all required context must be discovered, and the information should be up-to-date. As a result, we identify the following as the two main principles which establish the self-configuration and self-healing properties of service discovery:

1. *Discover services that match requirements (self-configuration)*

2. *Detect changes in service availability and attributes (self-healing)*

Since application designers are responsible for programming the logic for context inference, and specifying the actions to be taken by the applications, it is essential to clarify *service discovery functionalities* so that application designers are unambiguous on what exactly the service discovery system is capable of. If a chosen service discovery system is found to be missing one or two functionalities, it would be vital for the applications to absorb them as part of their own functionalities, so that the overall result still adheres towards satisfying to some extent the "invisible" property of pervasive and autonomous systems. This allows context-aware application to automatically execute tasks for or on behalf of the user, without troubling the user.

To ensure a clear definition of service discovery functionalities, we classify service discovery tasks into four main *functions*: Configuration Discovery, Service Registration, SD Discovery and Configuration Update. The term "configuration" refers to the entities in the system, Manager, User and Registry, and the services they offer. The Configuration Discovery, Service Registration (for Registry based architectures) and the SD Discovery functions are required for entities in the system to gather knowledge on the availability of nodes and services; thus they contribute towards self-configuration (principle 1). Meanwhile the Configuration Update function is required for the system to detect changes in service availability (services can suddenly disappear, or their attributes change). Thus this function contributes towards self-healing (principle 2). Each of the four functions can be accomplished using several different *methods*. We briefly summarize each of the function below (italics indicate the methods):

1. **Configuration Discovery** - This is the function that realizes the "self-configuration" aspect of service discovery systems. Here, Registries are setup, and identities of entities (e.g. Registries or cluster members) in the system are discovered automatically, without a system administrator. There are two sub-functions of Configuration Discovery:

   (a) Registry auto-configuration - Allows the system to self-configure one or more Registries through (a) *Registry election* algorithms, or (b)*Registry reproduction*, where a parent Registry spawns a child Registry. The Registry election or reproduction is done based on some criteria such as resource superiority, load threshold, service type or location. Registry auto-configuration is done on the fly, without supervision.

   (b) Entity discovery - Allows entities in the system to discover a Registry or cluster members automatically through (a) *active discovery*, where nodes initiate the discovery by sending announcements, or (b) *passive discovery*, where nodes discover the required entities by listening for announcements. In some systems, discovery via active and passive methods is integrated with the underlying routing protocol to optimize bandwidth utilization (such as in mobile ad-hoc networks).

2. **Service Registration** - This function allows Managers to register their services at a Registry, once the Registry is discovered via the Entity Discovery function. Registration methods include (a) *unsolicited registration*, where entities request the Registry to register their services and (b) *solicited registration*, where Registries request new entities to register their service descriptions (SD). The Registry keeps a cache of available SDs, and updates them according to requests from the Managers.

3. **SD Discovery** - This function allows Users to obtain SDs that satisfy their set of requirements. Users may cache the discovered SDs to reduce access time to the service, and reduce bandwidth utilization by avoiding multiple queries. There are two sub-functions in SD Discovery:

   (a) Query - This is a pull-based model where Users initiate (a) *unicast query* to a Registry, or (b) *multicast query*. The query specifies the requirements of the User. The Registry or Manager that holds the matching SD replies to the query.

   (b) Service notification - This is a push-based model, where Users receive (a) *unicast notification of new services* by the Registry, or (b) *multicast service advertisements* by Managers.

Table 3.1: Service discovery functionalities

Autonomous Property	Function	Subfunction	Method
Self-configuring	Configuration Discovery	Registry auto-configuration	(a) Registry election, (b) Registry reproduction
		Entity discovery	(a) active discovery, (b) passive discovery
	Service Registration		(a) solicited registration, (b) unsolicited registration
	SD Discovery	Query	(a) unicast query, (b) multicast query
		Service notification	(a) Registry notification, (b) multicast service advertisement
Self-healing	Configuration Update	Configuration Purge	(a) leasing, (b) advertisement TTL
		Consistency Maintenance	(a) pull-based polling for update by Users, (b) push-based update notification by Registry to Users, (c) push-based update notification among Registries

4. **Configuration Update** - This function monitors the node and service availability, and changes to the service attributes. Therefore, this is the function that implements "self-healing" aspects of service discovery. There are two sub-functions in Configuration Update:

   (a) Configuration Purge - Allows detection of "disconnected" entities through (a) *leasing* and (b) *advertisement time-to-live (TTL)*. "Disconnection" here includes connectivity breakdown (such as when nodes move out of range), nodes that have suddenly disappeared from the network due to failures or power outage, nodes that have been tagged by the application as unreliable, etc. Since the result of all these scenarios are end of communication, we abstract these incidents as "disconnection". In leasing, the Manager requests and maintains a lease with the Registry, and refreshes the lease periodically. The Registry assumes that the Manager who fails to refresh its lease has left the system, and purges its information. With TTL, the User monitors the TTL on the advertisement of a discovered Manager. The User assumes the Manager has left the system if the Manager fails to advertise before its TTL expires, and purges its information.

   (b) Consistency Maintenance - Allows Users and Registries to detect updates on cached SDs. Updates can be propagated using (a) push-based *update notification*, where Users and Registries receive notifications from the Manager, or (b) pull-based *polling for updates* by the User to the Registry or Manager for a fresher SD. (c) In a multiple Registry topology, push-based *update notifications among Registries* can be done to achieve consistency.

Table 3.1 summarizes service discovery functions and the implementation methods for each function. Every service discovery system implements the functions according to its own design rationale. We will discuss how selected state of the art systems implement service discovery functionalities in Section 3.6.

3.5 Operational Aspects of Service Discovery

In Section 3.3, we illustrated the different architectures and logical topologies, so that the system can be optimized for scalability, robustness and resource constraints. In this section, we explore in more detail the different operational considerations for designing service discovery systems and list solutions found in the wider distributed system paradigm to suit the service discovery context.

The operational environment influences the design rationale of service discovery systems. For example, a stable, wired office environment, with good system administration may not require too much emphasis on fault-tolerance towards communication and node failures. However, in the context of a less controlled environment such as the home, it becomes a necessity, because home owners are not restricted in how they manage their appliances (unplugging, moving). In a wireless, mobile environment, the system becomes even more vulnerable to certain communication and node failures.

We simplify the operational aspects into five main categories: system size, lossy environment, resource constraints, security and system heterogeneity.

1. **System size** - We define "system size" in terms of two dimensions: distance and the number of nodes. Small sized systems such as *Personal Area Networks (PAN)* and *Local Area Networks (LAN)* contain a limited number of nodes, and do not require a high degree of scalability. Large systems such as *Metropolitan Area Networks (MAN)* and *Wide Area Networks (WAN)* including the Internet require a scalable service discovery system. Scalability measures include setting up multiple Registries, whether in a tree or mesh topology, and applying query optimization and load balancing techniques to conserve bandwidth.

2. **Lossy environment** - Service discovery systems in wireless and mobile networks must assume that they will operate in a lossy environment with communication and node failures. *Communication failures* include message corruption, message loss and link failures. Message corruption is due to interference, noise or multipath fading. Message loss is due to loss of signal caused by physical obstacles, collisions, bandwidth limitations, etc. Link failures, especially in ad-hoc networks, are caused by mobile nodes losing radio contact with the destination node. *Node failures* include crash failures and interface failures. Crash failures are caused when nodes abruptly disappear from the system due to energy depletion, pulled out without warning, and overloaded processors (nodes simply stop communicating). Interface failures mean receiver and transmitter failure. Therefore, service discovery systems should be fault-tolerant. Some examples of fault-tolerant mechanisms in service discovery systems include redundant and replicated Registries, caching of alternate services, primary-based recovery protocols such as Registry monitoring and Registry backup [21], retransmissions and acknowledgments for reliable transmission and containment of unreliable behaviors by blacklisting suspicious nodes.

3. **Resource constraints** - Nodes with hardware constraints are *resource-lean* (low memory, processing power and energy). Systems with resource-lean nodes require resource-aware service discovery functions. One solution is to delegate more tasks to more powerful nodes. In systems with *low bandwidth availability*, cross-layer dependencies such as service

discovery with routing knowledge, and efficient query processing among Registries (e.g. through DHTs) can help conserve bandwidth. Furthermore, load balancing techniques help scale Registry-based architectures so that Registries do not overload.

4. **Security** - A secure service discovery system must support *confidentiality, message integrity and availability* [32]. Methods to address these concerns include authentication of communicating entities, access control so only a select few are able to communicate, protection of sensitive service attributes (e.g. location) by hiding the value, data integrity, so that communicating entities can detect when data are tampered during transit, and detection and blacklisting of malicious nodes (including authorized entities). The challenge for a secure service discovery system is to maintain self-configuration of the system, because the owner of the devices will most probably be required to provide authentication and access control. Security also consumes resource due to encryption algorithms. Most service discovery systems assume participating nodes are secure by delegating security to the application layer. However, full fledged deployment of a service discovery system will eventually require some secure measures integrated into the service discovery functions [33, 34]

5. **System heterogeneity** - Nodes in heterogeneous systems contain different types of network connectivity (e.g. Ethernet, 802.11 a/b/g, IRDA), and a variety of network stacks (e.g. transport, routing, addressing). A service discovery protocol that abstracts away as much as possible the lower-layer protocol stacks, and can perform its functions with minimum dependencies allow easier deployment in a heterogeneous environment.

Figure 3.4 summarizes the five operational aspects and the respective design solutions. By taking the operational aspects into consideration, it is possible to design a system that addresses more than one type of operational issue. For example, an architecture with replicated and redundant Registries supports a large and lossy system. State of the art systems usually base their design rationale on their own set of priorities for the design aspects, hence causing tradeoffs.

3.6 State of the Art

Having described the nature of service discovery from the point of view of (a) the architecture, (b) the functionality and (c) the operational aspects, we now investigate how existing service discovery systems fit into this mould.

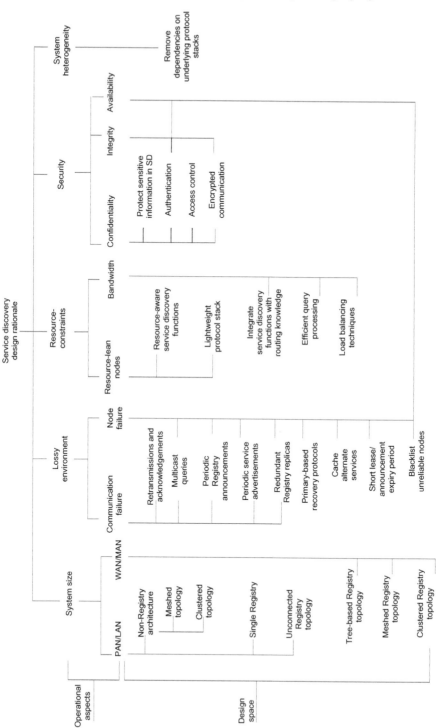

FIGURE 3.4: Summary of operational design aspects and solutions, tailored for service discovery. The design rationale for a service discovery system depends on its own relevant set of operational aspects.

We provide summaries of selected state of the art service discovery systems. We choose to describe these systems because of their popularity in the type of network and the size of system that they support. For ease of understanding, we will maintain the terms Manager, User and Registry to represent the protocol-dependent entities, even though the original papers use slightly different terms.

We divide the systems into two categories, based on the targeted system size: (1) *small systems*, which include LAN and PAN, with limited number of nodes, and (2) *large systems*, which include WAN and MAN, with a large number of nodes. The system size has the most influence on the design decisions in existing state of the art, where they implement similar service discovery functions and methods, and Registry topology (e.g. in large systems, the Registry-based architecture is chosen, where Registries are replicated, and arranged in either tree or mesh topology).

Unless mentioned specifically to support ad-hoc networks, the systems work in infrastructure-based wireless networks, and, by default, also work on wired networks.

3.6.1 Small Systems

Small systems are not usually concerned with scalability issues. The architecture type can be Registry, non-Registry or a combination of Registry with multicast query capability for Users (for resilience against single point of failure problems). Furthermore, as bandwidth utilization is not a critical issue in small systems, a strong support for consistency maintenance can be implemented, provided that the nodes in the system are not frequently moving in and out of the system.

1. Jini [35] - Jini was developed by Sun Microsystems, and is implemented using the Java programming language. Jini is a Registry architecture, where the Registry is called the *Lookup Service*. The Manager registers its service at the Registry, by uploading the service proxy code. The data stored are typically a part of a structured distributed shared memory, using tuple spaces, implemented through JavaSpaces [36]. The User queries the Registry for services matching its requirements, and receives the proxy code of the service. The User also requests the Registry to notify it if similar services register in the future. The Manager maintains a lease with the Registry, where it periodically refreshes the lease to indicate its continuous existence. The Registry automatically purges the information on the Manager that failed to refresh its lease. If the Manager updates its service description, it publishes an event to the Registry. The Registry propagates the event to interested Users. The use of Java allows code mobility and operating system flexibility for Jini devices. However, Jini uses the Java Virtual Machine (JVM) and Java Remote Method Invocation (RMI), and depends on TCP/IP for

reliable communication. These technologies cause dependencies on the underlying protocol stacks.

2. UPnP [37] - Universal Plug and Play (UPnP) was developed by Microsoft, and is based on the Simple Service Discovery Protocol [38]. UPnP is a non-Registry based architecture. The User is called the *Control Device*, and the Manager is simply called the *Device*. Service Description is described in XML. The Manager sends multiple multicast messages periodically to announce its presence and its services. The User also sends multicast queries to request services matching its requirements. The Manager sends XML documents to the User. The XML document provides the device and service descriptions along with URLs to view the user interface of the Manager. The User controls the remote service through the Simple Object Protocol (SOAP) [39] and XML parsing of action requests. The Manager updates its service through General Event Notification Architecture Base (GENA) [40]. GENA publishes notifications to subscribers. The Manager periodically sends multicast announcements, which is used by the interested User to monitor the continued existence of the Manager. The non-Registry based architecture eliminates single point of failure issues, and supports mobility. However, it increases network traffic due to extensive use of multicast messaging. Like Jini, dependencies on IP technologies cause network dependencies.

3. SLP [41] - The Service Location Protocol (SLP) was developed by the IETF SvrLoc working group. It is a combination of Registry and non-Registry architecture. The User is called the *User Agent*, the Manager is the *Service Agent* and the Registry is the *Directory Agent*. When a Registry exists, the Manager registers its Service Description at the Registry and the User queries for services matching its requirement. SLP provides filters that allows attribute and predicate string search. When the Manager updates its Service Description, it re-registers at the Registry. The User has to query the Registry periodically if it wants to discover the update. When the Registry is unavailable, the User can send multicast queries to discover the Manager. The Manager also periodically refreshes its registration by re-registering its data. If the Manager fails to refresh its registration on time, the Registry removes the data, and assumes that the Manager is no longer available in the system. A typical implementation of SLP depends on reliable TCP/IP. SLP provides basic service discovery functions, with limited consistency maintenance support. Unlike Jini, the combination of Registry and non-Registry based architecture reduces single point of failure issues.

4. Bluetooth SDP [42] - Bluetooth was developed by the Bluetooth Special Interest Group, an industry consortium consisting of companies like

Ericsson, Nokia and IBM. It is meant for low power, short range (within 10m), wireless (ad-hoc) radio system devices (PAN network) operating in the 2.4GHz ISM band. The Bluetooth Service Discovery Protocol (SDP) depends on the underlying connectivity; thus we first describe how devices establish their connectivity. Bluetooth devices periodically sniff for nearby Bluetooth devices and form a personal area network called piconets which has a maximum of 8 members. The member that initiates communication acts as the master of the piconet. Additional devices are supported by reusing addresses of a silenced existing member on a new member. Groups of piconets communicating with each other are called scatternets. The Bluetooth SDP [43] is a non-Registry based architecture. The Manager runs an *SDP server*, while the User runs an *SDP client*. Services are divided into classes. Each service is represented by a *service record*. The User sends a query for a particular service type, and receives a response from the Manager if it offers a matching service. The User detects that the Manager is no longer available when it does not receive a response to a request. The Bluetooth SDP has simple basic functions, where there is no leasing or subscription to receive updates. The tight dependency on the underlying protocol layers makes the Bluetooth SDP unsuitable for stand-alone deployment.

Other state of the art that fall in the "small systems" category include FRODO [44], Salutation [45], Konark [46] and DEAPspace [47]. FRODO and Salutation has an architecture similar to SLP, where it is both Registry and non-Registry-based. FRODO is a single Registry architecture for the home environment, where the Registry is elected. FRODO emphasizes robustness and resource-awareness, where guarantees of service delivery is offered, and entities perform service discovery based on their resource limitations. Unlike the Registries in SLP and Jini, the Registries in Salutation can query each other, forming a meshed Registry topology. Konark is similar to the non-Registry based UPnP architecture, but targets ad-hoc networks specifically. Users and Managers send multicast service advertisements and queries. Konark also uses HTTP, SOAP and XML like UPnP for service descriptions. However, Konark reduces multicast network traffic by using a gossip protocol for advertising services, instead of the more conventional periodic muticast advertisements used in UPnP. DEAPspace is targeted for a smaller system than in Konark or UPnP. DEAPspace is built for a single-hop ad-hoc network, similar to Bluetooth SDP. However unlike Bluetooth SDP, Users in DEAPspace cache the service descriptions of all Managers within their vicinity, and periodically send multicast messages containing the list of cached services. Managers that find their services missing, or nearing their expiry periods in the advertised lists, readvertise their services sooner than previously scheduled. We do not explore these protocols further in this paper because of their architectural and functional similarities to the four state of the art systems that we have described in this section.

3.6.2   Large Systems

Service discovery for large systems is designed with multiple Registry architecture because it reduces network traffic, thus increasing scalability. Since large systems are deployed over long distances, multiple, replicated Registries are available. To support a large number of nodes, Registries can do load balancing, and reproduce child Registries to help reduce load. To allow Registries to query and update each other efficiently, Registries are arranged in a logical mesh or tree topology.

1. Ninja SDS [48] - The Ninja project by the University of California, Berkeley developed the Service Discovery Service (SDS). SDS is a Registry architecture, where services are registered by Managers and discovered by Users through queries. The Registry in SDS is called the *SDS server*. For the purpose of scalability, Registries are organized into multiple shared tree-like hierarchies, so that tasks can be shared among several Registries. When a Registry is overloaded, it spawns a nearby node as a new Registry, which then becomes a child of the overloaded Registry. The new Registry is allocated a portion of the network extent, and, thus, a portion of the load. Security is also supported by SDS, where service discovery functions are wrapped around steps to allow authentication, authorization and data integrity. For service queries and Service Descriptions, SDS uses an XML-based query and description language. SDS is implemented in Java and requires the use of Secure Remote Method Invocation (Secure RMI) to perform secure communication, hence it requires substantial resources.

2. INS/Twine [49] - INS/Twine was developed at Massachusetts Institute of Technology. Like SDS, the architecture is based on the Intentional Naming System [50], where a number of Registries, called *resolvers*, map queries to destination addresses, and also distribute service information. Unlike tree-like Registries structures in SDS, INS/Twine Registries have a mesh-like topology, where Registries are peers with each other. A Manager is simply referred to as a *resource*. The Manager advertises its Service Description to the nearest Registry. INS/Twine is built on top of a distributed hash table (DHT), such as Chord [51]. The Registry extracts prefix subsequences of attributes and values in the Service Description, into *strands*. The Registry then computes hash values for each of these strands, which constitutes numeric keys used to map resources to resolvers. To avoid being overwhelmed with registrations, Registries in INS/Twine use keying mechanisms to limit registrations. The service information is stored redundantly in all Registries that correspond to the numeric keys. When a User queries the nearest Registry, the Registry splits the query similar to how the Service Description is split. The Registry then queries other Registries that are identified by one of the

longest strands. The query is further processed by the Registry, which returns the matching service information.

3. Jxta [52] - Jxta was developed by Sun Microsystems. It is a combination of Registry and non-Registry based architecture. Registries in Jxta are known as Rendezvous peers. Managers (simply known as *peers*) send multicast advertisements to make their presence and services known to the network. Registries that receive the advertisements cache the service information. A User can send multicast queries, and Managers and Registries with matching service information respond to the queries. Each Manager periodically refreshes its service advertisements. Users and Registries purge service information when the Manager fails to refresh the advertisements at the expected time. When a service changes, the Manager sends another advertisement (either immediately, or at the next periodic refresh time) so that Users and Registries can detect the change. An additional entity called the *relay peer* acts as name resolver to map a service to its destination address. The relay peer stores routing information and relays messages across firewalls. In Jxta, Users, Managers and Registries can form groups, based on a certain criteria (such as location, service type, etc). An entity can only communicate with members of the groups that it has joined. Unlike SDS and INS/Twine, Jxta does not provide load-balancing techniques to unburden overloaded Registries. It also provides limited consistency maintenance support.

4. Ariadne [53]- Ariadne was developed by INRIA Rocquencourt in France. It targets mobile ad-hoc networks, comprised of at least 100 nodes, and integrates routing with service discovery. Like FRODO, the protocol uses an election algorithm to elect Registries. The node with the highest number of neighbors and the smallest number of Registries within its vicinity is selected as a new Registry. Registries periodically announce their presence to nodes within their vicinity. Managers register their service descriptions to Registries within a limited number of hops, and Users query known Registries for services. WSDL is used for service description, and for specifying Quality of Service parameters such as availability of service within a particular time, and resource capacity of the Manager (e.g. memory, energy). If the Registry does not have the service description matching the User's query, it selectively forwards the query to other Registries, based on the distance between the sending and receiving Registries. Registries share each other's cached information (called as profiles) by using Bloom filters [54] to reduce memory and bandwidth utilization. Updates on service descriptions are not propagated among the Registries. Only when the number of false cache hits reaches a threshold, the Registry requests for replacements of the profiles.

Other state of the art that are categorized as "large systems" include Superstring [55], GloServ [56] and CDS [57]. The Registries in Superstring are a set of distributed query resolvers, arranged in a tree-based topology, like in SDS. Also, like in INS/ Twine, the resolvers route amongst themselves using a distributed hash table routing structure. A variant of Superstring, called SuperstringRep [55], uses a Bayesian reputation model to compute the reputation score of an entity. The reputation score reflects the quality of service, thus the trustworthiness of the entity. GloServ is a tree-based Registry architecture that uses DNS-like hierarchy and RDF [17] for describing services. GloServ requires administration (services grouping and managed Registries), and assumes a stable environment, compared to the other systems described in this section. CDS uses a distributed hash-based Registry (resolver) system. Existing hash algorithms are used to distribute service information and queries to the Registries. CDS also does dynamic load balancing by clustering the Registries. Each cluster shrinks and expands according to the number of registrations and queries it supports (cluster size is determined individually by each Registry). Users select and query a Registry in the cluster. As with miscellaneous small systems, we do not explore these state of the art systems further in this paper.

3.7   Taxonomy of State of the Art

In this section, we analyze our selected state of the art systems. Our aim is not to conclude which system is the best because such a deduction would not be beneficial for all types of scenarios and requirements. We mainly aim to present the advantages and limitations of these systems in terms of how they cope with the five operational issues (system size, lossy environment, resource constraints, security and system heterogeneity), and how capable they are in self-configuring and self-healing by looking into their functional implementations.

3.7.1   Taxonomy of State of the Art Solutions to Operational Aspects

Figure 3.5 shows a summary of our analysis on the solutions provided by state of the art systems for the operational design issues. The shaded columns for each system expose the issues that the system considers and the solutions. We also show which system provides the most support for each design issue by the number of shades for the issue across the systems.

For small sized, mobile PAN (less than ten nodes) the non-Registry architecture (UPnP and Bluetooth SDP) is the most suitable. This is because the number of nodes is small, and service discovery tasks will be accomplished faster than if a Registry is required to be setup and maintained. For LAN (tens to several hundred nodes), the Registry-based architecture would be more suitable, to help conserve bandwidth. The Registry can either be statically deployed (Jini, SLP), or dynamically elected (Ariadne, Jxta), depending on the degree of fault-tolerance required. For large systems, scalability is the primary concern. Registries are scoped according to location or services (Ariadne, SDS, Jxta), and arranged in a tree (SDS), or mesh (Ariadne, Jxta, INS/Twine) topology to optimize query processing and conserve bandwidth.

State of the art systems provide fault-tolerance for a lossy environment by implementing on-the-fly Registry setup (Ariadne, SDS), or multiple replicas of the Registry (as can be done in Jini, SLP, INS/Twine and Ariadne) to provide redundancy in the face of single point of failure problems caused by mobile Registries. However, redundancy increases design complexity and consumes additional resources. Registries can also be monitored by other nodes for node crash failures (SDS, Jxta and Ariadne). The non-Registry based architecture of UPnP is the most robust against crash failure and message loss, because Users and Managers can hear each others' multicast queries and announcements directly. However, the tradeoffs are scalability and conservation of resource consumption. A Registry-based architecture, with the ability for nodes to multicast queries when the Registry disappears (SLP, Jxta), increases robustness against message loss and crash failure, while also increasing scalability and resource consumption. To provide reliable transmission, TCP is used in Jini, SLP and UPnP. None of the state of art protocols address message corruption (assume lower protocol layers will address this problem), and detect and recover from Byzantine failures.

To support resource-lean nodes, a Registry-based architecture that delegates heavier tasks to more powerful nodes allows resource-lean nodes to use less memory, and energy. None of the selected systems here provide explicit resource-awareness. Service discovery for large systems provide load balancing techniques so that Registries are not overloaded; SDS and INS/Twine allow overloaded Registries to spawn another to take over a portion of their tasks. For conserving bandwidth, some systems use efficient replication and query processing such as by using DHTs (INS/Twine) and Bloom filters (Ariadne).

To integrate nodes into different types of connectivity (e.g. wired Ethernet to wireless 802.11b), nodes are assumed to have the necessary interfaces to the different networks, or have connectivity via access points. SDS and Jxta abstract away underlying protocol stacks (transport layer and beyond), therefore providing more flexibility for deployment over different types of connectivity and protocol stacks. Service discovery functions in these systems are self-sufficient, with minimum network layer assumptions (multicast and unicast capabilities required). Systems built for the ad-hoc networks (Ariadne and Bluetooth SDP) integrate routing knowledge with service discovery, and become more dependent on the network stack. Some systems explicitly require a certain type of technology, such as TCP and IP (Jini, UPnP, SLP and INS/Twine). INS/Twine also uses the integrated INS as the underlying name mapping framework.

Most systems depend on higher layers in the protocol stack to provide authentication, authorization, privacy and data integrity. Additional steps are required in between service discovery functions. For example, once the Manager discovers a Registry, it decrypts the message and verifies the signature in the message to authenticate the Registry (as is possible in SLP, Jxta and SDS). These are intermediate steps, before service registration. Users who discover the service can only access the service if they are authorized by a capability manager (as in SDS), by applying for group membership (possible in Jxta), or if allowed by the Manager (can be performed by security applications in all systems). The service discovery systems discussed here do not support detection and blacklisting of an authenticated and authorized entity that has turned malicious. As mentioned earlier, one protocol that does reputation-based service discovery is SuperstringRep [55]. Security measures consume substantial resources, increase complexity of the system extensively and reduce self-configuration of the system. Due to these reasons, a full-fledged secure service discovery system is yet to be deployed successfully.

3.7.2   Taxonomy of Service Discovery Functions and Methods

Once architectural decisions are made on how to address the operational design issues, the service discovery functions are designed. The service discovery functions implement self-configuring and self-healing aspects of service discovery. The more exhaustively the system implements the functions, the more capable it is to self-configure and self-heal.

Some functions may be provided by different protocol layers (such as network and application), which is useful for conserving resources, mainly memory and energy. However in such systems, the dependencies on different protocol layers mean they are not easily portable, and may be *less effective* [58, 31]. For example, Bluetooth SDP cannot easily replace SLP on a printing device. Meanwhile, SLP expects the User application to poll for changes in services, and if the application fails to do so (overlooked by application developer, retransmission limit, etc), the system will not be effective in conveying changes in service information to interested parties.

We show in our previous work [58, 31] that the functional differences impact system performance such as responsiveness, efficiency and resource consumption. Briefly, these performance outcome depend on how exhaustive the implementation of the function is. For example, SLP and Jxta implement both unicast and multicast queries, therefore when Registry disappears, services can still be discovered by Users through multicast queries. In Jini, SDS, INS/Twine and Ariadne, Registries must recover before the service can be discovered, causing slower responsiveness.

Operational Issues			Design Solutions		State of the art solutions to operational issues							
Issue		Specifics			Jini	SLP	UPnP	Bluetooth SDP	Jxta	SDS	INS/Twine	Ariadne
System size	1	Small systems (PAN/LAN)	Non-Registry architecture	Meshed topology								
				Clustered topology								
			Registry architecture	Single Registry								
				Unconnected topology								
	2	Large systems (MAN/WAN)	Multiple Registry architecture	Meshed Registry topology								
				Tree-based Registry topology								
				Clustered Registry topology								
Lossy environment	3	Communication and node failure		Retransmissions and acknowledgements	TCP	TCP	TCP					
				Multicast queries								
				Periodic Registry announcements								
				Periodic service advertisement								
				Redundant Registry replicas								
				Primary-based recovery protocols								
				Cache alternate services	Appl	Appl	Appl	Appl	Appl	Appl	Appl	Appl
				Short lease/announcement expiry period	Appl	Appl	Appl	Appl	Appl	Appl	Appl	Appl
				Blacklist unreliable nodes								
Resource constraints	4	Resource-lean nodes		Resource-aware functions								
				Lightweight protocol stack								
	5	Bandwidth constraint		Integrate routing and service discovery								
				Efficient query processing								
				Load balancing techniques								
Security	6	Confidentiality, integrity and availability		Remove sensitive information from SD	Appl	Appl	Appl	Appl	Appl	Appl	Appl	Appl
				Authentication	Appl	Appl	Appl	LM	Appl		Appl	Appl
				Access control	Appl	Appl	Appl	LM			Appl	Appl
				Encrypted communication	Appl	Appl	Appl	LM			Appl	Appl
				Blacklist malicious nodes								
System heterogeneity	7	Heterogeneous protocol stack		Abstract away underlying protocol stacks	TCP/IP	TCP/IP	TCP/IP	Bluetooth			INS/IP	Routing
Operational issues that are given priority by the system					1,3	1,3	1,3	1,6	1,2,3,6,7	2,3,6,7	2,3,5	2,3,5

Appl: The solution can be provided by the application layer
LM: The underlying Link Manager in the Bluetooth protocol stack provides authentication and encryption
RMI/JVM: Jini is supported by the security features in JVM and RMI/TCP: Reliable communication is provided using TCP
TCP/IP, Bluetooth, Routing, INS/IP: The system depends on the listed underlying protocol stack

FIGURE 3.5: Taxonomy of state-of-the-art solutions to operational aspects. Shaded service discovery systems support the proposed solutions. *Appl* means the solution to the operational aspect is supported by the application layer. Some systems depend on solutions provided by the underlying protocol stacks, such as TCP, IP, Bluetooth and ad-hoc routing protocols.

Service Discovery Functions		Methods	State of the art functional implementation							
			Jini	SLP	UPnP	Bluetooth SDP	Jxta	SDS	INS/Twine	Ariadne
Configuration Discovery	Registry auto-configuration	Registry election			N/A	N/A				
		Registry reproduction			N/A	N/A				
	Registry or cluster discovery	Passive discovery				N/A				
		Active discovery	*		N/A	N/A				
Registration		Solicited registration			N/A	N/A				
		Unsolicited registration			N/A	N/A				
SD Discovery	Query	Unicast query				**				
		Multicast query								
	Service notification	Service notification by Registry			N/A	N/A				
		Multicast service advertisement								
Configuration Update	Configuration purge	Leasing expiry								
		Advertisement TTL expiry								
	Consistency maintenance	Poll for updates (by the User;)	Appl	Appl	Appl	Appl	Appl	Appl	Appl	Appl
		Notification of updates (by the Manager / Registry)								Appl
		Update among Registries			N/A	N/A				

* Jini only does active discovery when the node initializes (powers on)

** Bluetooth SDP depends on the underlying Bluetooth network to detect neighboring nodes, so that it can query through unicast

N/A: the method is not relevant for the non-Registry architecture

Appl: the method can be supported by the application layer, by using the handles provided by the service discovery protocol

FIGURE 3.6: Taxonomy of state-of-the-art functional implementation. Configuration Discovery, Service Registration and SD Discovery functions offer self-configuring capability while the Configuration Update function offers self-healing capability. The extent of the implementation of each function impacts how able these systems are to self-configure and self-heal. Additionally, the choice of implementation method impacts other considerations, such as the efficiency, responsiveness and resource consumption.

Figure 3.6 summarizes the functional capabilities of state of the art systems. From this table, we can intuitively deduce the relative capabilities of each system to self-configure and self-heal. The effectiveness of self-configuration and self-healing increases with the number of methods the system implements for each function. We now give a detailed analysis.

For the Configuration Discovery function, the Registry reproduction method in INS/Twine and SDS requires the first set of Registries in both systems to be manually deployed by a system administrator, unlike the more dynamic Registry election method in Ariadne. For discovering Registries and cluster members, systems that do periodic passive and active discovery (SLP, INS/Twine and SDS) have higher responsiveness than systems that implement only one of the methods. Passive and active discovery are especially useful to allow the system to recover from failures that cause network partitioning.

In the Service Registration function, none of these selected systems allow the Registry that receives messages from unknown Managers to solicit registrations. This method allows the Registry to speedily recover purged information of a Manager, after communication failures. The rest of the systems allow only unsolicited registration, after a Registry is discovered.

For the SD Discovery function, UPnP and Jxta allow both multicast queries and multicast service advertisements. The combination of these two methods gives the highest probability for successfully discovering a service, even after message loss and temporary node failures (e.g. mobile nodes getting temporarily disconnected). Among Registry systems, the probability of discovering services is increased if the Registry can notify the Users of newly registered services matching the requirements of the Users (Jini).

For the Configuration Update function, Jini uses leasing for Configuration Purge, where the Registry can request Managers to lengthen or shorten their lease period, according to the Registry's processing capability. The rest of the systems require Users and Registries to monitor the advertisement TTL of Managers to detect defunct services. Leasing is more efficient in terms of bandwidth and resource utilization, compared to periodic multicast advertisements. For Consistency Maintenance, the state of the art systems provide handles to allow the application on the User to query the Registry or Manager periodically for updates. Only Jini and UPnP allow the Registry or the Manager to update the User directly on changes in the SD. In large systems (Jxta, INS/Twine, SDS and Ariadne), Registries achieve consistency by updating each other on changes in the cached SDs. Unlike small systems, updates are not propagated each time an SD changes, but in bulk (when a threshold is reached), thus providing weaker consistency maintenance (but necessary to conserve bandwidth).

3.8   Conclusion

We analyze the field of service discovery by first characterizing service discovery as a third generation name discovery system that solves the limitations of legacy naming systems for pervasive computing. We describe the different architectures and the main functions of service discovery that allows services to be discovered, and changes in service availability and attributes to be detected. We then classify the main operational design aspects for service discovery, and compare the state of the art solutions to these aspects.

Our state of the art analysis presents how these systems cope with the five operational aspects (system size, lossy environment, resource constraints, security and system heterogeneity), and implements the four functions which offers self-configuration (Configuration Discovery, Service Registration and SD Discovery) and self-healing (Configuration Update). By presenting the advantages and limitations of the state of the art, we hope to enlighten system and application designers so that they can choose the service discovery system that best suits their scenarios and requirements.

There are still several interesting directions in which future research on service discovery can be taken.

- The semantics of device, service and attribute names still require much attention, to improve the context of a discovered service. For a truly unattended system deployment, different service discovery systems should adhere to a single, standardized method for describing services. This is important to ensure that applications that rely on discovered services can make the correct inference on the usage of the service.

- Service discovery systems should ensure that a service is discovered and accessed by authorized entities only. However, authorization should be dynamically allocated and revoked, as time progresses, and the requirement or capability of the entity changes. A service discovery system that assigns, monitors, revokes and reassigns access to services is necessary for establishing a truly self-protecting system.

- For a service discovery system in the pervasive environment to mature (as DNS has done in the Internet), applications that use service discovery need to be actively developed and promoted. One major hindrance to achieving this objective is the lack of agreement by manufacturers of devices and applications on a standard service discovery platform. A service discovery system that unifies well-known service discovery protocols is a step towards this objective.

- Existing service discovery architectures for wide-area networks focus more on scalability issues (such as bandwidth efficiency, and supporting a large number of nodes). More work has to be done to produce a large-scale service discovery design, which is also evaluated against communication and node failures, such as done in smaller systems [59, 60, 31]. A scalable and robust architecture is especially important in mobile ad-hoc networks, because nodes are easily moved, uncertain wireless connectivity, low bandwidth availability, and energy constraints.

We conclude by stressing that a service discovery system is the medium for propelling the power of computing beyond the realm of personal computers, such that information and services are accessible anywhere, and anytime.

References

[1] Murch, R.: Autonomic Computing. Prentice Hall (March 2004)

[2] Ganek, A.G., Corbi, T.A.: The dawning of the autonomic computing era. IBM Systems Journal **42**(1) (2003) 5–18

[3] Weiser, M.: The computer for the 21st century. **265**(3) (September 1991) 94–104

[4] Satyanarayanan, M.: Pervasive computing: Vision and challenges. IEEE Personal Communications **8**(4) (August 2001) 10–17

[5] Schilit, B., Adams, N., Want, R.: Context-aware computing applications, Santa Cruz, CA, US, IEEE Computer Society (1994) 85–90

[6] Dey, A.K.: Understanding and using context. Personal Ubiquitous Comput. **5**(1) (2001) 4–7

[7] Kidd, C.D., Orr, R., Abowd, G.D., Atkeson, C.G., Essa, I.A., MacIntyre, B., Mynatt, E.D., Starner, T., Newstetter, W.: The aware home: A living laboratory for ubiquitous computing research. In: CoBuild '99: Proceedings of the Second International Workshop on Cooperative Buildings, Integrating Information, Organization, and Architecture, London, UK, Springer-Verlag (1999) 191–198

[8] Cheverst, K., Davies, N., Mitchell, K., Friday, A.: Experiences of developing and deploying a context-aware tourist guide: The guide project. In: MobiCom '00: Proceedings of the 6th annual international conference on Mobile computing and networking, New York, NY, USA, ACM (2000) 20–31

[9] Bettstetter, C., Renner, C.: A comparison of service discovery protocols and implementation of the service location protocol. In: Proceedings of 6th EUNICE Open European Summer School: Innovative Internet Applications, University of Twente (September 2000) 101–108

[10] Vanthournout, K., Deconinck, G., Belmans, R.: A taxonomy for resource discovery. Personal Ubiquitous Comput. **9**(2) (2005) 81–89

[11] Richard, G.G.: Service advertisement and discovery: Enabling universal device cooperation. **4**(5) (September-October 2000) 18–26

[12] Helal, S.: Standards for service discovery and delivery. IEEE Pervasive Computing **1**(3) (2002) 95–100

[13] Ververidis, C.N., Polyzos, G.C.: Service discovery for mobile ad hoc networks: A survey of issues and techniques. In: IEEE Communications Surveys and Tutorials, IEEE Computer Society (2008) to appear

[14] W3C Working Group Note: Web services architecture. http://www.w3.org/TR/ws-arch (February 2004)

[15] McGuinness, D.L., van Harmelen, F.: Owl web ontology language. http://www.w3.org/TR/owl-features (February 2004)

[16] Christensen, E., Curbera, F., Meredith, G., Weerawarana, S.: Web service description language. http://www.w3.org/TR/wsdl (March 2001)

[17] RDF Core Working Group. Homepage at http://www.w3.org/RDF/ (2004)

[18] Srinivasan, R.: Rpc: Remote procedure call protocol specification version 2 (1995)

[19] Gosling, J., Joy, B., Steele, G.: The java language specification. Homepage at http://java.sun.com/java.sun.com/newdocs.html (1996)

[20] Needham, R.: Names. In Mullender, S., ed.: An Advanced Course In Distributed Systems, Wokingham, England, ACM Press/Addison-Wesley Publishing Co. (1993) 315–326

[21] Tanenbaum, A.S., Steen, M.V.: Distributed Systems: Principles and Paradigms. Prentice Hall PTR, Upper Saddle River, NJ, USA (2002)

[22] Coulouris, G.F., Dollimore, J.: Distributed Systems: Concepts and Design. fourth edn. Addison-Wesley Longman Publishing Co., Inc., Boston, MA, USA (2005)

[23] Birrell, A.D., Levin, R., Schroeder, M.D., Needham, R.M.: Grapevine: An exercise in distributed computing. Communications of the ACM **25**(4) (1982) 260–274

[24] Lampson, B.W.: Designing a global name service. In: Proceedings of the fifth annual ACM symposium on Principles of distributed computing (PODC '86), New York, NY, USA, ACM Press (1986) 1–10

[25] Mockapetris, P., Dunlap, K.J.: Development of the domain name system. In: SIGCOMM '88: Symposium proceedings on Communications architectures and protocols, New York, NY, USA, ACM Press (1988) 123–133

[26] Peterson, L.L.: The profile naming service. ACM Trans. Comput. Syst. **6**(4) (1988) 341–364

[27] Bowman, M., Peterson, L.L., Yeatts, A.: Univers: An attribute-based name server. Software-Practices and Experiences **20**(4) (1990) 403–424

[28] Chadwick, D.: Understanding X.500 The Directory. Chapman & Hall, London (1994)

[29] Howes, T., Smith, M., Good, G.S.: Understanding and Deploying LDAP Directory Services. Macmillan Technical Publishing, Indianapolis, Indiana (1999)

[30] Object Management Group: OMG. The Common Object Request Broker: Architecture and Specification, Rev 1.2., OMG Document Number 93-12-43. (December 1993)

[31] Sundramoorthy, V., Hartel, P.H., Scholten, J.: On consistency maintenance in service discovery. In: 20th IEEE Int. Parallel & Distributed Processing Symp. (IPDPS 2006), Los Alamitos, California, IEEE Computer Society Press (April 2006) 10pp in CD–ROM

[32] Avizienis, A., Laprie, J.C., Randell, B., Landwehr, C.: Basic concepts and taxonomy of dependable and secure computing. In: IEEE Transactions on Dependable and Secure Computing. Volume 1., Los Alamitos, California, IEEE Computer Society Press (Jan-Mar 2004) 11–33

[33] Elkhodary, A., Whittle, J.: A survey of approaches to adaptive application security. In: SEAMS '07: Proceedings of the 2007 International Workshop on Software Engineering for Adaptive and Self-Managing Systems, Washington, DC, USA, IEEE Computer Society (2007) 16

[34] Merwe, J.V.D., Dawoud, D., McDonald, S.: A survey on peer-to-peer key management for mobile ad hoc networks. ACM Comput. Surv. **39**(1) (2007) 1

[35] Sun Microsystems: The Jini Architecture Specification, version 2.0. (June 2003)

[36] Sun Microsystems: JavaSpaces Service Specification , version 2.0. (June 2003)

[37] Microsoft: Universal Plug and Play Architecture, V1.0. (Jun 2000)

[38] Goland, Y., Cai, T., Leach, P., Gu, Y.: Simple service discovery protocol, version 1.0. (2000)

[39] Gudgin, M., Hadley, M., Mendelsohn, N., Moreau, J., Nielsen, H.: Simple Object Access Protocol (SOAP) V.1.2, Part 1: Messaging Framework. (June 2003)

[40] Cohen, J., Aggarwal, S., Goland, Y.: General Event Notification Architecture Base: Client to Arbiter. (June 1994)

[41] Guttman, E., Perkins, C., Veizades, J.C., Day, M.: Service Location Protocol, V.2, RFC-2608. Internet Engineering Task Force (IETF) (December 2003)

[42] Bray, J., Sturman, C.F., Mandolia, J.: Bluetooth 1.1 Connect Without Cables, 2nd Edition. Prentice Hall (December 2001)

[43] Bluetooth SIG: Specification of the Bluetooth System, Core, Vol. 1. (Feb 2001)

[44] Sundramoorthy, V., Speelziek, M.D., van de Glind, G.J., Scholten, J.: Service discovery with FRODO. In: 12th IEEE Int. Conf. on Network Protocols (ICNP), Berlin, Germany, Computer Science Reports, BTU Cottbus (Oct 2004) 24–27

[45] The Salutation Consortium Inc: Salutation Architecture Specification (Part 1), version 2.1. (1999)

[46] Helal, S., Desai, N., Verma, V., Lee, C.: Konark-service discovery and delivery protocol for ad hoc networks. In: Proc. IEEE Wireless Communications Networking Conf. (2003)

[47] Nidd, M.: Service discovery in Deapspace. **8**(4) (2001) 39–45

[48] Czerwinski, S., Zhao, B., Hodes, T., Joseph, A., Katz, R.: An architecture for a secure service discovery service. In: Proceedings of ACM/IEEE International Conference on Mobile Computing and Networking (MobiCom'99), Kluwer Academic Publishers (1999) 24–35

[49] Stoica, I., R.Morris, D.Karger, M.F.Kaashoek, H.Balakrishnan: A scalable peer-to-peer lookup service for internet applications. In: Proceedings of the 2001 ACM SIGCOMM Conference. (2001)

[50] Adjie-Winoto, W., Schwartz, E., Balakrishnan, H., Lilley, J.: The design and implementation of an intentional naming system. In: Proceedings of the 17th ACM Symposium on Operating Systems Principles (SOSP), ACM Press (December 1999) 186–201

[51] Stoica, I., Morris, R., Liben-Nowell, D., Karger, D.R., Kaashoek, M.F., Dabek, F., Balakrishnan, H.: Chord: a scalable peer-to-peer lookup protocol for internet applications. IEEE/ACM Transactions on Networking (TON) **11**(1) (2003) 17–32

[52] Gong, L.: Jxta: A network programming environment. IEEE Internet Computing **5**(3) (May-June 2001) 88–95

[53] Sailhan, F., Issarny:, V.: Scalable service discovery for manet. In: 3rd IEEE International Conference on Pervasive Computing and Communications (PerCom 2005), IEEE Computer Society (March 2005) 235–244

[54] Bloom, B.H.: Space/time trade-offs in hash coding with allowable errors. Communication of the ACM **13**(7) (1970) 422–426

[55] Wishart, R., Robinson, R., Indulska, J., Josang, A.: Superstringrep: Reputation-enhanced service discovery. In Estivill-Castro, V., ed.: 28th Australasian Computer Science Conference (ACSC2005). Volume 38 of CRPIT., Newcastle, Australia, ACS (2005) 49–58

[56] Arabshian, K., Schulzrinne, H.: Gloserv: Global service discovery architecture. In: First Annual International Conference on Mobile and Ubiquitous Systems: Networking and Services (Mobiquitous), Los Alamitos, CA, USA, IEEE Computer Society (August 2004) 319–325

[57] Gao, J., Steenkiste, P.: Rendezvous points-based scalable content discovery with load balancing. In: Networked Group Communication, New York, NY, USA, ACM Press (October 2002) 7178

[58] Sundramoorthy, V., van de Glind, G.J., Hartel, P.H., Scholten, J.: The performance of a second generation service discovery protocol in response to message loss. In: 1st Int. Conf. on Communication System Software and Middleware, New Delhi, India, IEEE Computer Society Press (Jan 2006)

[59] Dabrowski, C., Mills, K., Elder, J.: Understanding consistency maintenance in service discovery architectures during communication failure. In: Proceedings of the Third International Workshop on Software and Performance, ACM Press (July 2002) 168–178

[60] Dabrowski, C., Mills, K., Elder, J.: Understanding consistency maintenance in service discovery architectures in response to message loss. In: Proceedings of the 4th International Workshop on Active Middleware Services, IEEE Computer Society (July 2002) 51–60

Chapter 4

Managing Distributed and Heterogeneous Context for Ambient Intelligence

José Viterbo, Markus Endler, Karin Breitman

Departamento de Informática, Pontifícia Universidade Católica (PUC-RJ), 22453-900 Rio de Janeiro, Brazil

Laurent Mazuel, Yasmine Charif, Nicolas Sabouret, Amal El Fallah Seghrouchni, and Jean-Pierre Briot

Laboratoire d'Informatique de Paris 6 (LIP6), Université Paris 6 - CNRS, 75015 Paris, France

Abstract

Practical realization of applications for Ambient Intelligence (AmI) poses several challenges to software developers, many of them related to heterogeneity, dynamism (i.e., mobility) and decentralization. In this chapter, we focus on approaches for decentralized context reasoning and for semantic mediation, since we understand that these are some of the main challenges for enabling interactions among heterogeneous and context-aware entities in open and dynamic AmI environments.

4.1 Introduction

The *anywhere/any time* paradigm is becoming the new challenge to the conception, design and release of the next generation of information systems. New technologies, like Wi-Fi networks and 3rd generation mobile phones, are offering the infrastructure to conceive information systems as ubiquitous, that is, systems that are accessible from anywhere, at any time, and with (almost) any electronic device. However, the use of such ubiquitous access to information systems requires new conceptualizations, models, methodologies and support technologies to fully explore its potential.

In this context, mobility introduces new accessibility scenarios and increases complexity. New issues, such as how to enable users to retain their ability to cooperate while located in different workplaces, the role of context and location in determining cooperation, the support for *ad hoc* cooperation in situations where the fixed network infrastructure is absent or cannot be used, are beginning to arise. The approaches and technologies for supporting these new ways of working are still under investigation. Nevertheless, a particularly interesting trend explores the Ambient Intelligence paradigm, a multidisciplinary approach that aims at the integration of innovative technologies that support user activities through specific services of the environment, which are provisioned with minimal user intervention. Essentially, an Ambient Intelligence system should be aware of the presence of a person, perceive the needs of this person and be able to adapt to the needs of the users in a relaxed and unobtrusive manner [20].

Ambient Intelligence (in the following, abbreviated as AmI) requires new environments for software development and deployment, where large quantities of different devices and sensors need to be integrated, building a programmable and auto-configurable infrastructure. Several projects, e.g., Gaia, CoBrA, CHIL, etc., have developed prototypes of such environments, but usually with focus only at specific use cases, user tasks or application domains. Hence, most researchers have come up with pragmatic, problem-specific solutions, which are difficult to generalize and port to other applications. However, we believe that in a few years, nearly every public and private space will be equipped with sensors and smart appliances that are able to automatically adapt to the preferences and demands of the local user(s) and as such provide special context-specific services to them. Such systems will be open, i.e., these spaces will potentially serve any user with a communication device (e.g., a smartphone with powerful computing and multimedia capabilities), which will be the unique digital interface of the user with the ambient services and with the devices of other users. Thus, openness entails that both software agents responsible for user devices (e.g., agents for assisting the user), and agents responsible for smart spaces (e.g., agents that control the devices of a room according to its current use), must be prepared to interact with an *a priori* unknown set of other software entities.

In this chapter, we present existing technologies and current proposals toward the integration of heterogeneous entities within an Ambient Intelligence system. Practical realization of applications for AmI poses several challenges to software developers, many of them related to heterogeneity, dynamism (i.e., mobility) and decentralization. We make no claim of a complete or exclusive treatment of the subject. In fact, there are also several other related challenges [55], for example, identification of user intent, knowledge acquisition, negotiation, etc. But since these issues are a complex subject on their own right, in this work we will discuss only challenges related to context reasoning, distribution, interoperability and heterogeneity issues.

Context reasoning for AmI is very complex due to the dynamic, imprecise and ambiguous nature of context data, the need to process large volumes of data and the fact that reasoning needs to be performed in a decentralized, cooperative way among several entities of the system, e.g., entities representing spaces, devices or users. Decentralization takes the form of physical distribution of computing and sensor devices, of context providers and consumers, of entities responsible for reasoning and brokering, of applications and of users who may potentially engage into a spontaneous collaboration.

Besides, AmI spaces are intrinsically heterogeneous at several levels: At the infrastructure level, they include a wide range of appliances and gadgets with very specific data and control access protocols, and which are typically interconnected through different kinds of (wireless) networks with specific protocols and QoS parameters. Also the providers and the types of context information are usually very specific for each space and device, as well as their representations and models, making it difficult to achieve a common representation for different entities of the ambient context. A similar problem of heterogeneity can be identified at the level of services due to the very different kinds of ambient control functions provided, combined with the lack of standardized interfaces for service access. Finally, heterogeneity problems are found also at the level of knowledge representation and modeling, where systems may employ very different kinds of knowledge bases, descriptions and reasoning techniques. Hence, even if two elements are conscious of the same concrete fact, there is the problem of alignment of their knowledge representations.

In this chapter, we will focus mainly on decentralized context reasoning and on semantic mediation, since we understand that these are some of the main challenges for enabling interactions among heterogeneous and context-aware entities in open and dynamic environments of Ambient Intelligence. In the following subsection we describe a simple scenario, which highlights several problems related to AmI, such as location-specific context-awareness, ontology based distributed reasoning, heterogeneous knowledge basis and semantic mediation.

4.1.1   Scenario

Silva is a Brazilian professor and researcher who works at PUC-Rio. He is visiting LIP6 with several other researchers. Their purpose is to have joint workshops related to a collaboration project. Silva carries with him his smartphone and his notebook, both executing the Campus middleware services dedicated to collecting and interpreting context information and for collective reasoning with other ambient services and applications. The devices also host some context-aware applications that support the platform's self-configuration to adapt to different situations, according to user's preferences and environment conditions.

When Silva arrives at LIP6, his Wi-Fi and GPS enabled smartphone (SMP-1) connects to the network, and using the current GPS data, queries a location service to find out that its user (Silva) is at LIP6. It then determines that this university is a partner institution of PUC-Rio; obtains the IP address of the Ambient management service at LIP6 and registers with it, indicating the user's identity and preferences.

The Ambient management service registers SMP-1 and determines that it belongs to Silva, a visiting professor from PUC-Rio. The system verifies that Silva is involved with the collaboration project and sets a workspace for him, communicating with a service running on Silva's notebook (NTB-1) to configure it to grant access to the proper network directories and services. This system also informs other project members at LIP6 about Silva's arrival.

A personal agenda application running on SMP-1 contacts the context infrastructure to be notified about the beginning of each event involving the whole project team, based on the project schedule and the location. Another application on SMP-1, the Configuration manager, requests to be notified whenever Silva is in a room in which an *activity* has started, so that it may set the smartphone to vibe-mode, and as soon as the activity ends, switch it back to the ring mode.

Notice that when this application interacts with the Ambient's local context provider, there could be a semantic mismatch between the terms "activity," used in the device's ontology, and the terms "meeting" or "class," used in the Ambient ontology. Due to this semantic mismatch, Silva's application would not get the expected response from the Ambient Service and would issue a request for semantic mapping from a mediation service, which would try to identify equivalence or subsumption among the concepts and adjust the local ontology to reflect this new classification. Hence, we identify that the main requirements of AmI are context-specific reasoning capabilities (i.e., to enable the spaces and the interacting computing entities to "understand what is going on") and the ability to adapt services/behaviors to the current situation and user preferences. As the AmI environment is an open system, reasoning is inherently distributed.

Due to the intrinsic characteristics of Ambient Intelligence systems, ontological and distributed context-reasoning using multi-agent systems seems to be the most suitable development paradigm (i.e., each agent interacts with other agents to reinforce and complement its own knowledge about the context). However, the main problem with distributed reasoning is that heterogeneous knowledge bases and models have to be mapped (i.e., aligned, mediated), which leads to the problem of identifying and resolving semantic mismatch of knowledge representations.

4.1.2 Outline

The rest of this chapter is organized as follows: next section presents several fundamental concepts for AmI which are dealt with in this chapter. Section 4.3 discusses the related work on context awareness for AmI. In Section 4.4 we review the main approaches to deal with the ontology alignment problem. Section 4.5 presents the Campus approach for dealing with context and semantic heterogeneity in AmI. Section 4.6 concludes the chapter.

4.2 Fundamental Concepts

4.2.1 Ambient Intelligence

Ambient Intelligence (AmI), i.e., "intelligent" pervasive computing, builds on three recent key technologies [2]: Ubiquitous Computing, Ubiquitous Communication and Intelligent User Interfaces. Ubiquitous Computing is the integration of microprocessors into everyday objects like furniture, clothing, white goods, toys, even paint. Ubiquitous Communication enables these objects to communicate with each other and the user by means of *ad hoc* wireless networking. Intelligent User Interfaces enable the inhabitants of an AmI environment to control and interact with the environment in a natural (voice, gesture) and personalized way (preferences, context).

AmI aims at making use of those entities in order to provide users with an environment, which offers services when and if needed. One great challenge of such environments is how to adequately address the heterogeneity and dynamic nature of users, services and devices. Key issues of the development of AmI are context-awareness and reasoning and how to identify and activate the appropriate service within a continuously changing multitude of services [39]. The ultimate goal is to make the ambient services more *intelligent* and adaptive to the specific needs of their users.

4.2.2 Context Awareness

Context awareness is the ability of a system to sense the current environment and autonomously perform appropriate adaptations in regard to its optimal operation, general behavior and user interaction. When a user enters a new context, it is desirable that the applications on his devices be able to adapt to the new situation, and the environment be able to adapt its services to the presence of the new user.

There exist several definitions for context and context-awareness, but one of the most referenced one can be found in [19]: *"Any information which can be used to characterize the situation of an entity. An entity is a person, a place or an object which is considered relevant for the interaction between a user and an application, including the user and the application.*" In an attempt to classify context, Chen and Kotz [16] identified four basic types of context: computational context (i.e., state of resources at the device and of the network), user context (i.e., persons, places and objects), physical context (e.g., luminosity, noise, temperature) and temporal context (e.g., hour, day, period of the year). Abowd et al. [1] proposed the notions of *primary context* (localization, identity, activity and time) and of *secondary context*, where the latter one can be deduced from the former one and may be used for making adaptation decisions at a higher level of abstraction.

Conceptually, context provisioning can be organized in three layers [33]: data acquisition and distribution, interpretation and utilization. Before raw context data acquired from sensors and devices can be utilized, it must be interpreted and evaluated with respect to its accuracy, stability and reliability. The interpretation layer may also combine context data from different sources to enhance its reliability or completeness. For applications to be able to understand, describe and manage context-aware adaptations, it is necessary to have a context model, which can be defined at the application or the middleware layer. Strang and Linnhoff-Popien [62] identified and compared six types of context models: attribute-value pairs, schema-based models, graphic models, logic-based models, object-oriented models and ontology-based models. The author's main conclusion is that the object-oriented and the ontology-based models are the most complete and expressive ones, and hence are the most suited for modeling context for ubiquitous computing.

4.2.3   Ontology

Ontology has not only the advantage of enabling the reuse and sharing of common knowledge among several applications [58], but also of allowing the use of logic reasoning mechanisms to deduce high-level contextual information [68]. Therefore it has been widely adopted over other conceptual models, such as taxonomy, relational database schema and OO software models, for representing context information in ubiquitous systems.

A taxonomy is a set of terms arranged in a generalization-specialization (parent-child) hierarchy because they are much more expressive [34]. A controlled vocabulary simply lists a set of terms and definitions. A taxonomy may or may not define attributes of these terms. A relational database schema defines a set of terms through classes, attributes and a limited set of relationships among those classes. An OO software model defines a set of concepts and terms through a hierarchy of classes and attributes and a broad set of binary relationships among classes. Constraints and other behavioral issues may be specified through methods on the classes (or objects).

An ontology can express all of the preceding relationships, models and diagrams as well as n-ary relations, a rich set of constraints, rules relevant to usage or related processes and other differentiators including negation and disjunction [25]. Table 4.1 summarizes the benefits of the adoption of ontology.

Table 4.1: Benefits of adopting formal ontology to model ambient knowledge in Campus.

- Ontologies are semantically richer, i.e., have greater expression power than taxonomies, entity relationships or OO models;
- Conceptual knowledge is maintained through complex and accurate representations above and beyond hierarchical approaches;
- Ontologies are formal — OWL DL ontologies map directly to Description Logic (a dialect of first order logics);
- Formal ontologies in the OWL DL standard can be verified/classified with the aid of Inference Mechanisms, e.g., RACER and FaCT:
  - consistency checks;
  - classification;
  - new information discovery;
- OWL ontologies use a XML/RDF syntax that allows them to be automatically manipulated and understood by most resources on the Internet;
- Ontologies capture and represent finely granulated knowledge;
- Ontologies can be used to reduce ambiguity so as to provide a model over which information can be freely shared and acted upon by autonomic managers;
- Ontologies are modular, reusable and code independent — ontology driven applications are specified separately from the ontology itself. Changes to the ontology should not impact the code or vice versa;
- Ontologies can be combined with emerging rule languages, such as SWRL.

4.2.4   Context Reasoning

Reasoning is necessary in context aware systems to deal with the intrinsic imperfection and uncertainty of context data, and also to infer secondary context data. Henricksen and Indulska [28] have characterized four kinds of context imperfectness: unknown, ambiguous, imprecise and erroneous. The main tasks of reasoning are to detect possible errors, make estimates about missing values, determine the quality and validity of the context data, transform context data into meaningful information and infer new, implicit context information that may be relevant for the applications. Reasoning is also fundamental for any kind of context-oriented decision-making, e.g., system adaptations according to user-provided or learned decision rules.

According to [46] reasoning for context-aware systems can be approached from four main perspectives: the low-level perspective, which includes basic tasks such as data pre-processing, data fusion and context inference, usually performed by the sensors or the middleware, the application-oriented perspective, where the application can use a wide variety of reasoning methods to process the context data, the context monitoring perspective, where the main concern is a correct and efficient update of the knowledge base as the context changes and, finally, model monitoring perspective, where the main task is to continuously evaluate and update learned context classifiers/interpreters and their models, also taking into account user feedback. Although Nurmi and Floren give an interesting perspective on context reasoning, we understand that instead of four perspectives, these are in fact complementary tasks, which should be present in every approach for reasoning in context-aware systems.

For context reasoning, several approaches have been adopted: ontological reasoning, rule-based reasoning, distributed reasoning and probabilistic reasoning [7]. Instead of presenting and comparing the general reasoning approaches, which are very well surveyed in Bikakis et al. [7], in this paper, we will focus only on ontological and distributed reasoning approaches. On the one hand, ontologies offer high expressiveness and the possibility to develop a formal context model that can be shared, reused, extended to specific domains, and on the other hand, distributed reasoning is a direct requirement that arises from the open, dynamic and heterogeneous nature of AmI.

In the next section, we review the main approaches to deal with the contextual reasoning and ontological representations for AmI. Section 4.4 will focus on ontology alignment and semantic mapping between concepts, to deal with semantic heterogeneity in AmI. Section 4.5 will present our proposition to tackle both issues within the Campus framework.

4.3 Ontological Representation and Reasoning about Context

In this section we survey several research works that deal with ontological representation and reasoning about context for Ambient Intelligence. We first present the main criteria used for comparison and a proposed taxonomy; then we present each work with respect to each criteria; and finally, we classify the systems according to our taxonomy and discuss their suitability for implementing Ambient Intelligent environments.

4.3.1 Evaluation Criteria and Taxonomy

There is much research work on middleware systems that support context modeling and ontological reasoning about context. However, as expected, each one is based on a different notion of context, uses ontologies in a different way, has specific goals and approaches for context-specific reasoning and handling heterogeneity, and is targeted at specific applications or use scenarios. In this section, we will compare the works in regard to the following criteria:

4.3.1.1 Types of Context.

Which types of context information are collected, processed and distributed by the system (e.g., system context, location, physical context, user role, preferences, etc.). This information will give an idea of the framework's usefulness, scalability and practical feasibility.

4.3.1.2 Ontologies.

Which ontologies are used, and for which purpose? What sorts of concepts and relationships are represented? Is the ontology extensible? How are context instances updated and persisted, etc.? This criterion assesses the system's expressiveness and flexibility.

4.3.1.3 Inference/Reasoning Techniques.

What kind of reasoning is supported? What sorts of higher-level context is inferred? Does the work consider uncertainty of the inferred context? This aspect determines the expressive power, reliability, completeness and preciseness of the systems reasoning, as well as its practical applicability.

4.3.1.4 Knowledge Management.

Is the knowledge base static, or do most of the facts in the knowledge require continuous updates? Does the system handle decentralized or heterogeneous knowledge bases, and if so, do they handle evolving knowledge models (ontologies)? If heterogeneity is supported, what is the basic mediation or semantic alignment technique employed and how powerful is it? The evaluation with regard to this aspect will give insight on how well the system is suited to deal with the inherently dynamic, decentralized and unpredictable nature of Ambient Intelligence.

4.3.1.5 Architecture.

Is the system based on a centralized, fully decentralized or hybrid architecture, with respect to the knowledge bases, the reasoning process and the mediation/brokerage support? By discussing this aspect, we have an idea on the system's scalability, reliability and of the implicit execution overhead related to the distributed interactions.

Although there are many possible means of classifying the context systems, we believe that the following aspects are the most relevant for assessing their suitability for developing open and heterogeneous Ambient Intelligence environments. Hence, we will use them as the basis for our taxonomy.

1. Centralized *versus* decentralized knowledge base;

2. Static *versus* dynamic (or extensible) set of context providers;

3. Main goal of context reasoning: enhance reliability of context information, derive higher-level context facts, or both;

4. Means of handling heterogeneous knowledge bases, if any.

In the following subsections we summarize and analyze the most representative middleware systems with regard to the presented criteria, and in Subsection 4.3.10, classify each system according to the proposed taxonomy.

4.3.2   Gaia

Gaia provides a generic computational environment that integrates physical spaces and their ubiquitous computing devices into a programmable computing and communication system [52]. It is similar to traditional operating systems in that it manages the tasks common to all applications built for physical spaces [50]. Each space is self-contained, but may interact with other spaces. Gaia provides core services, including events, entity presence (devices, users and services), discovery and naming. By specifying well-defined interfaces to services, applications may be built in a generic way so that they are able to run in arbitrary active spaces. Gaia uses CORBA to enable distributed computing. Gaia is a mature project. The first prototypes were implemented in 2002 and several applications for active-classrooms have already been developed.

4.3.2.1   Types of Context.

The Gaia Context Infrastructure allows applications to obtain a variety of contextual information. Various components, called Context Providers, obtain context from either sensors or other data sources. These include sensors that track people's locations, room conditions (for example, temperature and sound) and weather conditions. Context Providers allow applications to query them for context information. Some Context Providers also have an event channel to asynchronously send context events. Thus, applications can either query a Provider or listen on the event channel to get context information.

4.3.2.2   Ontologies.

Gaia's context model is based on first-order predicates. The name of the predicate indicates the type of context that is being described (e.g., location, temperature or time), and its typed arguments describe the properties of the context. For example, if the predicate is "location," the first argument has to be a person or object, the second argument has to be a preposition or a verb like "entering," "leaving" or "in" and the third argument must be a locationID. The structures of different context predicates are specified in an ontology. Each context type corresponds to a class in the ontology, which also defines the corresponding arguments of the predicate. Moreover, Gaia uses ontologies to describe various concepts of an Ubiquitous Computing Environment, such as kinds of applications, services, devices, users, data sources and other entities. They also define all terms used in the environment and the relationships between different terms. These ontologies are written in DAML+OIL.

4.3.2.3   Inference/Reasoning Techniques.

Context Synthesizers are Gaia components that get sensed context data from various Context Providers, derive higher level or abstract context from these lower-level context data and provide these inferred contexts to applications. Whenever a Synthesizer deduces a change in the inferred context, it publishes the new information. Gaia adopts two basic inference approaches. Rule-based Synthesizers use pre-defined rules written in first order logic to infer different contexts. Each of the rules also has an associated priority, which is used to choose one rule when multiple rules are valid at the same time. However, if all the valid rules have the same priority, one of them is picked at random. Alternatively, some Synthesizers may use machine learning techniques, such as Bayesian learning and reinforcement learning, to infer high-level contexts. Past context information is used to train the learner.

4.3.2.4   Knowledge Management.

All the ontologies in Gaia are maintained by an Ontology Server. Entities contact the Ontology Server to get descriptions of other entities in the environment, information about context or definitions of various terms used in Gaia. The server also supports semantic queries to get, for instance, the classification of individuals or subsumption of concepts. The Ontology Server also provides an interface for adding new concepts to existing ontologies. This allows new types of contexts to be introduced and used in the environment at any time. The Ontology Server ensures that any new definition is logically consistent with existing definitions. Since the ontologies clearly define the structure of contextual information, different agents can exchange different types of context information easily. For example, Context Providers and Context Synthesizers can get the structure of contexts that they provide, while Context Consumers query the Ontology Server for the structure of the requested context, and then frame appropriate queries to Context Providers to get the context information they need.

4.3.2.5   Architecture.

The Gaia kernel consists of a Component Management Core that dynamically loads, unloads, transfers, creates, and removes any Gaia component or application. Each active space is self-contained but may interact with other spaces. For each space, Gaia manages its resources and services; provides location, context and event services; and stores information about it. Gaia provides a set of basic services to be used by all applications. Among them, the Space Repository stores information about all software and hardware entities in the space and lets applications browse and retrieve an entity on the basis of specific attributes. The Space Repository learns about entities entering and leaving the active space through the Presence Service, which detects and maintains soft state information about applications, services, devices and people in a active space. When the Presence Service detects that an entity is no longer available in an active space, it notifies the rest of the space that the entity left. In the context infrastructure, the Context Provider Lookup Service allows searches for different context providers. Providers advertise the set of contexts they provide in the form of a first order expression that describes the context provided. Applications can query the Lookup Service for a context provider that provides contextual information it needs.

4.3.3   CoBrA

Context Broker Architecture (CoBrA) is an infrastructure that supports agents, services and devices that interact in order to explore context information in active spaces [17, 18]. Its main component is an intelligent agent called *context broker*, which is responsible for providing a common model to represent context information, mediating the information exchanged between context providers and resource constrained context consumers, and inferring higher-level context information not directly available from sensors [17]. In addition, the context broker is capable of detecting and correcting inconsistent context data, and supports the enforcement of privacy policies defined by the users to control the sharing of their contextual information among other users. The proposed architecture is based on a central entity that was implemented as a FIPA-compliant agent using Jade.

4.3.3.1   Types of Context.

CoBrA has a context-acquisition module, which is a set of library procedures for acquiring contextual information from sensors, agents and the Web. This library includes procedures for collecting information from Smart Tag sensors (location) and environment sensors (temperature, sound, luminosity, etc.), but any other information can be added.

4.3.3.2   Ontologies.

The base ontologies used for representing context information are the Co-BrA Ontology (COBRA-ONT) and SOUPA. COBRA-ONT is a set of on-tologies for agents to describe contextual information and to share context knowledge. It defines concepts for representing actions, agents, devices, meet-ings, time and space. The SOUPA ontology, on the other hand, is a standard ontology for supporting pervasive and ubiquitous computing applications. It consists of vocabularies for expressing common concepts that are associated with person, agent, belief-desire-intention (BDI), action, policy, time, space and event, and also a set of vocabularies for supporting specialized domains of pervasive computing, such as smart spaces and peer-to-peer data manage-ment. The developer of a new system must design its specific ontology reusing some others that may be adequate.

4.3.3.3   Inference/Reasoning Techniques.

CoBrA's context reasoning is backed by the Jena rule engine, the Java Expert System Shell (JESS) and the Theorist system. The reasoning for in-terpreting context information uses two different rule-based systems. Jena rule-based reasoners are used for OWL ontology inferences and the JESS rule engine is used for interpreting context using domain specific rules. CoBrA supports also reasoning for maintaining a consistent context model by detect-ing and resolving inconsistent information, and the Theorist system is used for supporting the necessary logical inferences in that case. When a new context data is asserted into the knowledge base, the context broker first selects the type of context it attempts to infer (such as a person's location or a meeting's state). If such information is unknown, the broker decides whether it can infer this type of context using only ontology reasoning (Jena Rules). If logic infer-ence is required, the context broker attempts to find all essential supporting facts by querying the ontology model and asserts them into the Jess engine. Before asserting the new inferred information into the knowledge base, ontol-ogy reasoners are used to infer whether the context described by the instant data is consistent with the model defined by the ontology. If not, a Theorist assumption-based reasoning is used for resolving inconsistent information.

4.3.3.4   Knowledge Management.

The system provides a centralized (and homogeneous) model of context that all devices, services, and agents in the space must share. The knowledge of the context broker is represented as RDF statements and is stored in a persistent knowledge base. To acquire contextual information, all agents must send query messages to the context broker.

4.3.3.5   Architecture.

CoBrA has a centralized architecture, where a single context broker agent should be deployed and all computing entities must be aware of this broker from the beginning. Usually, a single context broker is sufficient to support a small-scale smart space. However, being the main service provider in the space, the context broker may become the bottleneck of the system and a single point of failure. A team of context brokers can be deployed to overcome this problem, as well as to improve system robustness through redundancy.

4.3.4   Semantic Space

Semantic Space [67, 68] is a context infrastructure developed to address three key issues. First, it aims to provide an explicit representation of the raw context data that is obtained from various sources in different formats. Furthermore, it provides means for the applications to selectively access a subset of context data through expressive context queries. Finally, it provides reasoning capabilities for inferring higher-level contexts. A prototype of the context infrastructure has been developed, and a prototype context-aware application was also implemented. The application, called SituAwarePhone, adapts mobile phones to changing situations while minimizing user distraction.

4.3.4.1   Types of Context.

In Semantic Space, *context wrappers* obtain raw context information from various sources such as hardware sensors and software programs and transform them into context data. Some context wrappers work close to the hardware sensors deployed in the prototypical smart space, gathering information such as user's location, environmental temperature, noise, and light, status of doors (open or closed) of rooms, etc. Software-based context includes the activity of the user, based on the schedule information from Outlook Web Access; the status of different networked devices (such as voice over IP or mobile phones), the status (idle, busy, closed) of applications such as JBuilder, Microsoft Word, and RealPlayer from their CPU usage; and weather information obtained by periodically querying a weather web service.

4.3.4.2   Ontologies.

Semantic Space uses the CONtext ONtology (CONON) for modeling context in pervasive computing environments [68]. Rather than completely modeling all sorts of context in different kinds of smart spaces, this ontology aims to be an extensible upper-level context ontology providing a set of basic concepts that are common to different environments. To characterize smart spaces, there are three classes of real-world objects (user, location and computing entity) and one class of conceptual objects (activity), which together form the skeleton of a "contextual-rich environment." Consensus domain ontologies such as friend-of-a-friend (FOAF), RCAL Calendar and FIPA Device Ontology were also integrated into CONON to model users, activities and device contexts, respectively.

4.3.4.3   Inference/Reasoning Techniques.

Two context reasoners are available, a description logic based reasoner and a first-order logic based situation reasoner, both implemented using Jena Semantic Web Toolkit to perform forward reasoning over the knowledge base. The description logic based reasoner was built to carry out ontology reasoning. The more flexible first-order logic based situation reasoner deduces a wide range of higher-level, conceptual context from relevant low-level context, such as user's activity. Semantic Space requires developers to write rules describing higher-level context information for each particular application based on its needs.

4.3.4.4   Knowledge Management.

In each smart space resides a *Context Knowledge Base*, which provides persistent context knowledge storage. It stores the extended context ontology for a particular space and the context data provided by users or gathered from context wrappers. The *Context Aggregator* is responsible for discovering context wrappers, gathering context data from them, and then asserting the gathered data into the context knowledge base. It updates the knowledge base whenever a context event occurs. The scope of contexts that the knowledge base manages may change depending on the availability of wrappers. When a context wrapper joins the smart space, the context aggregator adds the provided contexts to the knowledge base, and when the wrapper leaves, the aggregator deletes the contexts it supplied to avoid stale information.

4.3.4.5   Architecture.

The architecture is centralized around a *Context Aggregator* and a *Context Knowledge Base*. Developers can add new wrappers to expand the scope of contexts in a smart space or remove existing wrappers when the contexts it provides are no longer needed.

4.3.5   CHIL

The middleware infrastructure developed in the CHIL (Computers in the Human Interaction Loop) Project [60] provides mechanisms for service access, context modeling, control of sensors and actuators, directory services for infrastructure elements and services, as well as fault tolerance mechanisms. In general, this middleware infrastructure allows developers to focus on the service logic, rather than on the details of context processing and utility services, also providing a framework with several components that can be reused across different ubiquitous computing services. Mechanisms for modeling composite contextual information and describing networks of situation states are also available. The middleware has been implemented as a distributed multi-agent system where the agents are augmented with fault tolerance capabilities using the agent's capacity to migrate between hosts.

4.3.5.1   Types of Context.

The infrastructure can exploit numerous sensors for context acquisition, and new sensors can be plugged into the framework to provide information that may be used to compound derived contextual information or define situations that will trigger system's responses. Context information is obtained from sensors by software agents and made accessible to other agents of the system through the *Knowledge Base Agent*. Monitoring and control of sensors is performed through special *Proxy agents* that represent the sensors in the world of agents. Each proxy agent exposes a *universal virtualized interface* to the agent framework. A sensor specific driver is required to adapt the universal interface commands to the low-level capabilities of each particular sensor. This low-level driver is based on the control API offered by the sensor. Actually, three concrete proxy agents were implemented: one generic, one for microphones and one for cameras.

4.3.5.2   Ontologies.

The CHIL ontology aims to establish a general-purpose core vocabulary for the various concepts comprising a multi-sensor smart space and the context-aware applications associated [47]. It was modularized to allow different parts to be used in different contexts and applications. Separated namespaces are used so that developers may safely introduce new concepts locally in their module's name-space without interfering with other modules. Assuming that other modules use similar concepts that should be merged, the core module may provide a merged version of the concept. To globally put together all the modules, the ontology consists of a main OWL file, which imports all modules. Developers interested only in a subset of modules can define a main OWL file of their own that imports only the modules of interest. The main component is the core module *chil-core*, which introduces concepts of perceivable entities such as, for example, *Person*, *MeetingRoom*, *Table* or *Whiteboard*, as well as perceivable roles of such entities, such as the *Location* of a *Person* or the *ActivityLevel* of a *MeetingRoom*.

4.3.5.3   Inference/Reasoning Techniques.

The approach adopted by CHIL to infer high-level contexts is based on the notion of *networks of situation states*. According to this approach a situation is considered as a state description of the environment expressed in terms of entities and their properties. Changes in individual or relative properties of specified entities correspond to events that signal a change in the situation. The concept of *role* serves as a variable for the entities to which the relations are applied, thus allowing an equivalent set of situations to have the same representation. A role is played by an entity that can pass an acceptance test for the role, in which case, it is said that the entity can play or adopt the role for that situation. For example, in the scope of a meeting involving short presentations, at any instant, one person plays the role of "presenter," while the other persons play the role of "attendees." Dynamically assigning the role of "presenter" to a person makes it possible to select sensors to acquire images and sound of the current speaker. Detecting a change in some role allows the system to reconfigure the video and audio acquisition systems.

4.3.5.4   Knowledge Management.

The knowledge base was developed as a server accessible both locally and remotely through a unique interface. The server remote interface is programming language independent, so that client components may be written in a variety of programming languages. The knowledge base server API is tailored to OWL.

4.3.5.5   Architecture.

This architecture is centralized around some core agents, which are independent of the service and smart room installation. They provide the communication mechanism for the distributed entities of the system, control of the sensing infrastructure, and allow service providers to register their service logic into the framework. Besides, some agents that provide basic services, such as the ability to track composite situations, the control of sensors, access to the knowledge base, are tightly coupled with the installed infrastructure of each smart room.

4.3.6   SAMOA

SAMOA framework [8] supports the creation of semantic context-aware social networks, which consist of logical abstractions that represent groups of mobile users who are in physical proximity and share common affinities, attitudes and social interests. In particular, SAMOA lets mobile users create *roaming social networks* that, following user movements, at each instant reflect all nearby encounters of interest. Mobile users interested in creating social networks are called *managers*. They are responsible for defining the scope (i.e., radius) of discovery of their social network and the selection criteria. Other users located within the discovery boundaries are those *eligible* to become members of the *manager*'s social network. But only the users that are selected by the *manager* become *affiliated* with that social network.

4.3.6.1   Types of Context.

To support the creation of social networks in ubiquitous environments, SAMOA relies on geographical context information, e.g., a user's location and reciprocal proximity, user attributes and social preferences, and place descriptions. Users' location and proximity are determined either by the network cell (or the WiFi access point) the user is currently attached to, or by the number of network hops between users in an *ad hoc* network. The middleware provides graphic tools for specifying profiles of users and places.

4.3.6.2   Ontologies.

SAMOA models and represents context data in terms of semantic metadata. Places and users are the entities in the system. They are associated with profiles describing their characteristics. A *place profile* has an identification and an activity parts. The former includes a unique identifier, a name and a description of the physical place, and the latter includes all of the social activities that characterize the place, and which sorts of information members located in that place are expected to share. The *user profile* consists of an identification and a preference part. The identification part provides user naming information and describes user properties, such as age, gender and education, and the preference part defines the activities the user is interested in and, for each of these activities, the user's specific preferences. Besides *place* and *user profiles*, managers also have a *discovery profile* associated with each place, defining which preferences *user profiles* must match to join the manager's social network at that place. Preferences in *discovery profile* include desired client attributes for each activity. While activities and preferences in the *place profile* and in the manager's *discovery profile* are represented as classes, activities and preferences in a *user profile* are defined as instances.

4.3.6.3   Inference/Reasoning Techniques.

SAMOA exploits two semantic matching algorithms for analyzing profiles and inferring potential semantic compatibility among users. The first algorithm operates on *user* and *place profiles* to identify a first set of eligible members located within an area of interest around a place. Only those users whose profiles have activities that are semantically related to that of the *place profile* activities become *eligible members*. The second matching algorithm selects among the previously selected *eligible members* only those users whose attributes semantically match the preferences included in the manager's *discovery profile* for that particular place. Moreover, the matching algorithms perform also ontology reasoning to identify if the activity or preference in the *user profile* is an instance of a more generic activity or preference class, or an instance of a more specialized activity or preference class in the manager's *place* or *discovery profile*. SAMOA relies on the Pellet DL reasoner [59] for implementing both matching algorithms.

4.3.6.4   Knowledge Management.

No centralized database is kept in SAMOA. Place and discovery profiles are maintained and analyzed separately. The user's mobile devices keep their own user profiles. Some users that may become managers of a social-network keep on their devices discovery profiles associated with each place. Stationary devices may keep place profiles for each place. The manager communicates only the place profile to co-located users, preserving the privacy of its discovery profile. Similarly, users return their user profiles only to managers that provided places with activities of interest. In addition, keeping place and discovery profiles separate lets SAMOA distribute the overhead of the social-network extraction among all users, since the semantic analysis of the *place profile* is performed on user's devices, and semantic matching between *discovery* and *user profiles* is performed on manager devices.

4.3.6.5   Architecture.

The SAMOA middleware has totally distributed architecture organized in two logical layers: *the basic service layer* and the *social-network management layer*. The basic service layer provides facilities for naming, detection of co-located users and device communication. In this layer, the location/proximity manager (L/PM) lets SAMOA entities advertise their online availability by periodical broadcasts of advertisement messages. L/PM senses incoming advertisements and builds a table of "discovered" co-located users. The social-network management layer includes facilities for semantic-based social network extraction and management. In this layer, the *place-dependent social-network manager* (PSNM) creates and maintains a table that includes all members of the manager's social-network that are currently co-located with the manager. The *global social-network manager* (GSNM) keeps a record (in a dedicated table) of all place-dependent social networks previously formed at the visited places, i.e., the manager's global social network. In addition, the table stores the *place profile* and the *discovery profile* of the manager, which guided the selection of each member.

4.3.7   CAMUS

Context-Aware Middleware for URC (Ubiquitous Robotic Companion) System (CAMUS) is a context-aware infrastructure for the development and execution of a network-based intelligent robot system [36]. It was designed to overcome limitations of the ubiquity, context-awareness and intelligence that existing mobile service robots have. CAMUS gathers context information from different sensors and delivers appropriate context information to different applications. Moreover, CAMUS provides context-aware autonomous service agents that are capable of adapting themselves to different situations.

4.3.7.1   Types of Context.

In CAMUS, a *sensor framework* processes input data from various sources such as physical sensors, applications and user commands and transfers them to the *Context Manager* through an *Event System*. The Context Manager manages context information collected from the Sensor Framework. When context information in the environment is changed, the Context Manager transfers events to the Event System. The context represented includes the user context, environment context and computing device context. User context includes user profile, user's task information, user preference, etc. Environment context includes hierarchical location information, time, etc. Computing device context includes information about available sensors and actuators.

4.3.7.2   Ontologies.

The context model in CAMUS is represented as a four-layered space, where each layer has a different abstraction level. In the *common ontology layer* are modeled the ontology concepts that are commonly used in various applications. The common ontology provides the high-level knowledge description to context-aware applications. Generally, highly abstracted knowledge can be easily reused by various applications. The *domain ontology layer* comes below the *common ontology layer*. It provides the domain specific knowledge to context-aware applications. This layer is composed of the *infrastructure domain ontology* and a set of *specific domain ontologies* for the application. The *infrastructure domain ontology* is the schema of the context model that is represented and managed in the context-aware system. The *specific domain ontology* is about specific services, for example, a presentation service. The *domain ontology layer* provides the schema to the layer below, the *instance layer*, where instances of the ontology concepts are represented. Above the *common ontology layer* there is the *shared vocabulary layer*, where is defined a set of shared vocabulary (and their semantics) used in the *common ontology layer*.

4.3.7.3   Inference/Reasoning Techniques.

CAMUS context reasoning engine includes many different reasoners, which handle the facts present in the repository and produce higher-level contexts [26]. The reasoning service is used by some context mapping services and context aggregators. They invoke the reasoners through a fixed API, providing the reasoners with context data. All new inferred facts will be inserted into that context data for later queries. The use of a fixed interface for all kinds of reasoning engines makes it possible to add and handle different reasoners. Multiple reasoning mechanisms are available. Reasoners can infer high-level contexts using rules written in different types of logic like first order logic, temporal logic, description logic (DL), higher order logic, fuzzy logic, etc. Instead they can also use various machine learning techniques, such as Bayesian learning, neural networks, reinforcement learning, etc. The middleware defines wrappers for each reasoner type. Besides, a Racer server [27] provides ontology reasoning to infer subsumption relationships, instance relationships and consistency of context knowledge base.

4.3.7.4 Knowledge Management.

The application context model is stored in the CAMUS context storage through a Knowledge Base adaptor. Applications can refer and change application context through Jena APIs. Moreover, application context models can be updated and changed through the Jena rule engine and OWL reasoner depending on application-specific inference rules and subsumption reasoning.

4.3.7.5 Architecture.

CAMUS has a centralized architecture composed of three parts: Main Server, Service Agent Manager and Service Agents. The Main Server manages context information delivered from Service Agent Managers. It generates and disseminates appropriate events to applications according to the context changes. The Service Agent Manager provides the container where Service Agents are executed. A Service Agent is a software module that acts as a proxy to connect various external sensors and smart devices to CAMUS. It delivers information of sensors in environment to the Main Server, receives control commands from the Main Server, controls devices in the environment and conducts applications. The entities in the system communicate using PLANET, a lightweight and fault-tolerant communication mechanism which also supports the disconnected operations and asynchronous operations.

4.3.8 OWL-SF

The distributed semantic service framework, OWL-SF [44], supports the design of ubiquitous context-aware systems considering both the distributed nature of context information and the heterogeneity of devices that provide services and deliver context. It uses OWL to represent high-level context information in a semantically well-founded form. Devices, sensors and other environmental entities are encapsulated and connected to the upper context ontology using OMG's Super Distributed Objects technology [54] and communicate using the Representational State Transfer protocol [23]. Integrated reasoning facilities perform the automatic verification of the consistency of the provided service specifications and the represented context information, so that the system can detect and rule out faulty service descriptions and can provide reliable situation interpretation. A prototype of the system has been implemented and tested.

4.3.8.1 Types of Context.

OWL-SF uses Super Distributed Objects (SDOs) [54] to encapsulate context providers, which may be sensors, devices, user's interfaces (GUIs) or services.

4.3.8.2 Ontologies.

Each SDO that encapsulates context providers and service-providing devices is an OWL-SDO. This OWL extension adds new methods to a standard SDO which allow accessing the current state of an object as an OWL description. Each functional entity implemented as OWL-SDO has to be described using its own ontology containing terminological knowledge that enables the automatic classification of the object into appropriate service categories. The state of an object stores context values and is represented by an instance of a class in the ontology.

4.3.8.3 Inference/Reasoning Techniques.

Deduction servers (DSs) are specific OWL-SDO with an RDF inference mechanism and an OWL-DL reasoner. The rule-based reasoning process is provided by the RDF inference component and the deduced facts are used to trigger events to other SDOs and to process service calls. A subscription notification mechanism is used to monitor the SDO parameters to generate notifications whenever an observed parameter changes, triggering the deduction process to update the global ontology model accordingly. The RDF inference component is connected to the OWL-DL reasoner, which is responsible for classification and answering OWL-DL queries. The Racer system [27] is used as an OWL-DL reasoner.

4.3.8.4 Knowledge Management.

Besides providing deductive support, DSs are responsible for collecting the status of SDOs, published in the OWL format, and building an integrated OWL description accessible to the reasoning process. The semantic representation of each SDO is added to the internal database of the DS. This semantic representation consists of a set of instances augmented with rules. Facts deduced from rules are only used to change parameters and to call services but never modify the knowledge base.

4.3.8.5 Architecture.

OWL-SF is a distributed system and its functional architecture integrates two basic building blocks: OWL-SDOs and DSs. A system may be composed of multiple components of both types which can be added and removed dynamically at runtime. DSs use the SDO discovery and announcement implementation to become aware of new SDOs in the environment. Whenever a new SDO is discovered, its semantic representation is added to the internal database.

4.3.9   DRAGO

Distributed Reasoning Architecture for a Galaxy of Ontologies (DRAGO) is a distributed reasoning system, implemented as a peer-to-peer architecture in which every peer registers a set of ontologies and mappings, and the reasoning is implemented using local reasoning in the registered ontologies and by coordinating with other peers when local ontologies are semantically connected with the ontologies registered in other peers [57]. DRAGO is implemented to operate over HTTP and access ontologies and mappings published on the web.

4.3.9.1   Types of Context.

DRAGO does not implement a context layer, i.e., it does not have any service for context collection, storing or distribution.

4.3.9.2   Ontologies.

DRAGO considers a web of ontologies distributed among a peer-to-peer network. Each peer may contain a set of different ontologies describing specific domains of interest (for example, ontologies describing different activities of users in a university). These ontologies may differ from a subjective perspective and level of granularity. In each peer there are also semantic mappings defining semantic relations between entities belonging to two different ontologies. These semantic mappings are described using C-OWL [9]. To register an ontology at a peer the users specify a logical identifier for it, i.e., a URI, and inform a physical location of the ontology in the web. Besides that, it is possible to assign semantic mappings to the ontology, providing, in the same manner, the location of the mappings on the web. New peers may be added dynamically to the system, providing new ontologies and semantic mappings.

4.3.9.3   Inference/Reasoning Techniques.

The reasoning process may compare concepts in different ontologies to check concept satisfiability, determining if a concept subsumes the other (i.e., the latter is less general than the former), based on the semantic mappings relating both ontologies. In a set of ontologies interconnected with semantic mappings, the inference of concept subsumption in one ontology (or between ontologies) may depend also on other ontologies related to the previous ones through those mappings. Every peer registers a set of ontologies and mappings, and provides reasoning services for ontologies with registered mappings. Each peer may also request reasoning services from other peers when their local ontologies are semantically connected (through a mapping) with the ontologies registered at the other peer. The reasoning with multiple ontologies is performed by a combination of local reasoning operations, internally executed in each peer for each distinct ontology. A distributed tableau algorithm is adopted for checking concept satisfiability in a set of interconnected ontologies by combining local (standard) tableaux procedures that check satisfiability inside the single ontology. Due to the limitations of the distributed tableau algorithm, for a semantic mapping DRAGO supports three types of rules connecting atomic concepts in two different ontologies: *is equivalent*, *is subsumed* and *subsumes*. A Distributed Reasoner was implemented as an extension to the open source OWL reasoner Pellet [59].

4.3.9.4   Knowledge Management.

As each peer registers sets of heterogeneous ontologies and mappings, the knowledge base is totally distributed. When users or applications want to perform reasoning with a registered ontology they refer to the corresponding peer and invoke its reasoning services giving the URI to which the ontology was bound.

4.3.9.5   Architecture.

DRAGO aggregates a web of ontologies distributed amongst a peer-to-peer network in which each participant is called a *DRAGO Reasoning Peer* (DRP). A DRP is the basic element of the system and is responsible for providing reasoning services for ontologies using the semantic mappings registered. As these mappings establish a correlation between the local ontology and ontologies assigned to other DRPs, a DRP may also request reasoning services of other DRPs as part of a distributed reasoning task. A DRP has two interfaces that can be invoked by users or applications. A *Registration Service Interface* is available for creating/modifying/deleting registrations of ontologies and mappings assigned to them. A *Reasoning Service Interface* enables requests of reasoning services for registered ontologies. Among the reasoning services DRAGO allows to check for ontology consistency, build classifications, verify concepts satisfiability and check entailment.

4.3.10   Conclusion

In this section, we classify the surveyed systems according to our taxonomy (cf. Table 4.2), and discuss their suitability for implementing context-oriented ontological reasoning for Ambient Intelligence. The eight systems we presented not only have different features, but some of them have been developed with different purposes. Gaia, CoBrA, Semantic Spaces and CHIL offer middleware infrastructure for Smart Spaces; SAMOA is designed specifically to support applications that deal with social networks in ubiquitous environments; CAMUS provides an infrastructure for the development and execution of a network-based intelligent robot system; OWL-SF supports the design of generic distributed context-aware systems; finally, DRAGO provides reasoning about heterogeneous ontologies.

Table 4.2: Classification of middleware systems for context-oriented ontological reasoning.

	Set of context providers	Ontology update	Knowledge base	Main goal of reasoning	Handling of heterogeneous knowledge bases
Gaia	Dynamic	Dynamic	Distributed	Derive higher-level facts	No
CoBrA	Dynamic	Static	Centralized	Both	No
Semantic Spaces	Dynamic	Static	Centralized	Derive higher-level facts	No
CHIL	Static	Static	Centralized	Derive higher-level facts	No
SAMOA	Static	Dynamic	Distributed	Derive higher-level facts	No
CAMUS	Dynamic	Dynamic	Centralized	Derive higher-level facts	Shared vocabulary layer
OWL-SF	Dynamic	Static	Distributed	Both	No
DRAGO	—	Dynamic	Distributed	Classification	Semantic Mapping

Comparing the four frameworks for Smart Spaces, it may be said that Gaia is the only one that supports distributed knowledge bases and is the one that best deals with dynamic scenarios, allowing context providers to be added or removed dynamically and ontologies to be dynamically modified with regard to types of context and their properties. Despite not being tailored specifically for smart spaces, OWL-SF may be used for implementing such systems, as its singular characteristic is its support for distributed inference. Similar to Gaia, OWL-SF considers a distributed knowledge base. Hence, in each space, the aggregated context information will depend on the available providers, avoiding communication bottlenecks and allowing more efficient information processing and dissemination. The disadvantage of this approach is that context consumers cannot know beforehand which context information will be available at each space, and that it may happen that the necessary information may not be available.

While most systems have no mechanism to deal with heterogeneity of context representation through different spaces, CAMUS and DRAGO pay attention to this subject. CAMUS has its ontology structured in layers to provide a shared vocabulary as an approach to tackle the problem, while DRAGO is the only one that supports the inclusion of generic mappings between the ontologies. However, DRAGO does not provide a context infrastructure, i.e., it does not have any service for context acquisition and distribution. In fact, it is solely dedicated to support reasoning with heterogeneous ontologies.

AmI applications are composed of independent entities that act autonomously in an open-ended environment, driven by their own goals. In order to fulfill their tasks, collaboration with peers is often required. Different entities are very likely to employ different knowledge representations; therefore the ability to align such representations into a single one that can be shared by different applications is paramount to ensure communication. DRAGO architecture, presented in this section, relies on pre-defined mappings to align different ontologies. Nevertheless, in practical implementations of AmI it is not feasible to build in advance mappings of all possible pairs of different ontologies that may be needed. There are other techniques to overcome the barrier of heterogeneous representations in such conditions. The next section is dedicated to a survey of approaches that try to solve exactly this problem.

4.4 Approaches for Ontology Alignment

Entities acting autonomously in an open-ended environment will often require collaboration with peers to fulfill their goals. Because different entities are likely to provide separate ontologies, the ability to integrate the ontologies into a single representation is paramount to ensure overall communication. This need, often referred to the *ontology alignment* problem [35], consists in finding a set of equivalence between a set of nodes in ontology A and a set of nodes in ontology B (see Figure 4.1). More formally, the problem of ontology alignment can be compared to that of database schema matching. Given two schemas, A and B, one wants to find a mapping m from the concepts in A into the concepts of B in such a way that, for all $(a, b) \in A \times B$, if $a = \mu(b)$, then *b* and *a* have the same meaning. Several approaches have been proposed to perform such alignments. They can be organized into three categories [35]: structural methods (which rely only on the structure of the ontology and the nodes labels), instance-based methods (which compare the instances of each concept in the ontologies) and methods based on a reference ontology which acts as a mediator. This field is wide and complex,[1] but its application to the interaction of entities in ubiquitous environments leads to the specification of a sub-category of problems:

- The alignment process must be performed *on the fly* and in a *limited amount of time*. Indeed, in open systems, it is not possible to know in advance the nature of the entities that interact, which makes impossible to compute in advance the alignment of their ontologies.

- The entities that interact share common goals or common capacities. Thus, one can consider in most applications that the intersection of ontologies will not be empty. As a consequence, there always exists an acceptable alignment between two ontologies. However, one cannot take for sure that concepts will appear at the same level of specialization. For instance, one ontology can have a single class for the concept of *research paper*, while the other directly works with the sub-concepts *journal*, *conference_proceedings*, etc.

- The ontology alignment must be performed automatically (whereas a lot of work in this domain relies on semi-automatic approaches). As a consequence, entities must decide on alignments without the validation of a human expert. Thus, they must be able to evaluate the trust they have in the resulting alignment, e.g., by valuating the equivalence links depending on their ambiguity.

The next subsection presents the lexical alignment (a.k.a *anchoring*) that is used as a basis by all ontology alignment approaches. We then present the three main approaches for ontology alignment (structural, instance-based and mediation-based). We illustrate the advantages and drawbacks of each technique and a brief overview of the most significant work in each category. Subsection 4.4.5 then presents a brief overview of semantic similarity measures and how they can offer a new solution for ontology alignment.

4.4.1   Lexical Alignment

Lexical anchoring is, generally, the first processing step of ontology alignment tools. It is possible to differentiate several kinds of approaches, with advantages and drawbacks. First are classical Natural Language Processing

---

[1] See http://www.ontologymatching.org for a complete description.

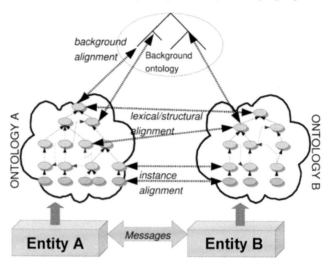

FIGURE 4.1: Ontology alignment is a set of equivalences between nodes of both ontologies. This schema presents the three classical solutions: alignment based on the structural properties, alignment based on the concepts instances and alignment based on a "background ontology."

tools, such as lemmatization (which constructs singular or infinitive forms of words, for instance, determining that *kits* is the plural of *kit*, *bought* is a derived form of *buy*), tokenization (which considers each word of a compound concept, like *long_brain_tumor subClassOf long_tumor* [3]) or suffix/prefix approach (which searches in a sub-part of the words. For instance, like *net* is an abbreviation of *network*, *ID* can stand for *PID*). However, these approaches have some limitations: the lemmatization can be ambiguous (out of the sentence context, *left* can be lemmatized either into *left:adjective* or *leave:verb*); the tokenization requires choosing the correct sub-concepts inference (is *brain_tumor subClassOf brain* a valid association?); and the prefix/suffix alignment is strongly dependent on the language (for instance, *hotel* should not match *hot*, nor can *word* be seen as an abbreviation of *sword*). For these reasons, the lexical anchoring has to be used with great care and to be completed and/or confirmed with other techniques.

A complementary approach of all these methods is the lexical distance measure, so called "edit distance" between two strings (Hamming distance or Levenhstein distance). For example, the Levenhstein distance is given by the minimum number of operations needed to transform one string into the other, where an operation may be an insertion, deletion or substitution of a single character. It is widely used for spell checking. The main advantages of edit distance are that it reproduces NLP approaches when words do not have too much complexity. For instance, the translation from plural to the singular form has a cost of 1 in most words (removing the trailing "s"). However, some drawbacks still remain, like *sword* is equivalent to *word*, which has a cost of 1 and could be wrongly accepted.

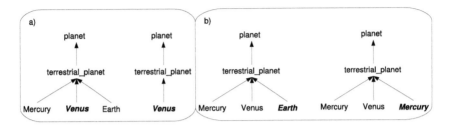

FIGURE 4.2: Structural alignment error example based on hierarchy analysis.

4.4.2 Structural Approaches

Structural approaches are based on the structural comparison of the two concepts graphs (in the meaning of graph theory). It relies on lexical anchoring as a first step for associating lexically-close labels from both ontologies. The complementary alignment pairs are obtained by an extended hierarchy comparison around these anchored concepts (e.g., in CATO [14], the authors make use of a specific algorithm for tree comparison (so-called TreeDiff) to find the largest common substructure between trees. The CATO system will be presented in more details in Subsection 4.5.4. More generally, such structural methods will match terms like PC and Personal Computers when sub-classes and properties describe the same concept (like ID, model, etc.). However, structural alignment may fail if the information is not classified using the same criterion or if the ontologies do not cover the same fields or instances. As Figure 4.2a shows, the concept "Venus" from the ontology to the right will be correctly aligned with the concept "Venus" from the ontology to the left, because they share lexically-close concepts in their whole hierarchical structures. But in Figure 4.2b, the concept "Mercury" from one ontology will be wrongly aligned with the concept "Earth" from the other ontology, because, although they do not have the same meaning, they also share lexically-close concepts in their whole hierarchical structures.

4.4.3 Instances-Based Approaches

The objective of these methods is to determine an alignment using common instances between the two ontologies. When the common instances are identified, the main idea is to suppose that the hierarchy declares these instances under the same concepts (maybe structurally or lexically different).

For example, in [30], the authors tried to align the category's hierarchy of Google and Yahoo. An instance is identified using the URL of websites. Regarding [63], the positive and/or negative matches of instances between two concepts allows them to compute subsumption alignment, in addition to equivalence alignment. For example, in Figure 4.3, if instances of the concept "a" of ontology A are classified as instances of the concept "c" of ontology B and the opposite is not true, then it is possible to deduce that the concept "c" is a super-class of the concept "a".

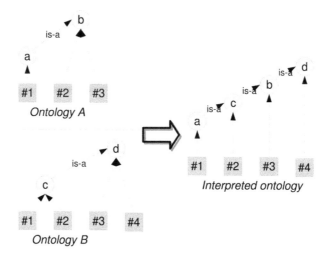

FIGURE 4.3: Instance-based alignment allows construction of subsumption alignment in addition to equivalence alignment.

However, the main drawback of this approach is the instances detection. For example, the work by Ichise et al. comparing Yahoo and Google hierarchy only generates 10% of common instances. Moreover, in van Diggelen work, it is difficult to conclude if instances intersection is not complete (i.e., if one class does not contain all instances of another class), even if it is just a problem of misidentification of concepts in one of the two ontologies.

4.4.4   Mediated Approaches

Mediated approaches are based on the use of a third ontology to mediate the alignment process (see for instance [4, 10]). The main advantage of these methods is to be more robust in case of ontologies that differ greatly either lexically or structurally, or when no instances are provided. For example, in [3], the authors align two ontologies with very different formalisms, which could not be done using a structural approach. Thus, one of the major pre-requisites (which is also the major limit of the approach) is to have access to a mediator-ontology with enough information to anchor concepts from the two initial ontologies on it. The anchoring stage is generally a lexical anchoring as presented in the previous section. After the anchoring stage, the two ontologies are represented by two set of concepts from the mediator ontology. For example, in Figure 4.4, following the subsumption relation allows the authors to find an alignment between "jeep" (ontology A) and "car" (ontology B), even if the two ontologies do not share any label. The main difficulty in me-diated approaches is to define a strategy to construct semantic paths between these two sets, using the structure of the mediator ontology.

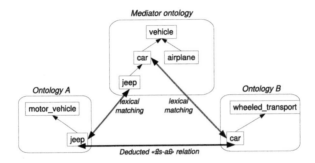

FIGURE 4.4: Mediated alignment approach.

4.4.5 Alignment Based on Semantic Similarity

Finding paths between concepts in an ontology is at the core of mediated approaches but it can also be used to complete lexical and structural align-ments. For instance, a given ontology concept may not be directly attached to an application-defined concept, as required for context interpretation. This case may happen, for instance, if the alignment was difficult, or if the ontol-ogy is large. Thus, we can use the *semantic similarity measure* on the entity ontology (as it is done on the background ontology in mediated approaches) to compute correct semantic paths and to valuate the strength of this path.

In this section, we first recall the general principle of semantic similarity and we then propose to use it within a mediated approach for aligning two entity ontologies.

4.4.5.1   Semantic Similarity

From a theoretical point of view, semantic similarity is a formula that allows users to evaluate the amount of common attributes shared by two concepts. Different kinds of approaches were proposed, based on a concepts hierarchy (e.g., [15]), on glosses from a dictionary (e.g., [6, 43, 48]), etc. In this paper, we will focus in similarity for ontology, since it is suitable with our initial problem of interaction between entities.

Work on semantic similarity with ontologies can be split in two major methods: the *edge-based* approach and the *node-based* approach. The edge-based approach [49] makes use of the shortest taxonomic path to define the semantic similarity between two concepts. The main idea behind it is intuitive: in a semantic taxonomy, the longer the path, the less semantically similar are the two concepts. Recent work in this area has focused on the issue of weighting the edges, which allows refining the value of a semantic link (e.g., [29, 31, 70, 71]).

The node-based approach [51] is the most used nowadays and is considered to be the most efficient for the semantic similarity ([15] provides a good survey on the subject). The weight of a node represents the information content of the concept. In other words, the more general a concept is (i.e., near from the root), the less information it contains. There exist different kinds of formulae that combine the information content of the two target nodes and their closest common parent. The *closest common parent* is the node that is the most specific in the set of common ancestors nodes (e.g., [32, 37, 38]).

4.4.5.2   Semantic-similarity Based Alignment

In Aleksovski's work (Section 4.4.4), the use of mediator ontologies is limited to a reduced set of patterns of paths, which are considered to be "semantically correct." Thus, two concepts can only be related with a binary relation (the alignment exists or it does not exist). In the framework of interacting entities, we suggest that it is necessary to have a solution to valuate the strength of an alignment. This weight will be useful to solve ambiguity and to propose more complex dialogue strategies, as proposed in [41].

The mediator ontology should be either WordNet [22] (especially for human/agent communication) or the ontology of a third mediator agent if this ontology contains some common concepts of the two other ontologies. A first lexical anchoring of terms within the mediator ontology is performed, using the edit distance of Levenshtein. Then, the system computes the set of all possible semantic scores between each concept from the set of anchored concept of the first ontology to the set of anchored concept of the second ontology.

The key problem in this approach is that a real ontology inherently contains a lot of different relation types. To tackle this problem, we have proposed a measure of semantic relatedness [42], which is more general than the similarity measure [51], to take into consideration the entire graph and not only the hierarchy. The preliminary evaluation of our measure, applied to human-machine communication, emphasized that our correlation factor with human judgment is approximately 20% better than other measures.

4.4.6   Conclusion

Ontology alignment is a key issue in open systems that require some form of distributed reasoning, such as in open Ambient Intelligence. While most middleware systems currently use a single ontology, openness and heterogeneity require distributed entities to be able to interpret information derived from other peers, based on an a priori unknown ontology. The three main approaches for ontology alignment we presented (structural, instance-based and mediation-based) all contain some limitation. While the structural methods are the most efficient ones, they require that the ontologies be very similar (e.g., two ontologies derived from a single initial specification). The instance-based approaches offer the consistency of Description Logic inference rules, but they work only if each concept is associated with a complete set of instances (e.g., document URIs). Mediation allows aligning very heterogeneous ontologies (even the knowledge representation formalisms can differ), but the background ontology is generally very large (i.e., larger than each one of the mediated ontologies).

4.5   The Campus Approach

In this section, we discuss Campus, a framework for the development of Ambient Intelligence applications [56]. Based on multi-agent systems technology, Campus provides an infrastructure to develop innovative context aware applications that accommodate mobile devices and environment sensor devices. The Campus architecture is intended as a configurable framework in which users can decide what services they want to enable in their environments, rather then a monolithic application. It is composed of three levels: the context-provisioning layer, the communication and coordination layer and the ambient services' layer, as illustrated by Figure 4.5. In a nutshell, the bottom level is responsible for offering basic middleware services and functionalities, such as providing context information and device discovery. The communication and coordination layer offers support for semantic interoperability, providing discovery, exchange and collaboration among hybrid entities, regardless of proprietary representations of information. Finally, the topmost layer provides application specific and ambient services and acts as a hotspot, i.e., allows users to extend the framework by plugging in specific services required by a particular user, environment, type of collaboration, of interest to their environment.

As shown in Figure 4.6, agents distributed through the two bottom layers

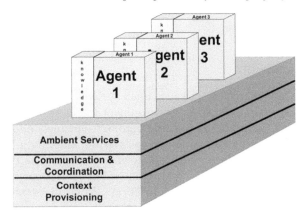

FIGURE 4.5: Abstract view of Campus architecture.

implement the main functionalities of Campus. In the context-provisioning layer, *Context Monitor Agents* (CMA) collect raw context data from various sources such as devices, sensors and applications, and make it available for interested entities. *Distributed Reasoning Agents* (DRA) infer and disseminate higher-level context information. In each smart space a *Local Knowledge Agent* provides persistent knowledge storage. It aggregates the context information obtained from context providers (i.e., CMAs and DRAs) available in that area, and builds a partial ontological view. The LKA may be queried by entities interested in finding context providers in that area. In the communication and coordination layer, a *Knowledge Interoperability Agent* (KIA) is responsible for semantic alignment of ontologies. It will provide this information to LKA whenever needed. In a further part of this section, the main features of Campus are discussed separately.

4.5.1 Context Types

In the Campus framework, context data comprises not only information about mobile devices, users' preferences and roles, description of institutional physical spaces, but also data collected from personal and smart spaces applications (e.g., appointments in a personal agenda, list of activities in an organizational scheduler, etc.). *Context Monitor Agents* (CMA) are responsible for collecting raw context data from various sources such as sensors, devices and user applications; interpreting it as context information according to a predefined ontology; and making this information available for interested entities.

Most information about mobile devices is acquired with the aid of the MoCA middleware [53], a component that supports the development of context-aware applications and services for mobile computing. MoCA provides efficient services to collect context information associated with mobile devices (e.g., CPU usage, available memory, battery level, etc.). This information comprises not only raw data related to the device's resources and the wireless links (currently, only IEEE 802.11), but also the symbolic location of each device, which is inferred from the RSSI values measured at the device with respect to all the Wi-Fi Access Points in the device's vicinity. On the other hand, much context data is obtained from applications and files that store personal and organizational information. For example, a CMA running on the notebook of professor Silva could make available information about Silva's agenda and preferences, and also about Silva's notebook. A CMA agent running on a fixed computer at LIP6 could collect data about the schedule of activities, worker profiles, etc.

4.5.2   Ontologies

The Campus upper ontology serves as a knowledge base for the framework implementation, i.e., provides the necessary semantics to allow high-level exchanges, including brokering, negotiation and coordination amongst software entities. It contains precise definitions for every relevant concept in the framework, e.g., it defines that context providers and services are described by a tuple containing its name, a parameter list, a capability list, the communication port number and protocol. Of course, the concepts of name, parameter list, capabilities list, port and protocol are also defined in the ontology. This ontology serves as a static model of our domain and will be used as a basis upon which mediation services will try to reason and understand the information provided by entities in the environment.

4.5.3   Reasoning

In Campus, we propose a distributed reasoning mechanism to infer and disseminate higher-level context information, i.e., context information that may be deduced using data obtained from other context providers. In our approach, each reasoning element is called a *Distributed Reasoning Agent* (DRA). A DRA is able to deduce new knowledge reasoning about description logic rules. In such rules, atoms that depend on some context data compose the antecedent of the rule, while the consequent of the rule defines a new piece of context information. CMAs collect context information from several sensors or obtain it from applications and database files that contain user's preferences, device's descriptions or specific data, such as the list of activities scheduled for a set of rooms, and make it available for DRAs. Facts are deduced in runtime, described according to the respective ontology, and updated in the knowledge base. DRA implements also an event-based communication interface to where other entities subscribe their interest about a high-level context (defined by a rule). These entities are notified whenever the state context satisfies a given rule [64].

Since we consider a fully distributed environment, some context informa-

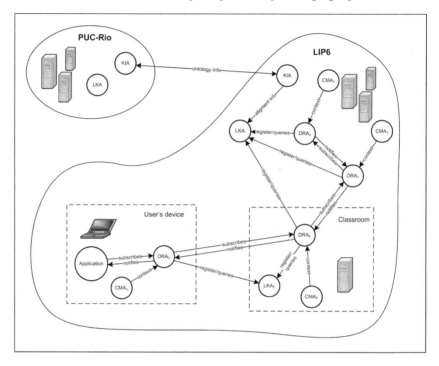

FIGURE 4.6: Campus multi-agent architecture.

tion necessary for processing the logical inference of a given rule may not be directly available from CMAs for a DRA. In such a case, the DRA will subscribe at other DRAs, which are capable of providing (by inference) the required piece of context information. In this sense, each DRA acts simultaneously as a provider, a reasoner and a consumer of context knowledge. This distributed approach brings several advantages. It allows the distribution of the high computational cost of the inferences process. As each DRA may have a different "context view," depending on the device on which it is running and its location, not all items need to be kept at a single database, avoiding a communication bottleneck. Besides, distributed inference may hide private data still revealing context information that may be inferred from it.

Figure 4.6 shows an example of this distributed interaction in the scenario where Mr. Silva enters a classroom to attend a meeting with the Campus team. In this case, Silva's smartphone executes DRA_U (user's DRA) that has access to the data obtained from the sensors at the smartphone (sound, luminosity, movement), its own location data and some administrative data available for Silva. As he enters the room, i.e., changes its location, DRA_U initiates a discovery process, and as a result, it detects the presence of LKA_E and then another reasoner, DRA_E, which is responsible for accessing the room's sensors, storing their context data and doing ambient-specific context reasoning. We assume that an application at the smartphone responsible for managing the ring tone has already subscribed at DRA_U to get control notifications for the ring tone adjustment according the following rule:

Device(?d) ∧ isLocatedIn(?d,?e) ∧ ClassroomInUse(?e) ⇒ InSilenceMode(?d)

While a CMA that wraps a location service delivers the smartphone's binary property "isLocatedIn" to DRA_U, the "ClassroomInUse" unary property is only made available by the DRA_E responsible for the classroom. This property is inferred from the following rule:

Environment(?e) ∧ hasScheduledActivity (?e, ?a) ∧ ClassActivity(?a) ∧
ActivityOncourse (?a) ∧ LecturerPresent(?e) ⇒ ClassroomInUse(?e)

Moreover, supposing that DRA_E has no direct access to the activities' time schedule and time reference, nor to the location data of every person in the institution, it has to rely on context knowledge inferred by another two agents: DRA_3 to monitor the "ActivityOnCourse" unary property and DRA_4 to obtain notifications of the "LecturerPresent" unary property. Then, DRA_3, running on an "activity manager" (i.e., fixed device running a CMA dedicated to monitor the time and the schedule of activities), will process the following rule:

Activity(?a) ∧ startTime(?a, ?t1) ∧ finishTime(?a, ?t2) ∧ presentTime(?t3) ∧
isLessThan(?t1, ?t3) ∧ isBiggerThan (?t2, ?t3) ⇒ ActivityOncourse (?a)

At the same time, DRA_4, running on a "location manager" which has access to location information from all mobile devices detected in the building, will process the following rule:

Device(?d) ∧ isLocatedIn(?d,?e) ∧ isCarriedBy(?d,?p) ∧ playsRole(?p, ?r) ∧
Lecturer(?r) ⇒ LecturerPresent(?e)

In fact the reasoning outcome of DRA_U will be the result of the cascading reasoning, with new context data being inferred initially by DRA_4.

It is straightforward to notice that for interactions among DRAs referencing different ontologies, it is necessary to provide an intermediating agent that has the ability to resolve – or at least, try to resolve – the semantic mismatch among nodes in the different ontologies. For example, when the DRA_U, which is a foreign entity, wants to obtain the "ClassroomInUse" context data, and this same information is represented as "RoomBusy" in DRA_E, then the intermediate agent will have to support the interaction between the entities identifying the identity of "ClassroomInUse" and "RoomBusy" in the scope of this particular application.

4.5.4   Ontology Alignment

In this section, we will consider the problem of ontology alignment (as defined in Section 4.4) in the Campus architecture. To enable the automation of the process, we coded the resource representations using W3C's OWL-DL ontology standard, for the reasons previously discussed in Subsection 4.2.3. Ontologies are expressive, formal, machine processable representations that fulfill the knowledge requirements of AmI applications.

Our past experience with semantic interoperability enabled us to provide CATO [11, 13, 14, 21], a solution that combines well known algorithmic solutions, e.g., natural language processing, the use of similarity measurements, and tree comparison, to the ontology alignment problem. We propose to incorporate CATO to the kernel of the Campus framework. The philosophy underlying CATO's strategy mixes syntactical and semantic analysis of ontological components. Its current implementation combines the lexical and structural approaches discussed in Subsections 4.4.1 and 4.4.2 respectively.

During the alignment, lexical and structural comparisons are performed in order to determine if concepts in different ontologies should be considered semantically compatible. A refinement approach is used that alternates between lexical and structural comparison between ontological concepts. The process begins when concepts from both ontologies go through a lexical normalization process, in which they are transformed to a canonical format that eliminates the use of plurals and gender flexions. The concepts are then compared, with the aid of a dictionary. The goal is to identify pairs of lexically equivalent concepts.

We assume that lexically equivalent concepts imply the same semantics, if the ontologies in question are in the same domain of discourse. For pairs of ontologies in different domains, lexical equivalence does not guarantee that concepts share the same meaning [45, 61]. To solve this problem, we adopted a structural comparison strategy. Concepts that were once identified as lexically equivalent are now structurally investigated. Making use of the intrinsic structure of ontologies, a hierarchy of concepts connected by subsumption relationships, we now isolate and compare concept sub-trees. Investigation on the ancestors (super-concepts) and descendants (sub-concepts) will provide the necessary additional information needed to verify whether the pair of lexically equivalent concepts can actually be assumed to be semantically compatible.

Lexical comparison is done during the first and second steps of the strategy. Structural analysis is done in the second and third steps of the strategy. The final result is a OWL document containing *equivalent class* statements (<owl:equivalentClass>) that relate the equivalent concepts from the two input ontologies. This is equivalent to a mapping between conceptual schemas. The proposed strategy is depicted in Figure 4.7.

4.5.4.1   First Step: Lexical Comparison

The goal of this step is to identify lexically equivalent concepts between two different representations, as presented in Subsection 4.4.1. We begin by assuming that lexically equivalent concepts are also semantically equivalent in the domain of discourse under consideration, an assumption that is not always warranted.

Each concept label in the first ontology is compared to every concept label present in the second one, using lexical similarity as the criteria. Filters are used to normalize the labels to a canonical format: (i) If the concept is a noun, the canonical format is the singular masculine declination; (ii) if the concept they represent is a verb, the canonical format is its infinitive. Besides using the label itself, synonyms are also used. The use of synonyms enriches the comparison process because it provides more refined information. For example, in the scenario proposed in Subsection 4.1.1 of this chapter, the "activity," "class" and "meeting" concepts were identified as synonyms in our database.

Lexical similarity alone is not enough to assume that concepts are semantically compatible. We also investigate whether their ancestors share lexical similarity. It is important to note that the alignment strategy in this step is restricted to concepts and properties of the ontology. As a result of the first stage of the proposed strategy, the original ontologies are enriched with synonyms and links that relate concepts that are known to be lexically equivalent.

4.5.4.2   Second Step: Structural Comparison Using *TreeDiff*

Comparison at this stage is based on the subsumption relationship that holds among ontology concepts, similarly to what was discussed in Subsection 4.4.1, not taking into consideration ontology properties and restrictions. Our approach is thus more restricted than the one proposed by Noy and Musen [45], that analyzes the ontologies as graphs, taking into consideration both taxonomic and nontaxonomic relationships among concepts.

Because we only consider lexical and structural relationships in our analysis, we are able to make use of well-known tree comparison algorithms. We are currently using the *TreeDiff* [65]. Our choice was based on its ability to identify structural similarities between trees in reasonable time.

The goal of the *TreeDiff* algorithm is to identify the largest common substructure between trees, described using the DOM (*Document Object Model*) model [5]. This algorithm was first proposed to help detect the steps, including renaming, removing and addition of tree nodes, necessary to migrate from one tree to another (both trees are the inputs to the algorithm).

The result of the Tree Diff algorithm is the detection of concept equivalence groups. They are represented as subtrees of the enriched ontologies. Concepts that belong to such groups are compared in order to identify if lexically equivalent pairs can also be identified among the ancestors and descendants of the

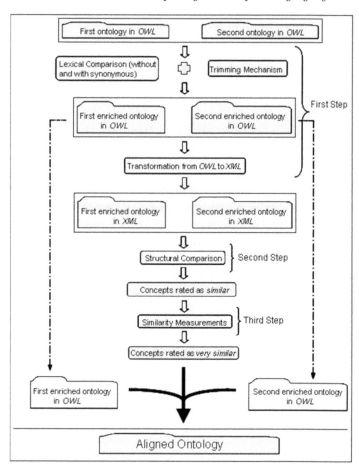

FIGURE 4.7: CATO ontology alignment strategy.

original pair. Differently from the first step, where we based our analysis and compared concepts that were directly related to one another, we are now considering the structural vicinity of concepts. Every concept in the equivalence group is investigated in order to determine lexically equivalent pairs, number of matching sons, number of synonymous concepts in the subtrees, available from the previous step, and ancestor equivalence.

4.5.4.3   Third Step: Fine Adjustments Based on Similarity Measurements

The third and last step is based on semantic similarity measurements, as discussed in Section 4.4.5.1. In CATO, concepts are rated as very similar or little similar based on pre-defined similarity thresholds. We only align concepts that were both classified as lexically equivalent in the second step, and thus rated very similar. Thus the similarity measurement is the deciding factor responsible for fine-tuning our strategy. We adapted the semantic similarity measurement strategies proposed in [40]. The similarity threshold is fixed by the users, and can be adjusted to enforce a firmer similarity policy.

During this step the "Activity" and "Class" concepts, from the visiting professor scenario, are aligned with the "Talk" and "Lecture" concepts belonging to the Campus upper ontology. Those concepts were rated equivalent during the second step. Their similarity level is calculated in the third step.

The final ontology, containing mappings between concepts imported from the two input ontologies, will provide a common understanding of the semantics represented by both input ontologies. This representation can now be shared by entities searching for information, seeking to discover or to compose with other AmI applications. Table 4.3 depicts a part of the output ontology for this example.

In Campus, the ontology alignment is implemented by the *Knowledge Interoperability Agent* (KIA), which is responsible for applying the CATO strategy for determining the equivalent classes between different ontologies, whenever foreign entities come to interact together. If an entity queries the context infrastructure looking for some context information represented using a different ontology, it will resort to KIA to obtain the equivalent class in the prevalent ontology.

Table 4.3: A short example of the code for ontology alignment.

```
<owl:Class rdf:ID="Activity">
    <rdfs:subClassOf rdf:resource="#CATO_Thing"/>
    <owl:disjointWith rdf:resource="#Event"/>
    <owl:equivalentClass>
        <owl:Class rdf:about ="Talk">
    </owl:equivalentClass>
</owl:Class>
<owl:Class rdf:ID="Lecture">
    <rdfs:subClassOf rdf:resource="Educational_Activity"/>
    <owl:equivalentClass>
        <owl:Class rdf:about ="Class">
    </owl:equivalentClass>
</owl:Class>
```

The main contribution of CATO's strategy is to combine well-known algorithmic solutions, such as natural language processing and tree comparison, to the ontology integration problem. CATO is fully implemented in Java and relies on the use of the JENA API. The use of the API helped us to focus on the alignment process, for it made ontology manipulation transparent. JENA reads and filters information from the tags of files written in an ontology language and transforms it to an abstract data model in which ontological concepts can be manipulated as objects.

4.6   Conclusion and Open Problems

Design and operation of open Ambient Intelligence Environments pose several huge challenges to the research community, which are caused mainly by the inherently heterogeneous, distributed and dynamic nature of these systems. In this chapter we first surveyed, analyzed and classified several middleware systems that propose partial solutions to the corresponding complex problem of distributed context reasoning. It turned out that only some of the systems support distributed context reasoning, and in fact only two tackle the problem of managing heterogeneous context knowledge. Then, we discussed the main general approaches for semantic alignment of ontologies, as this is a basic requirement for coping with heterogeneous knowledge representations. Finally, we presented our approach for distributed reasoning and semantic alignment in the scope of our ongoing effort to develop a multi-agent based framework Campus for development of Ambient Intelligence.

Within the Campus framework we focused on the provision of distributed context reasoning – using Distributed Reasoner Agents (DRAs) – and the construction of a software component responsible for the automatic alignment of ontologies – the Knowledge Interoperability Agent (KIA). Our strategy is based on the application of well-known software engineering strategies, such as lexical analysis, tree comparison and the use of similarity measurements, to the problem of ontology alignment. Motivated by the requirements of AmI applications, we proposed an ontology alignment strategy and tool that produces an ontological representation that makes it possible for such applications to share common understanding over information available on environment [69].

4.6.1   Discussion and Future Work

Building complex Ambient Intelligence environments requires the integration of several different context providers, which may be dedicated sensors, user's applications, databases monitors, etc. The inclusion of new types of context providers will require the implementation of new Context Monitor Agents with specific interfaces and functionalities.

The distributed inference of high-level context information brings the advantage of sharing the complexity of the reasoning process among several devices, allowing quicker response times. But on the other hand it requires efficient context dissemination to work efficiently. Aiming for efficient performances, the balanced distribution of DRAs among the devices that compose Ambient Intelligence typical scenarios is an issue to be investigated.

Any automated ontology alignment solution presents some degree of risk, in the sense that it cannot fully guarantee that the most adequate equivalence between concepts will be always identified. Limitations of the algorithms used, time to perform the computations and possible lack of information coded in the original ontologies may, in some of the cases, prevent the automated solution to identify answers that would be otherwise manually found. The success of the CATO approach depends on the volume and quality of the information coded in the input ontologies. The richer and more complete the information, the better the results. Conversely, if the input ontologies are poorly defined, incomplete or lacking, the ontology integration engine has little data to work upon, and thus is not likely to deliver adequate results.

To tackle such situations, an alternative solution may be the use of an instance-based approach, as presented in Subsection 4.4.3. Each implementation of a device, such as Dr. Silva's smartphone (SMP-1) or notebook (NTB-1) can be thus represented by an ontology, containing a set of classes, restrictions, properties (data schema), that corresponds to the internal knowledge representation of each device. The goal of the instance-based approach is the same, i.e., find matching classes across different ontologies.

The instance based approach uses a query probing technique that consists of exhaustively sending keyword queries to original ontologies [66]. Further analysis of the results using learning algorithms and statistical analysis provides indication of good matches. This approach can be generalized to any domain that provides a reliable substitute for an unique instance identifier. In the Geographic Information systems domain, for example, there are various geo-referencing schemes that associate a geographic object with a description of its location on the Earth's surface. This location acts as a universal identifier for the object, or at least an approximation thereof. We have successfully applied this approach to build mediators for Geographic Data Catalogs [11, 12, 24]. We are currently adapting the approach to be part of the Campus Framework kernel and help improve CATO results.

The CATO semantic adjustment makes use of the Maedche architecture [40], to confirm that a "very similar" rated alignment is semantically correct (and not only lexically and structurally). However, as stated by Maedche himself, the semantic measure used has two limits: 1) it only uses the taxonomic information from the ontology; 2) it does not consider that two different given edges in a taxonomy do not carry the same information content (as demonstrated in [51], see Subsection 4.4.5.1). We have proposed in [42] a new measure of semantic relatedness, which considers different weight for edges and different edges types. The preliminary evaluation of our measure shows that it increases approximately by 20% the correlation factor with human judgment. We currently try to integrate this measure in a refinement of the Maedche algorithm, in order to enhance the semantic adjustment in the CAMPUS framework.

References

[1] G. Abowd, A. Dey, P. Brown, N. Davies, M. Smith, and P. Steggles. Towards a Better Understanding of Context and Context-Awareness. In *Proc.of the 1st international symposium on Handheld and Ubiquitous Computing (HUC 99)*, pages 304–307, London, UK, 1999. Springer-Verlag.

[2] J. Ahola. Ambient Intelligence. Final report, IST Advisory Group, February 2001.

[3] Z. Aleksovski, M. Klein, W. ten Kate, and F. van Harmelen. Matching Unstructured Vocabularies using a Background Ontology. In *Proc. of Knowledge Engineering and Knowledge Management (EKAW)*, pages 182–197, 2006.

[4] Z. Aleksovski, W. ten Kate, and F. van Harmelen. Exploiting the structure of background knowledge used in ontology matching. In P. Shvaiko, J. Euzenat, N. Noy, H. Stuckenschmidt, R. Benjamins, and M. Uschold, editors, *Proc. of First International Workshop on Ontology Matching (OM-2006), co-located with the 5th International Semantic Web Conference (ISWC-2006), Athens, Georgia, USA*, CEUR Proceedings, November 2006.

[5] V. Apparao, S. Byrne, M. Champion, S. Isaacs, I. Jacobs, A.L. Hors, G. Nicol, J. Robie, R. Sutor, C. Wilson, and L. Wood. Document Object Model (DOM) Level 1 Specification. Recommendation, 1998.

[6] S. Banerjee and T. Pedersen. Extended gloss overlaps as a measure of semantic relatedness. In *Proceedings of the Eighteenth International Joint Conference on Artificial Intelligence*, pages 805–810, 2003.

[7] A. Bikakis, T. Patkos, G. Antoniou, and D. Plexousakis. A Survey of Semantics-based Approaches for Context Reasoning in Ambient Intelligence. In R. Bergmann, K.-D. Althoff, U. Furbach, and K. Schmid, editors, *Proceedings of the Workshop "Artificial Intelligence Methods for Ambient Intelligence" at the European Conference on Ambient Intelligence (AmI'07)*, pages 15–24, November 2007.

[8] D. Bottazzi, R. Montanari, and A. Toninelli. Context-Aware Middleware for Anytime, Anywhere Social Networks. *IEEE Intelligent Systems*, 22(5):22–32, 2007.

[9] P. Bouquet, F. Giunchiglia, F. van Harmelen, L. Serafini, and H. Stuckenschmidt. C-OWL: Contextualizing Ontologies. In *Proc. of the Second International Semantic Web Conference (ISWC-2003)*, volume 2870 of *Lecture Notes in Computer Science*, pages 164–179. Springer, 2003.

[10] P. Bouquet, L. Serafini, and S. Zanobini. Semantic Coordination: A New Approach and an Application. In Dieter Fensel, Katia P. Sycara, and John Mylopoulos, editors, *International Semantic Web Conference*, volume 2870 of *Lecture Notes in Computer Science*, pages 130–145. Springer, 2003.

[11] D. Brauner, M.A. Casanova, and R. Milidiú. Mediation as Recommendation: An Approach to Design Mediators for Object Catalogs. In Robert Meersman, Zahir Tari, and Pilar Herrero, editors, *OTM Workshops (1)*, volume 4277 of *Lecture Notes in Computer Science*, pages 46–47. Springer, 2006.

[12] D. Brauner, C. Intrator, J.C. Freitas, and M.A. Casanova. An Instance-based Approach for Matching Export Schemas of Geographical Database Web Services. In L. Vinhas and Antonio C. R. Costa, editors, *Proc. of the IX Brazilian Symposium on GeoInformatics. Porto Alegre : Sociedade Brasileira de Computao (GeoInfo)*, pages 109–120. INPE, 2007.

[13] K. Breitman, D. Brauner, M.A. Casanova, R. Milidiú, and A.G. Perazolo. Instance-Based Ontology Mapping. In *Proc. of the Fourth IEEE International Workshop on Engineering of Autonomic and Autonomous Systems EASe 2007*, pages 117–126. IEEE Computer Society Press, 2007.

[14] K. Breitman, C. Felicíssimo, and M.A. Casanova. CATO - A Lightweight Ontology Alignment Tool. In Orlando Belo, Johann Eder, Joo Falco e Cunha, and Oscar Pastor, editors, *CAiSE Short Paper Proceedings*, volume 161 of *CEUR Workshop Proceedings*. CEUR-WS.org, 2005.

[15] A. Budanitsky and G. Hirst. Evaluating WordNet-based Measures of Semantic Distance. *Computational Linguistics*, 32(1):13–47, 2006.

[16] G. Chen and D. Kotz. A Survey of Context-Aware Mobile Computing Research. Technical Report TR2000-381, Department of Computer Science, Dartmouth College, 2000.

[17] H. Chen. *An Intelligent Broker Architecture for Pervasive Context-Aware Systems*. PhD thesis, University of Maryland, Baltimore County, December 2004.

[18] H. Chen, T. Finin, and A. Joshi. An ontology for context-aware pervasive computing environments. *Special Issue on Ontologies for Distributed Systems, Knowledge Engineering Review*, 2003.

[19] A. Dey. Understanding and Using Context. *Personal and Ubiquitous Computing*, 5(1):4–7, 2001.

[20] K. Ducatel, M. Bogdanowicz, F. Scapolo, J. Leijten, and J.-C. Burgelma. Scenarios for Ambient Intelligence in 2010. Final report, IST Advisory Group, February 2001.

[21] C. Felicíssimo and K. Breitman. Taxonomic ontology alignment - an implementation. In *Proceedings of Workshop em Engenharia de Requisitos (WER 2004)*, pages 152–163, 2004.

[22] C. Fellbaum, editor. *WordNet: An Electronic Lexical Database*. MIT Press, 1998.

[23] R.T. Fielding and R.N. Taylor. Principled design of the modern Web architecture. *ACM Transactions on Internet Technology*, 2(2):115–150, 2002.

[24] A. Gazola, D. Brauner, and M.A. Casanova. A Mediator for Heterogeneous Gazetteers. In *Proc. of the XXII Brazilian Symposium on Databases (SBBD), Poster Session*, volume 1, pages 11–14. Sociedade Brasileira de Computao, 2007.

[25] A. Gómez-Pérez, M. Fernadéz-Peréz, and O. Corcho. *Ontological Engineering*. Springer-Verlag, London, 2004.

[26] D. Guan, W. Yuan, S.J. Cho, A. Gavrilov, Y.-K. Lee, and S. Lee. Devising a Context Selection-Based Reasoning Engine for Context-Aware Ubiquitous Computing Middleware. In J. Indulska, J. Ma, L. T. Yang, T. Ungerer, and J. Cao, editors, *UIC*, volume 4611 of *Lecture Notes in Computer Science*, pages 849–857. Springer, 2007.

[27] V. Haarslev and R. Möller. RACER System Description. In *Proc. of the International Joint Conference on Automated Reasoning (IJCAR'01)*, volume 2083 of *Lecture Notes in Computer Science*, 2001.

[28] K. Henricksen and J. Indulska. Modelling and using imperfect context information. In *Proc. Second IEEE Annual Conference on Pervasive Computing and Communications Workshops*. IEEE Computer Society, 2004.

[29] G. Hirst and D. St-Onge. Lexical chains as representation of context for the detection and correction malapropisms, chapter 13, in WordNet: An Electronic Lexical Database,. MIT Press, 1998.

[30] R. Ichise, H. Takeda, and S. Honiden. Integrating multiple internet directories by instance-based learning. In G. Gottlob and T. Walsh, editors, *IJCAI*, pages 22–30. Morgan Kaufmann, 2003.

[31] M. Jarmasz and S. Szpakowicz. Roget's thesaurus and semantic similarity. In N. Nicolov, K. Bontcheva, G. Angelova, and R. Mitkov, editors, *RANLP*, volume 260 of *Current Issues in Linguistic Theory (CILT)*, pages 111–120. John Benjamins, Amsterdam/Philadelphia, 2003.

[32] J.J. Jiang and D.W. Conrath. Semantic Similarity Based on Corpus Statistics and Lexical Taxonomy. In *Proc. on International Conference on Research in Computational Linguistics, Taiwan*, pages 19–30, 1997.

[33] G. Jones. Challenges and opportunities of context-aware information access. In *UDM '05: Proceedings of the International Workshop on Ubiquitous Data Management*, pages 53–62, Washington, DC, USA, 2005. IEEE Computer Society.

[34] M. A. Casanova K. Breitman and W. Truszkowski. *Semantic Web: Concepts, Technologies and Applications*. Springer-Verlag, 2007.

[35] Y. Kalfoglou and M. Schorlemmer. Ontology Mapping: The State of the Art. In Y. Kalfoglou, M. Schorlemmer, A. Sheth, S. Staab, and M. Uschold, editors, *Semantic Interoperability and Integration*, number 04391 in Dagstuhl Seminar Proceedings. Internationales Begegnungs- und Forschungszentrum fuer Informatik (IBFI), Schloss Dagstuhl, Germany, 2005.

[36] H. Kim, Y. Cho, and S. Oh. CAMUS: A Middleware Supporting Context-aware Service for Networkbased Robots. In *Proc. of the IEEE Workshop on Advanced Robotics and Social Impacts*, 2005.

[37] C. Leacock and M. Chodorow. Combining local context and WordNet similarity for word sense identication, chapter 11, in WordNet: An Electronic Lexical Database, pages 265–283. MIT Press, 1998.

[38] D. Lin. An Information-Theoretic Definition of Similarity. In Jude W. Shavlik, editor, *Proc. of the 15th International Conference on Machine Learning (ICML 1998), Madison, Wisconson, USA, July 24-27, 1998*, pages 296–304. Morgan Kaufmann, 1998.

[39] J. Lindenberg, W. Pasman, K. Kranenborg, J.Stegeman, and M.A. Neerincx. Improving service matching and selection in ubiquitous computing environments: a user study. *Personal Ubiquitous Computing*, 11(1):59–68, 2006.

[40] A. Maedche and S. Staab. Comparing Ontologies: Similarity Measures and a Comparison Study. Internal report, Institute AIFB, University of Karlsruhe,, 2001.

[41] L. Mazuel and N. Sabouret. Generic Command Interpretation Algorithms for Conversational Agents. In *IAT*, pages 146–153. IEEE Computer Society, 2006.

[42] L. Mazuel and N. Sabouret. Degré de relation sémantique dans une ontologie pour la commande en langue naturelle. In *Plate-forme AFIA, Ingnierie des Connaissances 2007 (IC 2007)*, pages 73–83, 2007.

[43] S. Mohammad and G. Hirst. Distributional measures of concept-distance: A task-oriented evaluation. In *Proceedings, 2006 Conference on Empirical Methods in Natural Language Processing (EMNLP 2006)*, Sydney, Australia, July 2006.

[44] B. Mrohs, M. Luther, R. Vaidya, M. Wagner, S. Steglich, W. Kellerer, and S. Arbanowski. OWL-SF - A Distributed Semantic Service Framework. In *Proc. of Workshop on Context Awareness for Proactive Systems (CAPS), Helsinki, Finland*, pages 67–78, 2005.

[45] N. Noy and M. Musen. The PROMPT suite: Interactive tools for ontology merging and mapping. *Int. J. Hum.-Comput. Stud.*, 59(6):983–1024, December 2003.

[46] P. Nurmi and P. Floreen. Reasoning in Context-Aware Systems, 2004.

[47] I. Pandis, J. Soldatos, A. Paar, J. Reuter, M Carras, and L. Polymenakos. An ontology-based framework for dynamic resource management in ubiquitous computing environments. In *Proceeding of the 2nd International Conference on Embedded Software and Systems (ICESS 2005)*, pages 1–8. Northwestern Polytechnical University of Xi'an, PR China, December 2005.

[48] S. Patwardhan and T. Pedersen. Using WordNet-based context vectors to estimate the semantic relatedness of concepts. In *Proc. of the EACL 2006 workshop, making sense of sense: Bringing computational linguistics and psycholinguistics together. Trento, Italy*, pages 1–8, 2006.

[49] R. Rada, H. Mili, E. Bicknell, and M. Blettner. Development and application of a metric on semantic nets. In *IEEE Transactions on Systems, Man and Cybernetics*, volume 19, pages 17–30, 1989.

[50] A. Ranganathan and R. Campbell. A Middleware for Context-Aware Agents in Ubiquitous Computing Environments. *Lecture Notes in Computer Science*, 2672:143–161, January 2003.

[51] P. Resnik. Using Information Content to Evaluate Semantic Similarity in a Taxonomy. In *Proc. of IJCAI*, pages 448–453, 1995.

[52] M. Román, C.K. Hess, R. Cerqueira, A. Ranganathan, R. Campbell, and K. Nahrstedt. A Middleware Infrastructure for Active Spaces. *IEEE Pervasive Computing*, 1(4):74–83, October-December 2002.

[53] V. Sacramento, M. Endler, H.K. Rubinsztejn, L.S. Lima, K. Gonalves, F.N. Nascimento, and G.A. Bueno. MoCA: A middleware for developing collaborative applications for mobile users. *IEEE Distributed Systems Online*, 5(10), 2004.

[54] S. Sameshima, J. Suzuk, S. Steglich, and T. Suda. Platform Independent Model (PIM) and Platform Specific Model (PSM) for Super Distributed Objects. Final adopted specification OMG document number dtc/03-09-01, Object Management Group, September 2003.

[55] M. Satyanarayanan. Pervasive computing: vision and challenges. *Personal Communications, IEEE*, 8(4):10–17, 2001.

[56] A.F. Seghrouchni, K. Breitman, N. Sabouret, M. Endler, Y. Charif, and J.-P. Briot. Ambient intelligence applications: Introducing the campus framework. In *Proc. 13th IEEE International Conference on Engineering of Complex Computer Systems (ICECCS 2008), Dublin*. IEEE Computer Society Press, 2008.

[57] L. Serafini and A. Tamilin. DRAGO: Distributed Reasoning Architecture for the Semantic Web. In Asunción Gómez-Pérez and Jérôme Euzenat, editors, *ESWC*, volume 3532 of *Lecture Notes in Computer Science*, pages 361–376. Springer, 2005.

[58] A. Shehzad, H.Q. Ngo, K.A. Pham, and S.Y. Lee. Formal modeling in context aware systems. In *Proceedings of the First International Workshop on Modeling and Retrieval of Context*, September 2004.

[59] E. Sirin and B. Parsia. Pellet: An OWL DL Reasoner. In Volker Haarslev and Ralf Mller, editors, *Description Logics*, volume 104 of *CEUR Workshop Proceedings*. CEUR-WS.org, June 2004.

[60] J. Soldatos, N. Dimakis, K. Stamatis, and L.Polymenakos. A Breadboard Architecture for Pervasive Context-Aware Services in Smart Spaces: Middleware Components and Prototype Applications. *Personal and Ubiquitous Computing Journal*, 2007.

[61] G. Stoilos, G. Stamou, and S. Kollias. A string metric for ontology alignment. In *Proceedings of International Semantic Web Conference (ISWC 2005)*, pages 624–637, 2005.

[62] T. Strang and C. Linnhoff-Popien. A context modeling survey. In *First International Workshop on Advanced Context Modelling, Reasoning and Management*, Nottingham, England, September 2004.

[63] J. van Diggelen, R. Beun, F. Dignum, R. van Eijk, and J. Meyer. Combining normal communication with ontology alignment. In *Proceedings of the International Workshop on Agent Communication (AC'05)*, volume 3859 of *LNCS*, 2005.

[64] J. Viterbo, M. Endler, and J.-P. Briot. Ubiquitous service regulation based on dynamic rules. In *Proc. 13th IEEE International Conference on Engineering of Complex Computer Systems (ICECCS 2008), Dublin*. IEEE Computer Society Press, 2008.

[65] J. Wang. An Algorithm for Finding the Largest Approximately Common Substructures of Two Trees. *IEEE Transactions on Pattern Analysis and Machine Intelligence*, 20(8):889–895, 1998.

[66] J. Wang, J.-R. Wen, F. H. Lochovsky, and W.-Y. Ma. Instance-based Schema Matching for Web Databases by Domain-specific Query Probing. In M. A. Nascimento, M. T. Ozsu, D. Kossmann, R.J. Miller, J. A.

Blakeley, and K. B.Schiefer, editors, *VLDB*, pages 408–419. Morgan Kaufmann, 2004.

[67] X.H. Wang, J.S. Dong, C.Y. Chin, S.R. Hettiarachchi, and D.Q. Zhang. Semantic Space: An Infrastructure for Smart Spaces. *Pervasive Computing*, 3(3):32–39, July-September 2004.

[68] X.H. Wang, D.Q. Zhang, T. Gu, and H.K. Pung. Ontology based context modeling and reasoning using OWL. In *Proceedings of 2nd IEEE Conf. Pervasive Computing and Communications (PerCom 2004), Workshop on Context Modeling and Reasoning*, pages 18–22, Orlando, Florida, March 2004. IEEE Computer Society Press.

[69] A. Williams, A. Padmanabhan, and M. Brian Blake. Local consensus ontologies for B2B-oriented service composition. In *Proceedings of the Second International Joint Conference on Autonomous Agents and Multiagent Systems (AAMAS'03)*, pages 647–654. ACM, 2003.

[70] Z. Wu and M.S. Palmer. Verb Semantics and Lexical Selection. In *Proc. of the 32nd. Annual Meeting of the Association for Computational Linguistics (ACL 1994)*, pages 133–138, 1994.

[71] J. Zhong, H. Zhu, J. Li, and Y. Yu. Conceptual Graph Matching for Semantic Search. In *Proceedings of the 10th International Conference on Conceptual Structures ICCS 2002*, pages 92–106, London, UK, 2002. Springer-Verlag.

Chapter 5

Dynamic Content Negotiation in Web Environments

Xavier Sanchez-Loro, Jordi Casademont, Jose Luis Ferrer, Victoria Beltran, Marisa Catalan and Josep Paradells
Wireless Networks Group, Department of Telematics, Technical University of Catalonia, Spain

Abstract

In the move towards a ubiquitous and device independent Web it is necessary to make a complete characterization of the delivery context, including the capabilities of the device. Current frameworks (CC/PP, UAProf 2.0) provide effective tools for achieving a device independent Web, although there are limitations in their specifications and current implementations. Many devices do not implement any tools to express their capacities. This chapter aims to investigate existing synergies among context acquisition, expression of device capabilities, delivery context characterization, content adaptation and application layer protocol optimization in web environments. Two main proposals are presented. The first is a proxy-based solution for the detection of device capabilities that allows the accurate acquisition and provisioning of the user's device context in web environments. The second is a collaborative gateway/proxy server interception solution for enhancing web browsing over cellular links that reduces HTTP overhead and uses the dynamic expression of capabilities to enhance content negotiation, and consequently content adaptation, by web servers and adaptation proxies alike.

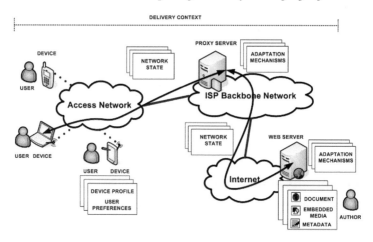

FIGURE 5.1: Elements involved in Web delivery.

5.1   Introduction

Web applications are deployed on many different devices and in many different environments. Mobile users access the Internet and the World Wide Web (WWW) via different wireless and cellular networks, each of them using different radio interfaces and protocols. Thus, their characteristics, services provided and performance vary widely. Users also access these networks using a wide spectrum of mobile devices, ranging from mobile phones and portable game consoles to laptops and Personal Digital Assistants (PDAs). These devices have significant differences in software, memory, computation power, networking, rendering capabilities, storage and battery power. Consequently, a contextualization or profiling of the user's device [1, 2, 3] is needed to provide pervasive and device-independent web access -with applications that automatically adjust their operation and presentation to changes in the user environment, network state and capabilities and configuration of the device. If this self-adaptation were based on accurate and up-to-date information about all the entities involved in the delivery context (e.g., the user's device and user agent, proxies, server, document, etc.), device characterization would allow web servers and adaptation proxies in the delivery chain to adapt and personalize content and application behavior to the real capabilities of the device. Furthermore, the WWW provides users with an infinite variety of contents, comprising almost every conceivable kind of media in different formats and languages. Most of these contents are designed for rendering on personal computers, needing adaptation for a proper rendering on other devices. This adaptation modifies the resulting content; so the content's original purpose can drastically change without some input from the author guiding the adaptation process [31]. So, content authors should provide metadata to allow content categorization in order to ease adaptation. From the content provisioning point of view, a high number of possibilities and architectures exist, from Content Delivery Networks to old plain web servers and chains of proxies. Moreover, a lot of networking and enterprise services (i.e., instant messaging and presence services, VoD, VoIP, etc.) use web applications as front-ends to configure account settings and even access their systems, increasing interaction modalities and communication possibilities. With the new trend in pervasive and ubiquitous services, complexity and diversity increases as new modalities of Human-Computer Interaction (HCI) appear [32]. This heterogeneity and diversity of possibilities of involved elements affects the user interaction with the applications and the delivery of web content. Figure 5.1 illustrates most of the elements involved.

The aim of this chapter is to investigate existing synergies among context acquisition, device capabilities expression, delivery context characterization, content adaptation and application layer protocol optimization in web environments. Two main proposals are presented. The first is a proxy-based solution for detecting device capabilities that allows the accurate acquisition and provisioning of the user's device context in web environments. This platform enables a database to be set up that is accessible from the Internet and that stores a complete and consistent profile of the user's different devices. This profile is expressed using UAProf and is compatible with available parsers. By this method we obtain a single profiling platform for different types of devices with a single profile vocabulary that can be used in multiple independent applications. The second is a collaborative gateway/proxy server interception solution for enhancing web browsing over cellular links that reduces HTTP overhead and uses dynamic expression of capabilities to enhance content negotiation, and consequently content adaptation, by web servers and adaptation proxies alike. The chapter is complemented with an introductory background on the Ubiquitous Web, focusing on protocols related to content negotiation and device capabilities expression, such as Composite Capabilities/Preferences Profile (CC/PP) [6] and UserAgent Profile (UAProf) [5].

5.2 Ubiquitous Web

Device and delivery context description is currently being addressed by different organizations and initiatives, especially in mobile environments. Some of the most relevant initiatives are those made by the World Wide Web Consortium (W3C) around the concept of the Ubiquitous Web, or as stated in [33]:

"The Ubiquitous Web will provide people with access whenever and wherever they find themselves, with applications that dynamically adapt to the user's needs, device capabilities and environmental conditions. Application mobility will allow users to seamlessly switch between devices whilst continuing to access the same applications. Device limitations will be circumvented through being coupled to other devices and by exploiting networked services as part of distributed applications. As users, we will be able to choose how we interact with these applications according to our current needs and the characteristics of the devices we are using. In one sense, the Web will disappear, as it becomes ubiquitous and taken for granted, thereby vanishing into the background of the global computing and network infrastructure."

To achieve this ambitious goal, W3C has created several working groups to develop different recommendations such as:

- Ubiquitous Web Applications Activity WG (former Device Independence Activity WG).

  - CC/PP (Composite Capabilities/Personal Preferences) 2.0
  - Device Independence Principles [31]
  - Content Selection for Device Independence (DISelect) 1.0 [10]
  - Device Independent Authoring Language (DIAL) [26]
  - Delivery Context: Interfaces (DCI) [12]
  - Core Presentation Characteristics [11]

- Semantic Web Activity WG.

  - OWL (Web Ontology Language) [34]
  - RDF (Resource Description Framework) [35]
  - Protocol for Web Description Resources (POWDER) [36]

- Multi-Modal Interaction WG

  - Multimodal Interaction Framework [32]
  - Multimodal Architecture and Interfaces [37]
  - InkML (Ink Markup Language) [38]

- Mobile Web Initiative Device Description WG:

  - Device Description Landscape [39]
  - Device Description Ecosystem [40]
  - Device Description Repository Requirements 1.0 [27]

  – Device Description Repository Core Vocabulary 1f [41]

- Web Accessibility Initiative (WAI) [42]

  – Web Content Accessibility Guidelines WG

  – User Agent Accessibility Guidelines WG

  – Authoring Tools Guidelines WG

In the mobile world, other related initiatives are UAProf from the Open Mobile Alliance (OMA), as the first massive deployment of CC/PP. UAProf vocabulary is being further expanded by 3rd Generation Partnership Project (3GPP) with components related to streaming media. Moreover, 3GPP is developing the Generic User Profile (GUP) [43] for mobile users profiling. GUP supports using UAProf vocabulary for describing device capabilities. OMA has also developed an application level protocol, the Mobile Location Protocol (MLP) [8], to obtain device location information independently from the network access technology in use.

Related to device characterization, there are other proposals such as WURLF [9], UPS [4] and FIPA Device Ontology [44]. WURLF is not endorsed by any standard organization but, although having drawbacks, it has some implantation. Universal Profile Schema (UPS) is a framework for content negotiation over the Internet and heterogeneous networks. It is based on CC/PP, expanding it with vocabularies to describe client-related and server-related profiles. The client profiles are divided in client profile and client resource profile. The server profiles are the document instance profile, resource profile and adaptation method profile. The FIPA Device Ontology is designed for agent-based applications and can be used as vocabulary for CC/PP-based applications.

Other initiatives, regarding content negotiation and adaptation for device independent service provisioning, are the Internet Content Adaptation Protocol (iCAP) [45] and OPES (Open Pluggable Edge Service) [46] and Transparent Content Negotiation [47]. iCAP and OPES allow to negotiate and order adaptation services by other proxies and servers. TCN provides mechanisms to transparently negotiate content with HTTP, but it is still in an experimental phase.

5.2.1   Related Concepts

This section will define some concepts related to the Ubiquitous Web for those readers unfamiliar with the subject.

- *Device Independence* is making the web accessible by any device under any circumstance and by all people. This means that the user's experience should be satisfactory regardless the device in use. This implies not only content adaptation for the successful rendering of documents on every type of device, but also adaptation in the modality domain to adapt user interaction with the Web, i.e. voice commands, text-to-speech. Device independence should be considered from different perspectives: user, author and delivery [31].

- *Capabilities detection* is the process where a server or proxy obtains the characteristics of a remote terminal device. There are many possible characteristics to be detected, ranging from hardware capabilities (i.e., type of CPU and screen size) and network characteristics (i.e., types of NICs) to software functionalities and browser characteristics. This detection is necessary to adjust and adapt the content to be served (like text, images, videos, etc.) to the capabilities of the remote client, in order to reach device independent user experience.

- *Content Negotiation* is the mechanism for selecting/adapting the appropriate content (HTTP representation) when servicing a request. The HTTP representation of entities in any response can be negotiated (including error responses). Web content negotiation is server-driven and is based on an interchange of client device capabilities.

- *Content Adaptation* is a process of selection, generation or modification that produces one or more perceivable units (media content) in response to a requested resource in a given delivery context. Content adaptation is critical to achieve the desired authoring principle of *"write once, deliver it anywhere"*[2], emerging as a inherent functionality in any context-aware system. There are different approaches and issues related to content adaptation which can be classified depending on: transparency [14]; location of adaptation process (client, server, intermediate proxy/middleware- collaborative or not, etc.) [48, 49]; affected domains (fidelity, modality, format, usability, etc.) [31, 49, 50, 51]; types of adaptation policies [50, 51]; rule-based, constraint-based and CDA [50], with implicit or explicit feedback; considered context dimensions (hardware and software capabilities, network state, user preferences, user's perceived QoS [51], etc.); distributed adaptation proxies architectures [49, 50, 51], etc.

- *Delivery Context* is a set of attributes that characterizes the capabilities of the access mechanism, the preferences of the user and other aspects of the context into which a web page is to be delivered. Thus, the client and intermediate proxies in the delivery chain add their perspective of the delivery context to the requests. Server and intermediate adaptation proxies adapt responses according to the perceived context. An illustrative list of possible entities and its characteristics is given in [2, 31].

5.2.2 Protocols Overview

The following section gives an overview of device capabilities and content negotiation protocols such as CC/PP and UAProf, giving also a small review of HTTP basic operation.

5.2.2.1 HTTP 1.1

HyperText Transfer Protocol [52] is an application level stateless protocol used to transfer web content (hypertext, images, objects, etc). It is based on a client-server architecture and a request-response paradigm. HTTP provides a simple content negotiation mechanism in order to select the most suitable content for the client. This negotiation mechanism is server-driven, so the server chooses the content, based on the information sent in the request of the client and the available web resources. The client specifies its preferences and capabilities in the HTTP headers sent in the request.

The following headers are those implied in content negotiation:

- *Accept*: this header specifies which MIME (Multipurpose Internet Mail Extensions) types the browser accepts in the response.

- *Accept-Charset*: it specifies which character sets the browser accepts in the response.

- *Accept-Encoding*: it specifies which encoding the browser accepts. Encoding is defined as those transformations applicable to the received content to restore the original information.

- *Accept-Language*: it specifies which human language the user prefers when reading text-based content.

- *User-Agent*: this header identifies the user agent or browser used by the remote client to interact with the server.

Figure 5.2 illustrates the use of these headers in a HTTP request. The information obtained about the device capabilities is too fragmentary. Besides, some browsers use a technique, known as browser cloaking, where they identify themselves as a different browser in order to assure that the server will deliver the requested content. In order to avoid these issues, some protocols, like CC/PP or TNC, have been created to enhance content negotiation and device capabilities expression. These protocols usually operate on HTTP extensions.

5.2.2.2 CC/PP 2.0

The proliferation of Internet-enabled devices supporting a variety of user-interface paradigms and modalities motivates the need to create presentations

GET / HTTP/1.1[CRLF]
Host: www.arrakis.es[CRLF]
Connection: close[CRLF]
Accept-Encoding: gzip[CRLF]
Accept: application/x-shockwave-flash,text/xml, application/xml, application/xhtml+xml, text/html;q=0.9, text/plain;q=0.8, image/png, image/jpeg, image/gif;q=0.2, */*;q=0.1[CRLF]
Accept-Language: en-us,en;q=0.5[CRLF]
Accept-Charset: ISO-8859-1, utf-8;q=0.7, *;q=0.7[CRLF]
User-Agent: Mozilla/5.0 (Windows; U; Windows NT 5.1; en-US; rv:1.6) Gecko/20040113 Web-Sniffer/1.0.20[CRLF]

FIGURE 5.2: HTTP request example.

of content that are optimized for specific rendering configurations. As a result, there is also a need for a standardized capabilities-and-content negotiation mechanism allowing clients accessing the web to assert their capabilities to the server serving the content. Composite Capabilities/Preferences Profile (CC/PP) 2.0 is a protocol-independent extensible framework that can be used for communicating any metadata information such as user location and preferences and device and document profiles. So CC/PP is an essential component to reach the goal of device independence. CC/PP is designed for sharing user profiles with other remote applications and servers in order to create presentations and applications optimized for the rendering device and the user preferences and context, like user location.

CC/PP uses RDF to describe location, capabilities and user preferences as profile attributes and components. A vocabulary or schema, defining different attributes and components with their structure and exact meaning, must be declared to describe a CC/PP profile without confusion. CC/PP doesn't define any standard vocabulary, but it defines a set of rules to create profiles using third-party vocabularies.

Being based on RDF, CC/PP is very flexible and extensible. It is flexible because it allows to describe new devices and properties using any vocabulary defined by third-parties like users, providers or manufacturer. It is extensible because a vocabulary can be extended introducing new attributes and/or components to describe new profile properties and characteristics. Also, profile serialization as RDF/XML document allows different network elements and devices to interchange CC/PP profiles across the web.

The flexibility of CC/PP is paradoxically its major disadvantage. In order to be compatible with most possible devices and maintain the solution scalability, there cannot exist so many different vocabularies as existing devices or manufacturers. So a viable approach could be defining some standard vocabularies or schemas to describe similar devices (like UAProf does). Besides,

```
<?xml version="1.0"?>
<rdf:RDF xmlns:rdf="http://www.w3.org/1999/02/22-rdf-syntax-ns#"
xmlns:ccpp="http://www.w3.org/2002/11/08-ccpp-schema#"
xmlns:example="http://www.example.org/vocabulary#">
<rdf:Description rdf:about="http://www. example.org/profile#MiProfile">
<ccpp:component>
<rdf:Description rdf:about="http://www. example.org/profile#Hardware">
<rdf:type rdf:resource="http://www.example.org/vocabulari
#HardwarePlatform"/>
<example:scrSize >320x200</example:scrSize >
<example:CPU>ARM</example:CPU>
</rdf:Description>
</ccpp:component>
</rdf:Description>
</rfd:RDF>
```

FIGURE 5.3: CC/PP profile example.

attribute definitions and meaning should be consistent between schema because some attributes can be defined in more than one schema (and as CC/PP stands now they could have very different meanings). So, different vocabularies should use the same semantic definition of a certain property. Some efforts are being invested in this area [11]. Even so, CC/PP does not propose any standard vocabulary to describe profiles, fact that difficulties semantic compatibility between profiles defined by different parties.

5.2.2.2.1 CC/PP Profile Composition. A CC/PP profile is composed by RDF/XML statements specifying location, device capabilities and user preferences. A CC/PP profile is divided hierarchically in two levels: a profile is composed of a number of *components* and each component is composed of one or more *attributes*. Components group characteristics in different categories (e.g., user context and preferences, device hardware platform, device software platform, network characteristics, etc.). This way, each component will contain certain attributes, one for each described characteristic. So Hardware component could contain attributes like type of CPU, RAM capacity, type of keyboard, screen size, and so on. All these attributes would be defined in a previously declared schema or vocabulary and referenced by XML namespaces.

Figure 5.3 shows an example of a profile described with RDF/XML composed just by one component and a pair of attributes. Default values for components can be defined in a schema. In order to get the default values of a component, a reference indicating where to find the default schema must be declared in the profile. Those attributes whose values are specified in the

```
GET /a-resource HTTP/1.1
Host: www.w3.org
Man: "http://www.w3.org/1999/06/24-CCPPexchange" ; ns=25
25-Profile: "http://www.ex.com/hw"," 1-CWccARHXxtYJE+rKkoD8ng=="
25-Profile-Diff-1: <?xml version="1.0"?>
<rdf:RDF xmlns:rdf="ttp://www.w3.org/1999/02/22-rdf-syntax-ns#"
xmlns:prf="http://www.wapforum.org/profiles/UAPROF/ccppschema-
20010430#">
<rdf:Description rdf:ID="MyDeviceProfile">
<prf:component>
<rdf:Description rdf:ID="HardwarePlatform">
<rdf:type rdf:resource="http://www.wapforum.org/profiles/UAPROF
/ccppschema-20010430#HardwarePlatform"/>
<prf:SoundOutputCapable>No</prf:SoundOutputCapable>
</rdf:Description>
</prf:component>
</rdf:Description>
</rdf:RDF>
```

FIGURE 5.4: CC/PPex request example.

profile will get these values, not the default values.

5.2.2.2.2   CC/PP Exchange Protocol (CC/PPex).   CC/PP profiles
are transferred in HTTP headers. There are two proposals which allow to
encapsulate CC/PP over HTTP. First there is CC/PPex, based on HTTP
Extension Framework, proposed by CC/PP WG and second there is OMA's
UAProf W-HTTP. However, CC/PP is independent from these two protocols,
and other profile transfer mechanisms may be used. Even so it is recommended
to support CC/PPex and, given that UAProf is the only real massive deployment of CC/PP, it is also recommendable to support transferring of CC/PP
profiles with W-HTTP.

CC/PPex is based on the introduction of new headers to manage the profile
transfer. CC/PPex recommends sending references to default profiles and the
differences of the device towards the default values, in order to minimize the
amount of data sent over the air. The headers defined in CC/PPex are (see
Figure 5.4 for an example of CC/PPex request):

- *Profile.* The Profile header field is a request-header field, which con-
  veys a list of references which address CC/PP descriptions. The Profile
  header field-value is a list of references. Each reference in the Profile
  header field represents the corresponding entity of the CC/PP description. A reference is either an absolute URI or a profile-diff-name. An
  entity of a CC/PP description which is represented by an absolute URI

exists outside of the request, and an entity of a CC/PP description which is represented by a profile-diff-name exists inside of the request (i.e., in the Profile-Diff header field). The profile-diff-name consists of a profile-diff-number part and a profile-diff-digest part. The profile-diff-number is the number which indicates the corresponding Profile-Diff header. The profile-diff-digest is generated by applying MD5 message digest algorithm and Base64 algorithm to the corresponding Profile-Diff header field-value. It is introduced for the efficiency of the cache table look up in gateways, proxies and user agent.

- *Profile-diff.* The Profile-Diff header field is a request-header field, which contains CC/PP description, usually describing changes between the referenced profile (in profile header) and the actual user device profile. The Profile-Diff header field is always used with Profile header in the same request. All profile information could be represented by absolute URIs in the Profile header. In this case, the Profile-Diff header field does not have to be added to the request. On the other hand, only one Profile-Diff header can contain all profile information. In this case, the Profile header includes only the profile-diff-name which indicates the Profile-Diff header.

- *Profile-warning.* The Profile-warning header field is a response-header field, which is used to carry warning information. When a client issues a request with the Profile header field to a server, the server inquires of CC/PP repositories the CC/PP descriptions using the absolute URIs in the Profile header field. If any one of the CC/PP repositories is not available, the server might not obtain the fully enumerated CC/PP descriptions, or the server might not obtain first-hand or fresh CC/PP descriptions. In this case the server should respond to the client with the Profile-warning header field if any one of the CC/PP descriptions could not be obtained, or any one of the CC/PP descriptions is stale.

5.2.2.3   UAProf 2.0

The User Agent Profile (UAProf) specification extends WAP 2.0 to enable the end-to-end flow of a User Agent Profile (UAProf), also referred to as Capability and Preference Information (CPI), between the WAP client, the intermediate network points and the origin server. It seeks to interoperate seamlessly with CC/PP. Thus, it uses the CC/PP model to define a robust, extensible framework for describing and transmitting CPI about the client, the user and the network that will process the content contained in a WSP/HTTP response. The specification defines a set of components and attributes that WAP-enabled devices may convey within the CPI. This CPI may include, but is not limited to, hardware characteristics (screen size, color capabilities, image capabilities, manufacturer, etc.), software characteristics (operating system vendor and version, support for MExE, list of audio and video encoders, etc.), application/user preferences (browser manufacturer and version, mark-up languages and versions supported, scripting languages supported, etc.), WAP characteristics (WML script libraries, WAP version, WML deck size, etc.), and network characteristics (user location, bearer characteristics such as latency and reliability, etc.). This specification seeks to minimize wireless bandwidth consumption by using a binary encoding for the CPI and by supporting efficient transmission and caching over WSP in a manner that allows easy interoperability with HTTP.

UAProf is considered the first massive scale deployment of a CC/PP-based solution. As it is noted above, UAProf defines a CC/PP-compliant vocabulary for describing profiles or CPIs.

5.2.2.3.1 Profile Resolution. Unlike CC/PP, UAProf not only proposes a standard vocabulary, but also a profile resolution mechanism. An origin server or proxy that needs to determine the correct values for CPI attributes must resolve the profile. This resolution process applies a collection of default attribute values and then applies appropriate overrides to those defaults. Because different network elements may provide additional (or overriding) profile information, the resolution process must apply this additional information to determine the final attribute values.

The User Agent Profile is constructed in three stages:

1. Resolve all indirect references by retrieving URI references contained within the profile.

2. Resolve each *Profile* and *Profile-Diff* document by first applying attribute values contained in the default URI references and by second applying overriding attribute values contained within the category blocks of that *Profile* or *Profile-Diff*.

3. Determine the final value of the attributes by applying the resolved attribute values from each *Profile* and *Profile-Diff* in order, with the attribute values determined by the resolution rules provided in the schema. Where no resolution rules are provided for a particular attribute in the schema, values provided in profile diffs are assumed to override values provided in previous profiles or profile diffs.

The value of an attribute with multiple descriptions is resolved as follows:

1. The description of the attribute within the Default tag is resolved first

2. Any other description of the attribute identified in a subsequent instantiation of the attribute overrides the default description

3. Where multiple descriptions of the attribute exist outside the default description block, the ultimate value of the attribute is determined by the resolution rules for that attribute. An attribute is associated with one of the following three resolution rules:

(a) *Locked.* The final value is determined by the first occurrence of the attribute outside the default description block. Subsequent occurrences are ignored.

(b) *Override.* The final value equals the last occurrence of the attribute with a description element.

(c) *Append.* The final value is a list of all the occurrences of the attribute.

5.2.2.3.2 Vocabulary. The schema for WAP User Agent Profiles consists of description blocks for the following key components:

- *HardwarePlatform.* A collection of properties that adequately describe the hardware characteristics of the terminal device. This includes the type of device, model number, display size, input and output methods, etc.

- *SoftwarePlatform.* A collection of attributes associated with the operating environment of the device. Attributes provide information on the operating system software, video and audio encoders supported by the device, and user's preference on language.

- *BrowserUA.* A set of attributes to describe the HTML browser application.

- *NetworkCharacteristics.* Information about the network-related infrastructure and environment such as bearer information, location, etc. These attributes can influence the resulting content, due to the variation in capabilities and characteristics of various network infrastructures in terms of bandwidth and device accessibility.

- *WapCharacteristics.* A set of attributes pertaining to WAP capabilities supported on the device. This includes details on the capabilities and characteristics related to the WML Browser, WTA, etc.

- *PushCharacteristics.* A set of attributes pertaining to Push specific capabilities supported by the device. This includes details on supported MIME-types, the maximum size of a push-message shipped to the device, the number of possibly buffered push-messages on the device, etc. Additional components can be added to the schema to describe capabilities pertaining to other user agents such as an e-mail client or hardware extensions.

- *MMSCharacteristics.* Contains properties of the device's MMS capabilities, such as supported content mime types and max size of an MMS message, etc.

Note this vocabulary is completely CC/PP compliant. This vocabulary has been extended by the 3GPP, adding components related to streaming capabilities and the Packet-switched Streaming Service (PSS). Also 3GPP provides a mapping to transfer profiles over Real Time Streaming Protocol (RTSP).

5.2.2.3.3   W-HTTP. Wireless Profiled HTTP is used to transport profile information over HTTP, providing a functional equivalent for the CC/PP exchange protocol (CC/PPex) but with slightly different semantics and syntax definition. The key difference is that CC/PPex uses HTTP extension framework (HTTPex) and W-HTTP is built on top of ad hoc HTTP headers. Thus, W-HTTP does not require support of any external framework like HTTPex, easing implementation.

- *x-wap-profile* equals to Profile header in CC/PPex.

- *x-wap-profile-diff* equals to Profile-diff header in CC/PPex

- *x-wap-profile-warning* equals to Profile-warning header in CC/PPex.

5.3   A Proxy-Based Solution for the Detection of Device Capabilities

Device characterization has already been considered by various standardization organizations, especially in the world of mobile communications. The most widespread proposals are the Composite Capabilities/Personal Preferences (CC/PP) specification [6] presented by the World Wide Web Consortium (W3C) and the User Agent Profile (UAProf) [5] specification presented by the Open Mobile Alliance. Although these frameworks provide good tools for attaining a device-independent web environment, there are limitations in their specifications and current implementations.

One such limitation in the case of CC/PP is its high flexibility, since it does not define or recommend a specific vocabulary. Indeed, it allows developers to define their own vocabularies. Although this flexibility is highly desirable, the presence of a potentially infinite set of different vocabularies without a univocally defined semantics hinders the adaptation of content and presentation because all the different vocabularies must be taken into account. Likewise, the implementation of generic web programming interfaces for accessing delivery context information [10, 12] is subject to the same difficulties. Currently, some work is underway to unify criteria to obtain common definitions and semantics for those attributes habitually used to build presentations [11]. Furthermore, W3C is working on an ontology for delivery context description [25], which involves defining use cases and requirements [27] so that device description repositories and APIs [22] can access this information.

One of the limitations of UAProf 2.0 is the fact that it is focused on mobile phones and there is a lack of perspective in the definition of the vocabulary that hinders the description of other kinds of devices. As we will explain later, we have nonetheless verified that UAProf 2.0 is able to properly characterize the web browsing capabilities of high computing power devices such as PCs or laptops and hand-held devices such as Pocket PCs and PDAs. However, for the other capabilities of the device, including the characteristics of peripherals and hardware/software, the vocabulary lacks completeness, although this limitation is only relevant for those web applications that are really ambitious in their ubiquity and context-awareness (such as a situated web application [1]) and thus require more exhaustive characterization, especially regarding peripherals. From a self-management point of view, in these cases the vocabulary should be expanded to include new components and attributes; otherwise, it would be too focused on web adaptation and not generic enough for other purposes.

Another problem with UAProf stems from the incomplete implementations made by manufacturers. Commercial UAProf browsers do not usually calculate or express the real differences between the device owned by the user (e.g., changes in configuration, installed applications, etc.) and the default profile of the model. Also, some profiles provided by manufacturers encounter problems when they are parsed by available libraries [23, 24] and do not properly fulfill the referenced XML schemas.

Finally, the most serious problem is that none of these frameworks have been implemented in browsers that run on devices other than mobile phones. For instance, PC browsers do not provide any context information. The problem is also widespread in the case of PDAs and other hand-held devices. Given the ubiquitous and changing environments from which users access information, it would be very desirable for all types of devices to provide information on their capabilities and characteristics.

In order to solve these problems, and until terminals can provide their characteristics, we present an implementation of a proxy-based application that accurately contextualizes a user's devices. The context information provided allows us to broadcast a complete and consistent profile of the user's device on the Internet. The profile is expressed using UAProf and is compatible with available parsers. Thus, we obtain a single profiling platform for different types of devices that has just one profile vocabulary.

5.3.1   System Description

We propose an intermediate solution that provides a mechanism for properly detecting the capabilities and characteristics of the device and the web browser in use that dynamically expresses a consistent and valid profile of the user's device across the delivery context. Additionally, the platform stores

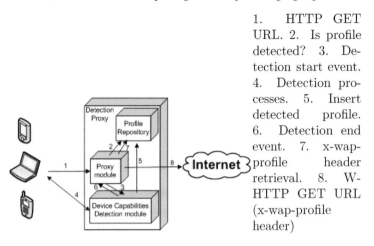

1. HTTP GET URL. 2. Is profile detected? 3. Detection start event. 4. Detection processes. 5. Insert detected profile. 6. Detection end event. 7. x-wap-profile header retrieval. 8. W-HTTP GET URL (x-wap-profile header)

FIGURE 5.5: Device capabilities detection process.

and announces the information collected using Wireless Profiled HTTP (W-HTTP) or the CC/PP Exchange Protocol (CC/PPex) [5, 6], in the form of profiles. A single profiling platform can therefore be used for all types of devices and a common vocabulary can be shared by all the profiles. The architecture of the system is divided into three modules (Figure 5.5): the detector, the profile repository and the proxy. Once the device's capabilities are detected, its profile is stored in the repository and can be retrieved by content servers and adaptation proxies. This profile is consistent with the present configuration of the device and complies with the vocabulary defined by UAProf 2.0.

The detection process is illustrated in Figure 5.5. First, the client requests a web page via HTTP and the proxy module checks whether the user's device has been previously detected. A negative response triggers the detection processes managed by the detection module. Once the profile is created, it is stored in the repository database and the proxy module is alerted to continue and request the web page using W-HTTP with profile headers.

Another option allows users to manage part of their own profile according to their preferences. In this case, the system announces several capacities that do not coincide with the ones detected. This feature is useful if the user wishes to cloak the capabilities of the device to force different content adaptations. One example would be the case of a user who prefers to download images smaller than the display resolution in order to reduce the download time.

The proxy module carries out the common functions of HTTP proxies, but it uses W-HTTP rather than HTTP and adds *x-wap-profile* and *x-wap-profile-diff* headers provided by the navigator to refer to the profile stored in the repository. Also, in the case of a device that already expresses differences by itself, the platform carries out a consistency test of the default profile and compares it with the characteristics detected by other methods. The final result is a customized profile for the device.

5.3.1.1 Detector Module

The aim of the detector is to obtain the maximum amount of information about the hardware and software characteristics of the device, as well as to completely characterize the browser to create a consistent, valid and parsable profile. The detector module executes different detection mechanisms depending on the user's device.

The module carries out all possible tests on the device then matches the results. This matching gives preference to active tests (Applet, JavaScript) over passive ones (HTTP analysis, CC/PP-UAProf negotiation).

5.3.1.1.1 Detection Process. The following is a simplified description of the steps involved in the detection process:

1. The detection module is in charge of identifying the device and the tests it supports. This is a critical step, since forcing a test on a device that lacks the proper support can interrupt interaction with the detector; as a result, the HTTP transaction that triggered the test would be lost and navigation interrupted.

2. The client is forced to access the web applications that are managing the detection processes that the device is going to execute.

3. The profile is created based on all the information collected and is then stored in the repository.

4. The proxy modifies or adds the *x-wap-profile* headers that refer to this profile.

5.3.1.1.2 Detection Mechanisms. The most critical issue in the whole process is the detection mechanism, as system accuracy depends on it. The system takes advantage of the facilities that the existing web technologies offer, which are executed on both the client and the server side. Passive tests are run on the server side, with no extra action on the client side. Active tests require an active role by the client browser. Most of these technologies are well known [2] and their potential will only be discussed briefly here. The implemented tests are based on the techniques described below.

- *HTTP negotiation*: this consists in analyzing the HTTP headers sent by the web browser. It allows limited characterization of the browser and an estimation of which operating system is being used, although browser identification is not entirely reliable due to widespread browser cloaking.

- *JavaScript active detection*: information data on the browser and the JavaScript interpreter can be obtained by examining the Navigator object and using several JavaScript functionalities and tricks. This technique has several drawbacks, since there are substantial differences and inconsistencies between the JavaScript interpreters of different commercial navigators.

- *Java 2 Standard Edition (J2SE) applet*: this technique is mainly suitable for devices that have no restrictions on their computation power. It allows the user to directly interrogate the system, thereby obtaining the characteristics of the operating system, hardware and particularly the Java Virtual Machine (JVM). This information is accessed through the interfaces provided by the Java language or by invoking the console of the system. The main problem of invoking the console of the system is that it is operating system-dependent, although depending on the information that is desired, and especially where low level information is concerned (e.g., the physical address of the device), it may be the only choice.

  The most significant advantage of using applets and similar multiplatform client-embedded technologies is that they allow a small portable detector element to be implemented, which in its turn allows the dynamic interrogation of the device system and also acts as an authentic CC/PP client during the detection process. This is a very interesting feature because channel use can be optimized by referencing a previously detected profile (or a default profile for this type of device) and CC/PP and UAProf resolution rules can be used to ensure that only actual differences are sent to the referenced profile. Since in ubiquitous surroundings connections are usually wireless, optimizing the use of the channel is critical and makes the detection process lighter and less intrusive.

  In addition to evaluating the capabilities of the devices, it is also useful to ascertain the potential of the network. Once this has been done, J2SE applets can also be used to carry out tests at the network level, directly involving the client in the process (e.g., to calculate estimations on channel bandwidth at HTTP level).

- *Expression of capabilities with W-HTTP or CC/PPex (UAProf or CC/PP compliant browser)*: in the case of UAProf-compliant web browsers, default profiles provide most of the information on the device. This static information is taken as valid except when inconsistencies with the other (dynamic) tests are detected. Although these types of browsers are the most desirable, the consistency and veracity of the information expressed in the default profile should be dynamically verified via other tests because of the previously mentioned limitations in the implementations.

Table 5.1: Detection processes supported by type of device.

Device	HTTP analysis	UAProf browser	JavaScript	Applet
PC	x	–	x	x
PocketPC/PDA	x	–	x	?
Phone	x	x	?	–

x: mechanism available in most if not all devices
?: mechanism available in some devices
-: mechanism not usually available for this type of device

Although it is an arbitrary classification, since it is not always clear into which category some devices fall, Table 5.1 displays the viability of each mechanism according to the type of device.

5.3.1.1.3 Detection Mechanism Selection. During the initial browser identification phase, the platform has to select the detection mechanisms that are applicable to the connecting device on the basis of various parameters provided by HTTP headers (generally the User Agent and *Accept* headers) sent by the browser. The parameters provide information that can be used to ponder the viability of each mechanism, such as browser family and version (Mozilla, Opera, etc.), the appearance of *x-wap-profile* and profile headers, OS identification (Windows, Linux Embedix, etc.), the appearance of proprietary headers, the MIDP profile and supported MIME types.

The parameters were selected empirically by testing various commercial devices: the Nokia N80 and N91, the Nokia 6600 and 7710, the Qtek 9000, the Sharp Zaurus PDA, the HP iPAQ 4100 and iPAQ 5550, the Motorola MPX200 and the Dell Axim X50, as well as various PC browsers on a laptop, including Mozilla Gecko and Firefox, Internet Explorer and Opera. Lists compiled from the User Agent strings used by the devices served as a further source of information. Combining and processing these parameters allows suitable detection mechanisms to be selected for the connected device. In case of doubt, the platform chooses the least selective and intrusive tests (i.e., the passive ones), which will not obstruct navigation.

The selection mechanism has the disadvantage of being based partly on the analysis of the User Agent header, since this forces the algorithm to be updated as new navigators and/or variations appear. A further difficulty lies in the identification of the navigator, a process that is compromised by widespread browser cloaking; this, however, should not be too much of a concern, since in these cases browsers usually supplant the browsers of the same types of devices that have similar browsing capabilities. The mechanism also depends on the completeness of the list of announced *Accept* header values, as each mechanism has an associated MIME type, although unfortunately most browsers send an abbreviated list of these values.

5.3.1.1.4   Manual Configuration.   If the user so desires, the proxy can be manually configured to advertise a profile that is different from the one detected. In the present implementation of the platform, cookies can control which profile should be advertised by the proxy or the user can statically configure the proxy to use a given one.

As stated earlier, this feature is useful if the user wishes to cloak the capabilities of the device in order to force different content adaptations by the content servers or the adaptation proxies, such as in the case of a user in a wireless environment who wishes to decrease download time by turning color images into grayscale images. The user would cloak the actual capabilities of the device by specifying that it only supports grayscale images, thus improving performance. User experience can also be enhanced, such as in the case of a user who prefers to read text with no images. In this case, not only would the new profile be announced, the HTTP *Accept* header would also be modified to reflect the absence of support for images and any other type of multimedia that is not textual.

However, manual configuration is a serious burden for system administrators and users and is therefore not advisable. In Section 5.3 we discuss how this burden can be overcome by automatically cloaking device capabilities and adjusting to changes in the network state to enhance content negotiation, adaptation and performance in a collaborative environment. This self-adapting dynamic content negotiation takes advantage of the proxy's own vision of delivery context.

5.3.1.1.5   Profile Consistency Test.   Since the system obtains context data from different sources, it is necessary to implement a mechanism that contrasts and merges the obtained data in order to build as truthful and consistent a profile as possible. This profile must be compliant with the selected vocabulary and therefore parsable by available CC/PP parsing tools [23, 24] (evaluated by the W3C). This consistency test allows us to complement and alleviate the deficiencies of the different methods. For instance, in the case of a device of high capabilities the data obtained by analyzing the HTTP *Accept* header is complemented by JavaScript detection of plug-ins and supported MIME objects and with information obtained by the applet.

As information about the same attribute can be obtained by different methods, discrepancies may appear. In order to solve this problem, the system assigns a classification of confidence to the methods that follows this order:

1. J2SE detection (dynamic)

2. JavaScript detection (dynamic)

3. Detection of expressed differences with UAProf default profile (dynamic)

4. HTTP header analysis (static)

5. Detection of UAProf default profile information (static)

The system prioritizes active detections because, unlike passive ones, they do not provide static information and the browser and the system can be interrogated directly. We consider J2SE to be the most reliable mechanism because it allows characteristics at a level lower than JavaScript to be dynamically consulted. It should be emphasized that active detections are prioritized over the expression of profile differences by the browser because we know how the information is obtained. However, when profile differences are expressed by a browser the reliability of active detections should be contrasted. However, none of the studied browsers expressed differences with the default profile. Completeness is the reason for considering this case.

The default profile has the lowest reliability for two reasons: firstly, it provides completely static information and does not consider the changes made in the model concerned, and secondly, in some cases it may not have been expressed correctly by the manufacturer. These profiles do not conform to the referenced schemas of the chosen vocabulary; therefore, they cannot be validated and, in the best of cases, are only parsable in partial form. The system constructs a final profile that is not only valid and conforms to the proposed vocabulary but is also parsable by existing tools.

5.3.1.2 Profile Repository

The profile repository module is divided into two databases: one stores those user profiles detected by the detector module and those manually configured by users, while the other stores the default profiles of detected models. Once a profile is detected and created it is stored in the user profile repository. Profiles are written in RDF/XML using the UAProf 2.0 vocabulary and are retrievable via HTTP. In future extensions, we will consider implementing access to profile repository information using Device Description Repository API [30]. These two databases can be decoupled from the proxy and detector modules and be distributed among various nodes or centralized in a central server and proxies can share the same profile databases.

The default profile database exists to improve performance in difference resolution and profile analysis. The referenced default profile is retrieved more rapidly, because it is retrieved locally rather than from an external server. Thus, if another user has the same device model, or if a second detection is performed, the default profile is ready to be used by the proxy and does not have to be retrieved from the Internet. The default profile database should therefore be deployed as a local database or near the proxy.

With reference to the J2SE detection process, having a small CC/PP client that is able to express differences with a default profile allows generic default profiles to be defined for devices such as hand-held devices and PCs. These profiles express typical values for capabilities usually found in these

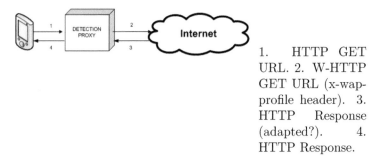

1. HTTP GET URL. 2. W-HTTP GET URL (x-wap-profile header). 3. HTTP Response (adapted?). 4. HTTP Response.

FIGURE 5.6: Proxy Web navigation.

types of devices. Thus, the system makes use of the profile difference-solving mechanism provided by CC/PP and UAProf from the very first detection. W-HTTP, which only expresses differences and not the whole detected profile, is used to reference these default profiles, thereby optimizing the channel utilization during the detection process.

5.3.1.3 Proxy

Once a user's device is detected and the user profile is created, the proxy module will be in charge of managing web navigation. Besides carrying out the habitual functions of HTTP proxies, it will also be in charge of presenting a valid profile for the device to the rest of the Internet, whether the device is UAProf-compliant or not. Thus, using W-HTTP, the proxy module will add UAProf *x-wap-profile* headers with references to the profile stored in the repository (a detected one or a manually configured one). Similarly, in the case of UAProf-compliant devices, the proxy module will modify *x-wap-profile* and *x-wap-profile-diff* headers expressed by the browser to reference the new profile (Figure 5.6). The navigation process, as seen in Figure 5.6, follows usual proxy behavior but includes *x-wap-profile* headers that reference the detected profile. A device-independent application can adapt content to fit the capabilities of the device and all devices navigating through the proxy will appear on the Internet as UAProf-compliant devices, expressing their capabilities and characteristics across the delivery context. This context expression facilitates content adaptation to context-aware servers and adaptation proxies.

An interesting feature of the proxy module is the possibility of it adding its own adaptation capabilities in the form of new supported MIME types, not only in the stored profile but also in the HTTP *Accept* header. Consequently, if a web server is not able to adapt content according to the expressed profile but can deliver those proxy-adaptable types of MIME objects, the proxy will adapt them to a device-displayable format.

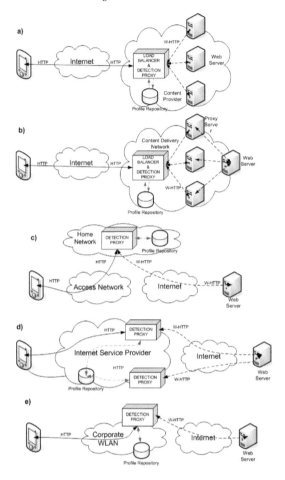

FIGURE 5.7: Proxy deployment configurations.

5.3.2   System Deployment

The proxy can be deployed in different ways depending on the service provider role and purpose. Figure 5.7 shows possible deployment options ranging from unfolding on a personal computer at the user's home to deployment on a content delivery network.

Figure 5.7a) shows the deployment of a detection proxy by a content provider with a cluster of web servers. The detection proxy is allocated at the load balancing point. This configuration reduces the number of redirection mechanisms that can be used for load balancing; HTTP redirection mechanisms and HTML rewriting cannot be used because HTTP requests should be transparently proxied if they are to be able to modify requests to incorporate profile headers. Therefore, the balancing mechanism should be done at the transport level. A cookie-based solution, which would map profile headers as cookies, could be devised to enable other load balancing mechanisms to be used at higher levels.

Figure 5.7b) shows a similar configuration, this time for a content delivery network, and it poses the same issue: load balancing mechanisms and redirections should be restricted to the transport level or managed by cookies.

Figure 5.7c) shows the deployment of a user's personal proxy. This configuration allows users to make use of the proxy in a ubiquitous manner wherever he or she accesses the web. The user deploys the proxy on a computer at home or in the office; the computer, which is permanently online, is managed by the user. In this way a user can access the Internet from any access network and still experience device-independent web browsing.

Figure 5.7d) shows deployment on an Internet service provider network, in which the profile repository is centralized and its data shared between different proxies allocated to different egression points on the Internet. All the operator's clients express their device capabilities when accessing the Web.

Figure 5.7e) shows deployment on a corporate LAN or WLAN, which is the detection proxy allocated at the office's proxy cache.

Section 5.4 explores how to deploy the detection proxy in a collaborative cellular environment in order to optimize web traffic by means of protocol reduction and self-adapting content negotiation and expression of capabilities.

Regarding the integration with other existing user privacy control, like P3P proxies[30], and content filtering proxies, like privoxy[29], the detection proxy could be chained to the privacy proxy node for a normal operation. However, it would be interesting integrating both systems in order to add a certain degree of privacy control regarding which context information the detection proxy provides to external servers and proxies.

5.3.3 Vocabulary

Regarding the vocabulary, UAProf 2.0 was chosen to describe the profiles over other possibilities (e.g., Universal Profiling Schemas [4, 14], Wireless Universal Resource File [9], etc.) for various reasons. It is the most widespread standard framework for expressing device capabilities on the Internet and it is a clear example of application of CC/PP. Furthermore, in addition to having an extensive definite vocabulary and wide-ranging support, it also defines attribute conflict resolution rules, which are essential in univocally resolving a profile and its differences. Unlike CC/PP with CC/PPex, it defines a transport protocol on HTTP (W-HTTP) that uses ad hoc HTTP headers and is not based on the HTTP extension framework. Using ad hoc headers eases protocol compatibility because it is not necessary for a complex framework like the HTTP extension framework to be supported.

The only aspect open to discussion is whether the defined vocabulary in UAProf allows devices other than mobile phones to be properly described. In the case of hand-held devices such as PDAs and PocketPCs their differences with mobile phones are minimal, so selective use of the vocabulary permits their complete characterization. In the case of high-computation devices such as PCs and laptops, the vocabulary allows their web browsing capabilities to be accurately described. Nevertheless, at the hardware and software levels, it only allows the most basic characteristics to be described and it requires extensions to describe peripheral accessories other than keyboards and screens. In all cases, this characterization is enough to allow device-independent web browsing.

5.3.3.1 Vocabulary Extensions

The implemented platform was extensively tested in different ubiquitous and device-independent environments. Some of these environments and applications needed information that was not included in UAProf vocabulary, so we extended the vocabulary to properly describe new attributes ranging from physical and temporal location information to dynamic network information.

In the case of many device-independent applications, some content and presentation adaptation rules are created specifically for a certain type of device and then customized to fit the actual characteristics of the model concerned [7]. For these types of applications, it would be desirable to have an attribute describing the type of device (PC, hand-held device or mobile phone, for instance). Thus, the HardwarePlatform component was extended with a new attribute (TypeOfDevice) to include this information. This attribute is fulfilled using the detection-mechanism selection process, whereby the proxy is able to properly identify the type of device (PC, hand-held device or mobile phone).

Some context-aware location-based applications were also tested. These applications work with physical and temporal location information obtained from different sources, so we investigated various ways of including this location information in the profile. UAProf preventively (for future use) defines attributes for the physical location of devices; so these attributes were tested. However, our requirements for these location-based applications meant that this information fell short; so we extended it by creating a new component (DeviceLocation) with attributes that included device location information such as the location server that provided the location information or the extended location information expressed in Mobile Location Protocol (MLP) format [8]. This location vocabulary allows enriched location information to be expressed, ranging from the location method used to the accuracy of the measurement.

In order to help and improve adaptation processes, certain dynamic calculations of network characteristics would be desirable. Following this idea, the proxy calculates the available bandwidth (up and downstream), the current bearer, etc. One vocabulary extension was designed to reflect these calculations and new attributes were added to the NetworkCharacteristics compo-

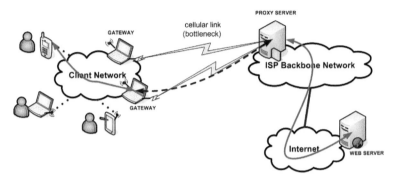

FIGURE 5.8: System architecture.

nent. These include information that is difficult to obtain behind a Network Address Translator (NAT) or outside the access network, such as the physical address and IP address.

5.4 Collaborative Optimization, Context Acquisition and Provisioning

Ubiquitous web access should take into account not only the capabilities of devices but also the characteristics of access and transport networks. Cellular networks such as the Universal Mobile Telecommunications System (UMTS) provide a wide coverage area and can provide access to the Internet to mesh, ad hoc and local area networks with no wired connection. Thus, a UMTS operator can use mesh or WLAN networks as a means of increasing indoor coverage area. Likewise, emergency networks or networks that are not supported by a wired infrastructure (e.g., in locations far from city centre, socioeconomic events, etc.) can use cellular networks to access the Internet. Figure 5.4 illustrates the proposed system architecture: the client access network (WLAN, ad hoc, mesh, etc.) is connected to the Internet through one or more gateways (GWs) and a Proxy Server (PS).

In these cases, the cellular link with the backbone network is usually an asymmetric link, e.g., General Packet Radio Service (GPRS) and High-Speed Downlink Packet Access (HSDPA), with less capacity in terms of bandwidth and delay than the client access network, becoming a potential network bottleneck. For this reason it is necessary to apply optimization techniques to improve access to Internet services such as web navigation. In addition, an enhancement on the uplink optimizes overall behavior and web page download time due to the Stop-and-Wait behavior of the HTTP protocol and the asymmetry between the uplink and downlink channels [13]. One technique is to remove most static content negotiation headers in HTTP requests over the cellular link and reconstruct them later on the wired side of the network. Headers must be restored reaching remote web servers. This topology provides some opportunities for further enhancing and optimizing web navigation, by configuring header values to take advantage of the content adaptation capabilities of servers and other proxies in the delivery chain.

On one level, header reconstruction involves some kind of mechanism to detect which current browser is in use and how it identifies itself and negotiates content. This could be something as straightforward as copying original header values sent by the device's browser but, with this necessity of browser capabilities profiling, we could take advantage of this opportunity to deploy a more sophisticated device capabilities detection and user profiling system. Such a system (see Section 3) allows enriched (i.e., more expressive) information about the user's device and browser capabilities to be provided and enhances navigation speed and the user's Quality of Experience (QoE) by the expression of dynamic capabilities, which enables adapted (and lighter) content and responses to be obtained from web servers and adaptation proxies alike.

Furthermore, this interception model allows PS to further toy with the content of HTTP headers sent by the browser to the remote server, and adds its own vision of the delivery context by modifying header and profile values according to the device's capabilities, access network state and cellular link condition. Consequently, different regeneration policies can be applied with different purposes. Dynamic policies could be applied to change header values depending on the available bandwidth of the cellular link, forcing remote web servers to deliver lighter or heavier content (e.g., plain text, HTML, DHTML or ActionScript documents) in order to mitigate the effect of network congestion and delay on the user's browsing experience. Hence, a web server with minimal intelligence can modify the content to deliver in order to better fit the user's device capabilities and delivery context characteristics.

5.4.1 Application Layer Optimization

The use of Application Layer Performance Enhancing Proxies (PEP) to address the challenges posed by the problematic characteristics of low-bandwidth wireless wide area network links is a widely studied field and its efficiency has been demonstrated [13, 16]. There are several techniques for performance enhancing at the HTTP layer such as extended caching [19], pipelining [13], compression and object adaptation [18, 19], and delta encoding [20]. The efficiency of link optimization at the HTTP layer by means of reducing the number and size of request headers has been demonstrated in several proposals [17, 18].

In [17], HTTP protocol reduction is limited to omitting the *Accept* header, and ignores other redundant content negotiation headers such as *Accept-*

Charset, *Accept-Language* and *Accept-Encoding*, which would otherwise further optimize performance. What is more, the results shown for protocol reduction are not detailed enough. An overall result for TCP and HTTP reduction techniques is given, including TCP connection overhead reduction and HTTP header reduction and compression, without differencing the contribution of each individual optimization mechanism. The client/intercept approach implemented relies heavily on caching solutions because it is aimed at optimizing web access to routine repetitive commercial applications, in contrast to the random surfing and ad hoc web browsing of our approach. Although the system does not force users to modify their browser software, it does require a new component, a local web proxy, to be installed in the client device in order to perform HTTP optimizations (compression, header reduction, caching and differencing). It also burdens users with the task of configuring their browser to use this local proxy. Thus, intercepting client requests, in contrast to hijacking connections, which facilitates the deployment and management of the system, is not done in an entirely transparent way. Besides, the approach followed in [17] for header restoring (copying original *Accept* header values) could be improved using an accurate device capabilities profiling system such as the one presented in Section 6.3 and further expanded herein. This complete device characterization enhances web browsing QoE by expanding content negotiation for devices with limited expression of capabilities in order to ease content negotiation and adaptation. This optimizes the amount of data transmitted and the response time as servers and proxies deliver lighter content that is specially tailored to fit delivery context characteristics, including device capabilities and network characteristics.

In contrast, the results shown in [18] for HTTP header reduction are obtained by performing GET requests directly from a modified browser software and omitting all HTTP headers. Consequently, the best results are achieved in minimizing delay and the amount of information exchanged, but this solution is not functional as web browsing is degraded by the omission of state, content negotiation, connection management, cache and session related headers. However, it does take into account the need for a profiling system to restore device capabilities, although it does not address the issue of how such a system might be built.

To summarize, these proposals do not explore header restoring strategies for enhancing and facilitating content negotiation and adaptation. Therefore, here we propose using intermediaries for the transparent management of HTTP protocol optimization and contextualization functions, without modifying browser implementation or configuration. This protocol reduction with posterior header restoring does not exclude using other optimization techniques such as the mentioned above to further enhance performance (i.e., delta encoding, extended caching, etc).

In this section, we see how the device capabilities detection platform described in Section 6.3 was restructured and distributed among different nodes (GWs and PS) to adapt it to the new collaborative environment. Optimization, analysis and restoration functions were added, and the following changes were made:

1. The device capabilities detection system's modules and functionalities were distributed between the GWs and the PS, with the aim of minimizing transactions over the cellular link.

2. Functions for header reduction and reconstruction of HTTP requests were implemented. Redundant headers were omitted on the GW side, and the User Agent was modified to give a univocal user and device identifier to the PS, which could then reconstruct most suitable headers according to the restoration policy applied.

3. Header restoration policies were designed and tested, including policies for ease device independence or for cloaking device capabilities based on network-state to influence content adaptation by web servers and proxies.

4. An available bandwidth estimation system based on Packet-Pair Probes [21] was implemented.

5. A coordination and collaboration protocol between GWs and PS was designed. In some scenarios several GWs collaborate with just one central PS (e.g., UMTS Assisted Mesh Network [28]).

The following section aims to investigate existing synergies among device capability expression, delivery context characterization, content adaptation and application layer protocol optimization.

5.4.2 System Description

The general system architecture is composed of one Gateway (GW) at one end of each link to be optimized, working collaboratively with a central Proxy Server (PS) on the wired side of the network. Figure 5.4 illustrates the unfolding of this architecture in a generic environment. On one side there is the client access network, which consists of one or more GWs to the cellular network. These GWs hijack HTTP-related TCP connections in order to perform HTTP protocol optimization and user and device profiling functions. Next is the link to be optimized, which is the most restrictive network section; with lower bandwidth and higher latency than the client access network, it forms the network bottleneck. On the other side of the cellular link is the central PS, which performs dynamic HTTP header restoring and profiling functions. The PS, being allocated to the ISP's backbone network, has no capacity restrictions. Once client requests reach the PS, content negotiation headers are restored with new values modified according to the restoring policy applied. Header values are inferred from information provided by the profiling system. The system is able to work with several GWs from the same client access network and the PS is also able to cope with multiple GWs from different client access networks.

Regarding channel characteristics, any link is susceptible to being optimized whenever request transmission time is the limiter rather than total optimization processing time, including header reduction at the GW and later header restoring at the PS. It should be noted that an enhancement of the response time increases with greater asymmetry between the uplink and the downlink channels. This is due to the HTTP's Stop-and-Wait behavior, by which a small reduction in uplink transmission time remarkably reduces overall download time because the server spends less time waiting for the arrival of requests and response delay decreases. As the results in Section 4.5 show, using the complete system the response time is enhanced by about 10% using HSDPA (approximately 1 Mbps in downlink and 384 Kbps in uplink).

5.4.2.1 System Modules

The system is built in modules that are distributed among GWs and the central PS. The modules collaborate to manage device capabilities detection and user profiling processes, the optimization operation and the HTTP header restoring function with enhanced content negotiation and capability expression, an enhancement that is obtained by complementing the expression of the device's capabilities and allowing intermediaries in the delivery chain (GW and PS) to add their point of view on delivery context to requests. Figure 5.9 illustrates the distribution of modules among each kind of node.

5.4.2.1.1 Gateway modules. The GW is composed of a transparent Java-based HTTP proxy and a JavaServlet container, which serves the web applications that provide the device capabilities detection service. They comprise the following modules.

1. Redirector: this module is in charge of hijacking HTTP-related connections and managing optimization and detection triggering. If there is an active profile for this user and device, it forwards client's HTTP requests to the header filtering module. If no active profile exists, it triggers the detection service.

2. Header Filtering: the HTTP optimizer manages the header reduction functions, modifying the *User-Agent* header value with a univocal user and device identifier. All the content negotiation headers are removed except for the Host and modified *User-Agent* headers before the request is forwarded to the PS. Other headers related to management of state, management, cache and session (i.e., cookies) are not removed so as not to degrade web application operation and consequently navigation experience.

3. Device Capabilities Detector: this is in charge of managing detection processes at one end of the link (from the access network up to the

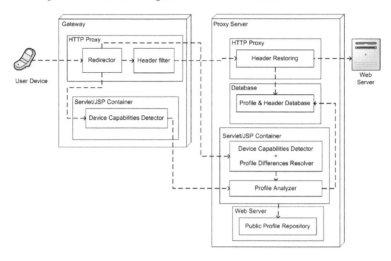

FIGURE 5.9: Deployment diagram.

cellular link) in order to obtain a complete device profile. This profile is then sent to the PS, which analyzes it and stores it in the repositories.

5.4.2.1.2 Proxy Server modules.   The PS is composed of several components: one HTTP proxy that manages header restoring procedures, an internal profile and header database, a servlet container serving the web applications associated with the device capabilities service and the profile analysis and, finally, a web server acting as a public profile repository for the Internet server. They comprise the following modules.

1. Header Restoring: this module restores content negotiation HTTP headers before forwarding requests to web servers on the Internet.

2. Profile and Header Database: this internal database stores the headers to be restored for each device and classifies them by executed analysis policy. It also acts as a private device profile repository for internal management of the complete system.

3. Device Capabilities Detector + Profile Differences Resolver: this module is in charge of managing detection processes at this end of the link (from the PS to the Internet). The detection processes of devices pointing to a profile hosted by a server on the Internet will be managed by this module rather than the GW detection module. The module is able to manage differences between the results of the detection processes and the content of profiles provided by manufacturers.

4. Profile analyzer: this module analyzes the content of detected profiles and infers header content according to the different profile analyzes and header restoring policies.

5. Public Profile Repository: the public profile repository serves detected and modified profiles in RDF/XML format (using UAProf as the vocabulary) to web servers and other context consumers on the Internet.

5.4.3   Header Restoring Policies and Context Provisioning

Pervasive and ubiquitous web access demands content adaptation for successful interaction with web applications and suitable rendering on any kind of device (e.g., PDAs, PCs and portable game consoles). These devices have very different hardware and software capabilities, including user interaction modalities that are different from traditional ones. Moreover, this ubiquitous access mechanism involves different wireless interfaces and networks, some with serious restrictions on channel capacity, available bandwidth and latency, which affect web interaction and content downloading. Hence, content adaptation and personalization is a critical factor in obtaining a functional user experience in any kind of environment, regardless of the access mechanism, and in accomplishing the desired author principle of "write once, deliver it anywhere" [2, 26].

W3C defines delivery context [2] as *"a set of attributes that characterizes the capabilities of the access mechanism, the preferences of the user and other aspects of the context into which a web page is to be delivered."* Therefore, each entity along the delivery chain should add its own context vision to web requests, so that content adaptation could be performed by content servers and/or adaptation proxies. Ideally, adaptation should be performed or managed by content servers so that the author's transformation guidelines can be fulfilled without it being too much of a burden [26] and QoE and adapted content fidelity can be enhanced to provide harmonized user experiences[1]. In order to reduce the load on origin servers, adaptation functions may be distributed on behalf of content servers among proxies and surrogates. Using protocols like the Internet Content Adaptation Protocol (ICAP) and Open Pluggable Edge Services (OPES), content servers can control the adaptation processes that are requesting proxies for specific adaptation tasks and still meet the quality criteria of the author by following his transformation guidelines.

However, a proxy that is not controlled by origin servers must either infer structural document information, the purpose of the document and the author's intentions or the original document must be overloaded with metadata

---

[1]A functional user experience that is sufficiently harmonized with the delivery context to meet the quality criteria of the author [2].

that give the author's own content adaptation guidelines. In any case, overhead increases significantly, performance decreases (due to the extra metadata that must be interpreted) and authors lose some degree of control over the adaptation process. Furthermore, client side adaptation is inefficient in terms of data traffic volume, and document overhead increases and usually demands too many resources from the client device.

Ideally, web content adaptation and personalization should be based on accurate and complete information about current device capabilities, user preferences and network characteristics that is dynamically expressed with HTTP content negotiation headers and profiling frameworks such as CC/PP and UAProf. At the moment other solutions such as WURLF [9] are based on identifying *User-Agent* strings to extract device capabilities information from a profile database. This approach has several drawbacks, such as the fact that configuration and personalization changes made to user devices are not reflected in the static information stored in the database. Device information is also limited to a known range of devices, namely mobile phones, which excludes other devices such as PPCs, portable game consoles and Tablet PCs, and is difficult to keep up to date. Furthermore, the approach converts a dynamic element, the user context, into a static element, leaving no way of adapting content dynamically to changes in the network state and/or device capabilities/configuration. If adaptation functions are allocated in content providers' servers and clusters and/or in third-party proxies along the delivery chain, these elements should be able to dynamically adapt content according to device capabilities, capabilities that are expressed by the client device itself and complemented by intermediate proxies (with or without adaptation functions) in such a way that all the entities in the delivery chain contribute their own vision of the delivery context. Thus, the proposed PS/GW collaboration system gives a privileged insight into a segment of the delivery chain with severe restrictions in available bandwidth, capacity and latency. Content can thereby be adapted bearing in mind the restrictions imposed by the cellular link.

The need to restore headers in order to provide a satisfactory browsing experience, with typical content negotiation and adaptation, can be used to take advantage of the synergy between device capabilities detection and application layer optimization and thus provide enhanced expression of capabilities and content negotiation mechanisms. Content negotiation information is dynamically modified by adapting context expression to the current network and device characteristics in order to enhance the user's perceived QoE. Different user profile analyzes and content negotiation header restoring policies can be designed with different enhancement philosophies and purposes in mind. Thus, the detected profile will be dynamically modified according to the restoring policy; this modified profile will be announced with W-HTTP, and content negotiation headers will also be modified to reflect the changes made to the announced profile.

Dynamic policies can serve two purposes. Firstly, they can modify header and profile values according to the state of the network in order to request lighter or heavier multimedia objects to mitigate the effect of network congestion on user browsing experience, i.e., if the network is congested users may prefer to see only text objects in order to speed up downloading time, so the user profile would be modified to appear as if it only supported text objects and only MIME text objects would be announced in the *Accept* HTTP header. Secondly, device capabilities information can be enriched in order to enhance content adaptation and personalization for the purpose of easing the device independence and/or the context-awareness of applications. Also, as explained above, some web pages use straightforward adaptation based on *User-Agent* string identification. In this case, server-driven web content adaptation is constrained to a limited range of profiled devices. For small hand-held devices like phones, this range can be increased by cloaking *User-Agent* strings so that the device appears to be another mobile device of similar characteristics, one profiled by the web application. Another possible method is to reflect client requests for the available adaptation capabilities of intermediate proxies on the cellular operator's network, e.g., by announcing adaptable MIME types, changing profile values, etc. Intelligent web servers can then better adjust content to device capabilities. For example, Google serves different search pages depending on whether the device is a PC, PDA or mobile phone. For mobile phones it also acts as a non transparent adaptation proxy using URL rewriting techniques, image adaptation, document repagination, eliminating embedded objects, etc. Thus, knowing this, we can configure header values in order to obtain different responses according to the user's needs, i.e., forcing Google to act as an adaptation proxy by cloaking the user's agent and device capabilities.

One of the policies implemented in the testbed aims to achieve device independence and thereby provide the most detailed information possible on device capabilities in order to allow web servers and intermediate proxies along the delivery chain to adapt content to device and browser capabilities. Thus, content negotiation header values are enriched with all the detected information about the capabilities of the device and a W-HTTP *x-wap-profile* header referencing the user's device enriched profile is added.

The other tested policy is designed to cloak device capabilities depending on the available bandwidth in the cellular link. In the case of bandwidth reduction, depending on the user's navigation preferences, PS modify headers and profile values to cloak the device's capabilities, profile and User Agent as if another device (with greater restrictions on size and hardware and software capabilities) were being used, in order to request lighter content. A bandwidth estimation system based on Packet-Pair Probes [17] monitors the cellular link on behalf of the PS and informs the PS when the bandwidth decreases considerably. This estimation service could be provided by third parties and operators. This restoration policy was tested against the Internet's most popular web page, the Google search page. It was chosen because Google servers adapt its search page's document structure, style (CSS), dynamism, language,

Table 5.2: Size in bytes of the Google search engine.

Device/Browser	Default headers	User Agent
Qtek 9000/MSIE 4.01	1875[1]	1875[1]
Qtek 9000/Minimo 0.2	1887[2]	1875[1]
UltraMobile PC (WinXP)/Mozilla 1.8.1	4075[1]	4075[1]
UltraMobile PC (WinXP)/IExplorer 7	4505[1]	4505[1]
UltraMobile PC (Debian 3.0)/Mozilla 1.8.1	4057[1]	4057[1]
UltraMobile PC (WinXP/Debian 3.0) /Opera 9	3785[1]	3795[1]
Nokia N91/default	1895[2]	1875[1]
Nokia 6329/Opera 8.00	1895[2]	3795[1]
Nokia 6329/ Default[m]	1539[3]	1895[1]
Fake mobile phone browser	1539[3]	3795[1]

[1]html [2]xhtml [3]wml [m]modified *Accept* header (wml prioritized)

pagination, format and encoding depending on the announced capabilities. Furthermore, for mobile phones Google acts as an adaptation proxy by reformatting web content to fit mobile phone characteristics and by using URL rewriting techniques to restructure document anchors and embedded object links. Therefore, we analyzed and tested Google's behavior to see the degree of context-awareness and adaptability of one of the most technologically up-to-date commercial web applications on the Internet.

5.4.3.1 Google Analysis

For the analysis (see Table 5.2), we requested the Google search engine page via different types of devices and models (mainly PCs, PDAs and mobile phones); we accessed it using different browsers for each hardware platform (i.e., Qtek 9000 was tested with IE MSIE 4.01, Mozilla Minimo 0.2 and Opera Mobile v 8.65) and we examined the differences between versions of the returning hyper-text document[2] (i.e., format, size, scripting structure, charset, encoding and style) and related multimedia objects (i.e., image size, format, quality and object size). We also searched for changes in application behavior to see when Google acts as an adaptation proxy. For the same model and browser, we also configured the existence and values of the differ-

---

[2]Google makes heavy use of transparent redirections and other balancing techniques that might not work properly with some User Agents.

ent content-negotiation related HTTP headers (*User-Agent, Accept, Accept-Charset, Accept-Encoding* and *Accept-Language*). W-HTTP profile headers (x-wap-profile and *x-wap-profile-diff*) and profile values were also taken into account. The procedure was carried out in the following order:

1. Complete requests (with all content negotiation headers)

    (a) With *x-wap-profile* header (when needed)

    (b) Without *x-wap-profile* header

2. Requests with just the *User-Agent* header (with and without other content negotiation headers)

    (a) Well-known *User-Agent* strings

    (b) Modified, unusual or new *User-Agent* strings

3. *User-Agent* + W-HTTP *x-wap-profile* header (with and without the rest of the negotiation headers)

    (a) Well-known *User-Agent* strings

    (b) Modified, unusual or new *User-Agent* strings

    (c) Varying *x-wap-profile* to refer to a modified profile

    (d) Adding *x-wap-profile-diff* to express differences with manufacturer profiles

When a user agent is not known, Google uses the existence (or not) of the *x-wap-profile* header and the content of the *Accept* list to infer whether it is a mobile device; in this case it does not act as an adaptation proxy. We also found it does not retrieve the referenced profile in the *x-wap-profile* header and ignores *x-wap-profile-diff*; therefore, it does not analyze profile information. Consequently, we deduce that the classification process is based mainly on user agent identification against some kind of profile database.

Regarding the adaptation process, Google serves different versions of its search engine page depending on the capabilities announced by browsers as traditional HTTP content negotiation demands. It delivers two visually different search engine pages, one for mobile or hand-held devices and one for PCs, but each of these visually equal documents shows considerable differences in behavior and in its internal foundation depending on the type of device. Hence, we discovered that Google classifies devices in three categories: PCs, hand-held devices and mobile phones.

In the case of devices without restrictions (e.g., PCs, laptops and UltraMobile PCs), Google adapts its search page to the announced capabilities. Thus, it adapts content to announced encoding (e.g., none, gzip, deflate); hypertext format (e.g., xhtml, html, wml); style (e.g., CSS); scripting (e.g., the use of different JavaScript/ECMA functions); image format (e.g., JPEG, GIF); human language and charset (e.g., ISO-8859-1, UTF-8).

In the case of mobile and hand-held devices (e.g., PDAs, mobile phones and portable game consoles), the Google search page also adapts to device rendering capabilities; it can, for example, adapt the size and format of images to screen size when image searches are conducted. However, the scope of the search engine application is expanded for mobile phones, as Google acts as an adaptation proxy on behalf of mobile devices. Web sites linked through the search results page are transparently proxied by Google using URL rewriting techniques. Thus, the content of the requested site is adapted to the hardware and software capabilities of the mobile phone.

This adaptation of third parties' web content is done in a very straight-forward manner. It is dependent on the static device capabilities information indexed by *User-Agent*, as in WURLF-based applications. Furthermore, adaptation is not transparent to users, as adapted documents clearly state that they are the result of the Google adaptation engine. Users have limited control over the adaptation procedure: they can disable image display and request the original documents. The following are some of the actions detected in the Google adaptation engine:

1. HTML is translated into xHTML.

2. Documents are repaged.

3. Embedded objects are removed (even in the case of supported MIME object types).

4. To reduce file size, images are resized and format and quality are otherwise modified.

5.4.4   Collaborative Device Capabilities Detection Service

Herein, the detection platform is distributed among GWs and a central PS. These nodes cooperate to obtain a complete profile of the same characteristics as in Section 5.3 but in a very different environment. This environment presents new challenges, especially regarding the number and size of detection-related HTTP transactions over the cellular link that must be minimized. Thus, pertinent management is performed to execute the detection processes on the most suitable network section, so that the detection process is as unobtrusive and agile as possible.

5.4.4.1   Use Cases

With this purpose in mind, two use cases are devised, depending on whether it is necessary to access Internet servers to complete the detection processes or they can be executed exclusively on the client access network.

1. Browser announces a reference pointing to a public profile on the Internet (external detection). The browser implements one of the device

Table 5.3: Size in bytes of detection-related transactions over each network section.

Device (Browser)	Client Access Network	UMTS	Internet
Ultramobile PC (Mozilla 1.7.12)	79385	2098	0
N91(default browser)	993	1188	24349
IPAQ h6300 (MSIE 4.01)	6692	1261	0

capabilities expression frameworks (CC/PP, UAProf, etc.) and provides a link to a default profile for this device. In this case it is necessary to access the Internet to retrieve the default profile and analyze it, so the PS will be in charge of executing profile resolution mechanisms and the remaining procedures of the detection process.

2. Browser does not announce any references pointing to a public profile (local detection). The browser does not implement any frameworks for the expression of device capabilities, or it expresses a complete profile without referencing a default profile on the Internet. In this case all the detection procedures will be executed on the client access network, and transactions will take place over the cellular link until the final profile publication on the PS.

It should be noted that both cases are independent of browser family and device type; it only matters if there are references to external profiles. The PS is in charge of managing the triggering of the detection service on a device. Thus, if a GW has not cached a device, it asks the PS to obtain the user and device identifier and the PS also indicates whether the GW should start the detection process or proxy the request.

As Table 5.3 shows, PDAs and PCs do not usually implement any capabilities expression frameworks; thus the detection processes are executed on the mesh network and the resultant profile is compressed and sent to the PS over the UMTS link. However, a UAProf-compliant device such as the Nokia N91 refers to a public profile on the Internet, and therefore the detection processes are delegated to the PS on the wired network section.

5.4.4.1.1   External Detection Use Case.   The sequence diagram in Figure 5.10 shows how the GW delegates detection management to the PS. The request is forwarded to the PS and it retrieves the default profile and resolves it against the differences expressed by the device and the information inferred from content negotiation headers. Once a complete profile is obtained, it is analyzed to infer the header values. The headers and the profile are stored

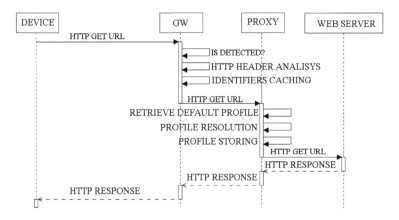

FIGURE 5.10: External detection use case sequence diagram.

on the private repository whilst the profile is stored on the public repository. Next, the original URL is requested with new content negotiation headers and subsequent requests are optimized.

5.4.4.1.2   Local Detection Use Case.   The sequence diagram in Figure 5.11 shows how the GW is in charge of most of the detection process, except for detection by profile analysis and publication, which are delegated to the PS. Next, the original URL is requested with new content negotiation headers and subsequent requests are optimized.

5.4.4.2   Device Identification Issues

One of the existing problems of a contextualization system such as the one proposed is how to univocally identify a device. Devices using multiple interfaces simultaneously or alternately have to identify them as part of the same device and not as different devices. In the testbed implementation, one device profile is created for each physical address in use, although all of the addresses are associated with the same user. Multihoming, whereby the same IP address is assigned to all the interfaces of the device, is only suitable if the IP address assignment is static and lasts for more than one session, and the Host Identity Protocol [15], which is based on public key interchange, cannot be executed by devices with limited computing power, battery life and memory (e.g., sensors/actuators, objects and phones) without severely degrading their performance, if it can be executed at all.

Devices using two different browsers is another case to consider. UAProf vocabulary and CC/PP framework specifications have some limitations and they do not allow collections of profiles or collections of components within the same profile. This makes context aggregation in one profile device diffi-

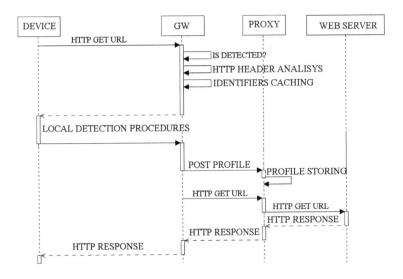

FIGURE 5.11: Local detection use case sequence diagram.

cult. Therefore, if a device uses more than one web browser, each browser is considered as a different device associated with the same user. Though several commercial devices such as mobile phones use more than one browser, the UAProf vocabulary does not allow collections of BrowserUA components for a specific device. To make the profile publication process easier, the system treats each browser as if it were a single device with just one browser component; the other components point at a common local default profile with the remaining device capabilities (hardware, software, etc.). In future research we will consider modifying the UAProf vocabulary and CC/PP structure to support collections of devices and profiles with collections of components.

5.4.5   Optimization Results

Tests were executed with the equipment listed below.

- Gateway. Linux-based laptop (Debian 4.0 distribution with 2.6.18 kernel version) using Intel ipw3945 802.11a/b/g chips

- Cellular link. HSDPA link, approximately 1 Mbps downlink and 384 Kbps uplink.

- Access Network. 802.11.b

- Client. Windows XP-based UltraMobilePC using Mozilla 1.7.12 as a web browser

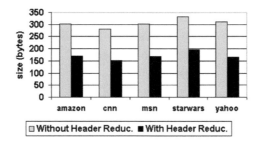

FIGURE 5.12: Average size of HTTP requests in uplink [bytes].

- Proxy Server. The PS is based on a set of software modules that run on a Linux OS Debian 4.0 distribution with 2.6.18 kernel version (PC with Intel P4 3GHz dual core processor and 2 GB of RAM).

The test methodology was as follows:

1. In order to avoid the effect of public server congestion, we deployed a virtual web hosting system [15] at the laboratory network that mirrored popular web pages (with the same hardware and software characteristics as the PS node). With this system, clients still accessed these pages through the Internet but the servers' work load was kept under control.

2. The first time a user requested a page, the PS retrieved the headers from the database, but in subsequent requests the header values were cached and the database was not accessed again.

3. For each GET request, the GW calculated the size of the original request and the size of the reduced one. Later, the average number of bytes saved (those that were not sent over the HSDPA link) on all GET requests was calculated.

4. Each page was requested 10 times, with optimizations disabled both on the GW and on the PS.

5. The same experiment is repeated, this time with the header reduction enabled at the GW and the header restoring enabled at the PS.

Figure 5.12 shows the reduction in request size obtained for different web pages. The differences depend on the implementation of the requested web application. This is due to the existence of cookies and similar dynamic headers and not changes in content negotiation headers, which are static. Header reduction mostly reduces request size by removing these static headers. These headers obviously vary between different browsers and may even vary between two browsers of the same family and version, depending on the plug-ins installed.

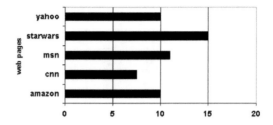

FIGURE 5.13: Average improvement in response time (%)

Figure 5.13 shows the average improvement in response time for the same pages, which is about 10%. The differences in the improvement in response time depend mainly on the number of objects, regardless of their type. The more objects there are to request, the more requests there are to optimize, and so improvement increases. However, it is demonstrated [13] that the downloading order of the objects of a web page affects the overall response time and that it is better to interleave small objects with large objects to keep the server transmitting rather than waiting. Differences in web document structure therefore affect the downloading order and, thus, the size of consecutive objects also affects the response time in conjunction with the number of objects that make up the web page.

5.5   Conclusion

In conclusion, pervasive and ubiquitous web access demands content adaptation for successful interaction with web applications and suitable rendering on any kind of device (e.g. PDAs, PCs and portable game consoles). These devices have very different hardware and software capabilities, including user interaction modalities that are different from traditional ones. Moreover, this ubiquitous access mechanism involves different wireless interfaces and networks, some with serious restrictions in channel capacity, available bandwidth and latency, which affects web interaction and content downloading. Hence, content adaptation and personalization is critical to obtaining a functional user experience in any kind of environment, regardless of the access mechanism. The self-adaptation process should be automated as this reduces the burden of content authoring on authors and content version management on web administrators. For this automatic adaptation and operational adjustment to be implemented it is critical that the capabilities of the terminal be known. Several proposals have been made but none have yet been fully implemented. Therefore, we propose a device capabilities detection proxy that provides us with a single solution that expresses device capabilities across the delivery context, with just one profiling platform and vocabulary for different types of device. This is a transitory solution while we advance towards a device-independent and ubiquitous web in which all devices express their capabilities in a reliable, dynamic and up-to-date manner.

Using optimization techniques based on request header reduction at the HTTP layer provides an excellent opportunity for deploying sophisticated device characterization and contextualization solutions that not only allow faster web downloading but also enrich content negotiation in order to enhance adaptation, thereby achieving more ubiquitous web access. The GW/PS collaborative interception model presented in Section 5.4 allows the PS to toy with the content of the HTTP headers sent by the browser to the remote server and thus add its own vision of the delivery context. In this paper, we have shown how PS can modify header and profile values according to its own privileged vision of the device capabilities, access network state and cellular link condition, and how to dynamically configure content negotiation HTTP header values and profile content to adapt web application behaviour and response to user's device capabilities and delivery context characteristics. We have also explained how to take advantage of synergies among context acquisition, content negotiation, device capability expression, delivery context characterization, content adaptation and application layer protocol optimization, and demonstrated how to build a collaborative system to improve performance, self-adaptability, ubiquity and the QoE of web applications in links with restrictions.

Acknowledgments

This work was supported in part by the i2CAT Foundation, Vodafone Research and Development.ES, and the Spanish government through the Ministry of Education and Culture and FEDER project TIC2006-04504.

References

[1] L. Suryanarayana and J. Hjelm, "Profiles for the Situated Web.", ACM 1-58113-449-5/02/0005. WWW 2002, May 7-11, 2002, Honolulu, Hawaii, USA.

[2] R. Gimson, R. Lewis and S. Sathish, "Delivery Context Overview for Device Independence.", W3C Working Group Note, 20 March 2006

[3] C. Canali, M. Colajanni and R. Lancellotti, "Distribution of adaptation services for ubiquitous Web access driven by user profiles.", Proceedings of the 11th IEEE Symposium on Computers and Communications (ISCC'06)

[4] T. Lemlouma and N. Layada, "Universal Profiling for Content Negotiation and Adaptation in Heterogeneous Environments." W3C Workshop on Delivery Context, W3C INRIA Sophia-Antipolis,France, 4-5 March 2002.

[5] User Agent Profile 2.0. Wireless Application Protocol WAP-248-UAPROF-20011020-a

[6] C. Kiss. "Composite Capability/Preference Profiles (CC/PP): Structure and Vocabularies 2.0." W3C Working Draft 30 April 2007. http://www.w3.org/TR/CCPP-struct-vocab2/

[7] Jordi Casademont, Ferran Perdrix, Martin Einhoff, Josep Paradells, Georg Dummer and Anne Boyer, "ELIN: a framework to deliver media content in an efficient way based in MPEG standards." IEEE International Conference on Web Services (ICWS 2005)

[8] OMA Mobile Location Protocol (MLP) Candidate Version 3.1, Open Mobile Alliance, revised 16 March 2004

[9] Luca Passani. "Wireless Universal Resource File (WURFL)." http://wurfl.sourceforge.net/

[10] R. Lewis, R. Merrick and M. Froumentin, Content Selection for Device Independence (DISelect) 1.0, W3C Working Draft 10 October 2006

[11] R. Gimson, L. Suryanarayana and M.Lauff, Core Presentation Characteristics: Requirements and Use Cases, W3C Working Draft 10 May 2003

[12] K. Waters, R. Hosn, D. Ragget, S. Sathish and M. Womer, "Delivery Context: Client Interfaces (DCCI) 1.0: Accessing Static and Dynamic Delivery Context Properties.", W3C Working Draft 04 July 2007

[13] C. Gomez , M. Catalan, D. Viamonte, J. Paradells and A. Calveras "Web browsing optimization over 2.5G and 3G: end-to-end mechanisms vs. usage of performance enhancing proxies." Wireless Communications and Mobile Computing. 12/9/2006

[14] T. Lemlouma. Architecture de Ngociation et d'Adaptation de Services Multimdia dans des Environnements Htrognes, PhD thesis, 9/5/2004. http://wam.inrialpes.fr/publications/2004/TheseLemlouma.pdf

[15] R. Moskowitz, P. Nikander and T. Henderson. "Host Identity Protocol." Internet draft. http://www.ietf.org/internet-drafts/draft-ietf-hip-base-10.txt

[16] J. Border, M. Kojo, J. Griner, G. Montenegro and Z. Shelby. "Performance enhancing proxies intended to mitigate link-related degradations." RFC 3135, June 2001.

[17] B. C Housel, G. Samaras and D. B. Lindquist. "WebExpress: a client/intercept based system for optimizing Web browsing in a wireless environment.", Mob. Netw. Appl., 3 (1999), 419-431.

[18] M. Catalan, C. Gomez, P. Plans, J. Paradells, A. Calveras, J. Rubio and D. Almodvar. "Extending Wireless Mesh Networks over UMTS: A proxy-based approach." Qshine'06, August 7-9, 2006, University of Waterloo, Ontario, Canada.

[19] R. Chakravorty, A. Clark and I. Pratt. "Optimizing web delivery over-wireless links: design, implementation and experiences." IEEE. Journal on Selected Areas in Communications 2005; 23(2): 402.

[20] J. Mogul, F. Douglis, A. Feldmann, Y. Goland, A. van Hoff and Marimba D. Hellerstein. "Delta encoding in HTTP." RFC 3229

[21] R. Prasad, C. Dovrolis, M. Murray and K. Claffy, "Bandwidth estimation: metrics, measurement techniques, and tools.", IEEE Network Magazine, Nov/Dec 2003

[22] J. Rabin, J. M. Cantera, R. Hanrahan and I. Marn "Device Description Repository API 1b." W3C Editors' Draft 31 August 2007 http://www.w3.org/2005/MWI/DDWG/drafts/api/latest

[23] N. Jacobs and J. Raj "JSR 188: CC/PP Processing." http://jcp.org/en/jsr/detail?id=188

[24] HP Labs. "DELI: A Delivery Context Library For CC/PP and UAProf." http://delicon.sourceforge.net/

[25] W3C Device Description Working Group. "Delivery Context Ontology." http://www.w3.org/2005/MWI/DDWG/wiki/Ontology

[26] K. Smith. "Device Independent Authoring Language (DIAL)." W3C Working Draft 27 July 2007. http://www.w3.org/TR/dial/

[27] D. Sanders. "Device Description Repository Requirements 1.0." W3C Working Draft 10 April 2006. http://www.w3.org/TR/DDR-requirements/

[28] J. Paradells, J.L. Ferrer, M. Catalan, W. Torres, M. Catalan-Cid, X. Sanchez-Loro, V. Beltran, E. Garcia, C. Gomez, P. Plans, et al. "Design of a UMTS/GPRS Assisted Mesh Network (UAMN)." 7th Wireless World Research Forum (WWRF) Meeting. Heidelberg, Germany. November 2006

[29] Privoxy. http://www.privoxy.org/

[30] Platform for Privacy Preferences (P3P) Project implementations. W3C. http://www.w3.org/P3P/implementations.html

[31] R Gimson. "Device Independence Principles." W3C Working Group Note 01 September 2003. http://www.w3.org/TR/di-princ/

[32] J.A. Larson, T.V. Raman and D. Raggett. "W3C Multimodal Interaction Framework". W3C NOTE 06 May 2003. http://www.w3.org/TR/mmi-framework/

[33] W3C The Ubiquitous Web Domain. http://www.w3.org/UbiWeb/

[34] D.L. McGuinness and F.van Harmelen. "OWL Web Ontology Language Overview." W3C Recommendation 10 February 2004. http://www.w3.org/TR/owl-features/

[35] F. Manola and E. Miller. "RDF Primer." W3C Recommendation 10 February 2004. HTTP://www.w3.org/TR/rdf-primer/

[36] P. Archer. "POWDER: Use Cases and Requirements." W3C Working Group Note 31 October 2007. http://www.w3.org/TR/powder-use-cases/

[37] J. Barnett, M. Bodell, D. Raggett and A. Wahbe. "Multimodal Architecture and Interfaces." W3C Working Draft 11 December 2006. http://www.w3.org/TR/mmi-arch/

[38] Y. Chee, M. Froumentin and S.M. Watt. "Ink Markup Language (InkML)." W3C Working Draft 23 October 2006. http://www.w3.org/TR/InkML/

[39] E. Nkeze. "Device Description Landscape 1.0." W3C Working Group Note 31 October 2007. http://www.w3.org/TR/dd-landscape/

[40] R. Hanrahan. "Device Description Ecosystem 1.0." W3C Working Group Note 31 October 2007. http://www.w3.org/TR/dd-ecosystem/

[41] A. Trasatti, J.Rabin Device and R. Hanrahan. "Description Repository Core Vocabulary 1f." W3C Working Group Note 11 February 2008. http://www.w3.org/2005/MWI/DDWG/Drafts/corevocabulary/latest

[42] W3C. Web Accessibility Initiative (WAI). http://www.w3.org/WAI/

[43] 3GPP TS 29.240. "3GPP Generic User Profile (GUP); Stage 3; Network." http://www.3gpp.org/ftp/specs/archive/29%5Fseries/29.240/

[44] FIPA Device Ontology Specification. 5/2002. http://www.fipa.org/specs/fipa00091/XC00091C.pdf

[45] J. Elson and A. Cerpa. "Internet Content Adaptation Protocol (ICAP)." RFC 3507.

[46] A. Rousskov and M. Stecher. "HTTP Adaptation with Open Pluggable Edge Services (OPES)." RFC 4236.

[47] K. Holtman and A. Mutz. "Transparent Content Negotiation in HTTP." RFC 2295.

[48] C. Canali, V. Cardellini, M. Colajanni and R. Lancellotti. "Performance comparison of distributed architectures for content adaptation and delivery of Web resources." Proceedings of the 25th IEEE International Conference on Distributed Computing Systems Workshops (ICDCSW'05).

[49] M. Butler, F. Giannetti, R. Gimson and T. Wiley. "Device independence and the Web." IEEE Internet Computing -September/October 2002 - pp 81-86

[50] Mohomed, I. Chin, A. Cai and J.C. de Lara. "Community-driven adaptation: automatic content adaptation in pervasive environments." Mobile Computing Systems and Applications, 2004. WMCSA 2004. Sixth IEEE Workshop on Volume , Issue , 2-3 Dec. 2004 Page(s): 124 - 133. 2004

[51] Wai Yip Lum y Francis C.M. Lau. "A Context-Aware Decision Engine for Content Adaptation." Pervasive Computing.

[52] J. Gettys, J. Mogul, H. Frystyk, L. Masinter, P. Leach and T. Berners-Lee. "Hypertext Transfer Protocol - HTTP/1.1." RFC 2616.

Chapter 6

The Road towards Self-Management in Communication Networks

Ralf Wolter

Cisco Systems, Duesseldorf, Germany

Bruno Klauser

Cisco Europe, Glattzentrum, Switzerland

6.1 Introduction

When reading the term "self-management", which picture do you associate with it? Think about it for a minute - before reading on! What did you see? Was it an illustration of the human body with the autonomous nervous system? Did you envision your heart and the fact that you only indirectly control it? Alternatively, did you imagine modern technology, for example an airplane with the "fly by wire" concept? Did you think of something else?

You might think these two examples have very little in common; however, they share exactly the same concept: trust and delegation are given to autonomous subsystems instead of micromanagement exercised by a single central authority. You trust that your autonomous nervous systems "knows" what is right for your body and in the same way, a pilot needs to trust that the commands he enters into the flight computer are translated into the correct actions and are properly and timely executed. Several proposals for network self-management are inspired by autonomous concepts in the biological area [1].

An example from our daily life comes from the car industry. When driving the car, we usually do not even realize the level of complexity that is nicely hidden from the driver and operated in the background. I only realized it when purchasing an old tractor, which was built more than 50 years ago. The simple procedure of starting the engine of this old tractor consists of six preparation steps, which have to be manually performed by the driver. Diagram 6.1 illustrates the flow.

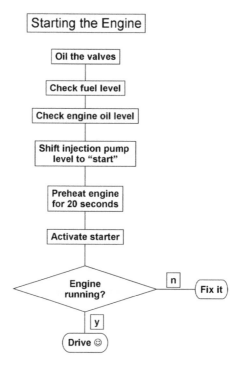

FIGURE 6.1: Flow of starting a diesel engine 50 years ago

How do you start the engine of your car in the 21st century? Your car probably has just one button "Start Engine", you press it, and the engine starts "by magic". All of the steps illustrated in the flow chart above (and many more) are still executed today; however, the user (i.e. the driver) is not aware of them and, better, does not have to be aware of them. This process is followed except in case of a problem that requires human intervention, in which case the responsible subsystem will bring this issue to the attention of the driver. Simple problems could be solved by the driver (for example, no fuel), while more complex technical issues might be beyond the competence or privileges of the driver, and will be flagged with a recommendation for maintenance as soon as possible. The technician in the garage however is quite familiar with these details and has access to detailed error reports and system diagnostics to help resolve the issue.

Note that the automotive illustration applies both abstraction and autonomy. An abstraction example is the "oil level indicator light" that keeps the driver from regular checking the engine's oil level. An autonomy example are preheating the engine, where the car senses the outside and engine temperature, automatically preheats if necessary until the right temperature is reached and only then applies the starter. Other examples are fully controlled air-conditioning systems where the user only selects the temperature; light sensors that provide input to automatically turn on the headlights in a tunnel and turn it off afterwards; or rain sensors that control the frequency of the windscreen wiper. The difference in these examples is that the car cannot fill in oil if the level reaches minimum (and therefore requires the driver's interaction) while it can control temperature and light.

A formal distinction could be defined as follows:

- Abstraction simplifies the user's interaction with a system by hiding complexity. It does not imply decision making by the system, other than notifying the user of exceptions or abnormal situations. Abstraction uses sensing only to identify a deviation from normal and does apply actions to return to the default state.

- Autonomy or self-management implies delegation for the system to make decisions on behalf of the user. In this case, sensing is applied as input to a state-machine, which takes concrete actions to return the system towards the default state.

6.2   Self-Management in Networks

If we translate the above automotive analogy to communication networks, we would like to apply the same principle: enable the network operator to make the right decisions while hiding the complexity, automatically perform the required actions in the background, and notify the operator and potentially the end-user only if something does not work properly. This has been implemented to a certain degree for PCs; however, there are huge opportunities for automation related to Information and Communications Technology (ICT) networks.

The promise of converged networks (i.e. when data, voice, and video are transported over a single infrastructure) was to simplify operations and reduce costs. Cost reduction has been achieved by converging to a common infrastructure instead of multiple parallel networks. At the same time, the level of complexity has grown exponentially: designing traffic and service classes, implementing redundancy and high-availability concepts, including fast-reroute functionality, securing the infrastructure against viruses and attacks, and trouble shooting and identifying the root cause of a problem are just some of the tasks network designers and operators encounter on a daily base. The configuration of network elements is done manually, by scripts, or by provisioning applications. Independently of the configuration method, network components as well as the Internet itself were initially designed to be "dumb" and therefore need to be told systematically what to do and how to do it.

Here is an access-control list example that only allows the primary NMS server (IP address 10.1.1.1) and secondary NMS server (IP address 10.2.2.2) to send SNMP requests to a router.

- Router# configure terminal

- Router(config)# access-list 10 permit 10.1.1.1

- Router(config)# access-list 10 permit 10.2.2.2

- Router# interface ethernet 1

- Router(config-interface)# ip access-group 10 in

If other protocols should be used for the communication between the NMS server and the router, such as NETCONF over BEEP, additional access lists are required. To avoid uploading of the routing table by the NMS application, some operators restrict SNMP read access, for example:

- snmp-server view noRouteTable internet included

- snmp-server view noRouteTable ip.21 excluded

- snmp-server view noRouteTable ip.22 excluded

These simple examples illustrate today's approach: the operator has full control at the micromanagement level and needs to configure all details; however, in general, there is either no link to a higher level or that link lies in the skills of a particular operator. For example, missing entries in an access-control list can easily result in an open back door that hackers might use to gain access to sensitive information.

What would be a better approach? Imagine devices could understand their "purpose in life", which means becoming context-aware? What about a server in a data center that "knows" the business priorities and automatically assigns resources according to it? Can you imagine a router that "understands" traffic patterns and blocks a denial-of-service attack immediately?

Related to the access control list described above, what if all devices have the knowledge that only one or two authorized NMS systems are allowed to send NMS requests, and these not only consist of SNMP requests, but also contain configuration requests as well as being the receivers for notifications. Depending on the location of the network element, it could apply traffic encryption for the communication with the NMS system etc. The idea of self-management is an area under discussion within academia and the industry for quite a while, [2] and [3] are examples.

At this point, the concept of a "system" needs to be introduced.

"A system is a set of entities, real or abstract, comprising a whole where each component interacts with or is related to at least one other component and they all serve a common objective." [1]

In the definition above, a system could be a single network element, such as a router, switch, intrusion detection system etc., or a complex structure of multiple network elements together with a central network management application. In this paper, we use the term "Self-Managing Network System" to describe the combination of one or multiple network elements in conjunction with a centralized or decentralized management application. In other words, multiple different components cooperate to achieve a common goal. A typical approach is to define the overall goal for the system, then derive sub-goals and link a set of entities to the sub-system. Assuming all sub-goals are completed, the overall goal is achieved as well. Unfortunately, the entities and sub-systems know nothing about the overall objective and therefore only perform or fail a defined task. They cannot adapt in case of undefined situations.

By adding the self-management dimension to the network systems approach, the sub-systems get a certain level of autonomy and awareness of the overall objective, and perform various actions to achieve the goal. A self-managing network system has the ability to monitor and adjust its behavior to achieve a defined business goal.

There are a number of requirements for self-managing network systems, starting with business objectives that are relevant for the full system, not just some subcomponents. Next, a translation mechanism should derive technical goals, policies, and rules from the overall business objectives. This should take place within in a trust model, where network management applications share information instead of acquiring them in isolation. A reality check from today's designs would be to count how many NMS applications repeatedly query the network elements instead of using a shared repository. It includes discovery of new devices as well as status changes, failures, and performance data. Implying a trust model between the applications and the network elements is the corner stone for development of self-management concepts.

Self-managing concepts would need to be applied at all levels of the network system, such as application servers, policy servers, management applications, routers, switches, blades, and so on. Another relevant aspect is sensing, where system components have the ability to sense their environment [4]. An event-condition-action concept could provide guidance towards the overall system as well as individual components how to act in certain situations, including a list of potential actions to solve unknown conditions. De Campos et al. describe a model for autonomic components guided by condition-action policies [5]. In addition to sensing the environment, at least systems, but ideally individual

---

[1] Wikipedia, http://en.wikipedia.org/wiki/System.

components as well, need to "understand" situations and be able to "learn" from them; this implies a dynamic concept for all methods applied.

The following statements are derived from IBM's Autonomic Computing manifesto [2] and have been adopted for networks. Instead of using IBM's term "Autonomic Computing System" we use the term "Self-Managed Network System", this could indicate an entire network, sub-parts of a larger network, or just individual network elements. IBM has published a large number of documents in the self* area, while the manifesto can be considered the vision for autonomic computing, [6] provides a practical toolset. Dobson et al. provides a status quo of autonomic communications in 2006 [7].

1. *Self-managed network systems needs to sense it's status and understand status changes, which relates to the overall goal it is expected to achieve as well as a detailed knowledge of its components, including resource consumption and capacity usage. This includes monitoring of the environment, including local and remote networks, through discovery, routing protocols, notifications from other elements or systems, and others.*

2. *Self-managed network systems need to perform some, or possibly all, of its initial configuration without human intervention (zero-touch deployment) and must be able to dynamically reconfigure under changing conditions. This should also include unknown and abnormal situations.*

3. *Self-managed network systems should run self-optimization functions regularly in order to optimize local functionality as well as overall network performance.*

4. *Self-managed network systems require self-healing functions, which includes pro-active diagnosis to predict failures early and activating redundant components, if applicable. If the system reaches the conclusion that an outage will happen soon that cannot be solved locally, it notifies neighbor systems for support.*

5. *Self-managed network systems require self-protection mechanisms. It must detect abnormal situations, identify the impacted functions, and ultimately protect itself against attacks to maintain operations as well as system security and integrity. It should also keep relevant traffic information for various actions, such as postmortem analysis, tracking the attacker's location, and setting up honey pots.*

6. *Self-managed network systems must leverage standards where applicable and propose new and innovative developments for standardization.*

7. *Self-managed network systems implement state-of-the-art technology to optimize resources and apply abstraction to simplify interactions and*

---

[2]http://www.research.ibm.com/autonomic/overview/elements.html

keeping complexity hidden. This could be achieved by service decomposition at all levels.

Moving from micromanagement in the old telephony and IBM SNA world or almost no control with the Internet's best effort approach towards self-monitoring and self-management combined with central or rather distributed policy rules and SLA monitoring is certainly a paradigm shift. What we gain from the cooperation of global policies, central servers, and network elements are "Self-Managing Systems".

Before developing the network self-management concepts any further, let us step back for a moment and find out if we really want communications networks to be self-managed, and if yes, to what degree? Remember, the initial idea of computing - and communications networks being no exception - was to help human beings to run their business better than before. Unvarying operations were no longer handled by humans but instead were processed fast and accurately by mainframes, which was a big achievement. With the emergence of pervasive computing, however, we make it our business to manage a plethora of computing entities. Instead of asking why the current solutions are not sufficient, the real question would be "Are you willing to give up micromanagement and instead trust the system to perform some operations autonomously?"

The next relevant question is: should network self-management be structured top-down, including modeling of the company's business objectives and processes, or is it better modeled in a bottom-up fashion?

Linked to the examples mentioned above, ask yourself if cars operate better if the driver still has to oil the valves manually before starting the engine or if the safety of a plane increases if the pilot is fully in control and potentially performs rollover maneuvers, just because the plane is capable of doing so. For the automotive and airline industry, we have a clear answer - why do we struggle to implement the same concept at communications networks? It looks like we still operate the networks with the aspect of "illusion of complete control" where many knobs and instruments pretend being in control of everything, while in fact we miss the big picture and highly relevant details due to information overflow. Have you ever noticed the frustration of an operator who gets flooded with thousands of events and as a result, deletes them all? Again, are you willing to give up control at the micromanagement level? How about giving up some level of control?

To ease answering the questions just raised, we propose a vision for self-managed network systems "Manage the Communication Infrastructure Intuitively". At the highest level, three major steps can help achieving the vision and all three build on top of each other. Starting with the technology and connectivity level, we should leverage the Internet Protocol (IP) ubiquitously. Over the last decade we have seen a large number of transport protocols (IBM SNA, DECnet, AppleTalk, IPX/SPX, ATM, and many others) and IP has succeeded them all. Second, we should design new solutions on the assumption of

"Internet connectivity everywhere"). Even though today there are still blind spots of coverage, they are literally reduced on a daily base. The third step addresses the operations level, which directly relates to self-management; it proposes using and managing the communication infrastructure intuitively.

A strong supporting argument for the three points above is a statement from Alfred North Whitehead, an English mathematician and philosopher (1861 - 1947):

"Civilization advances by extending the number of important operations which we can perform without thinking about them."

6.3   Defining Concrete Steps towards the Vision

The remainder of this article focuses on the subject of "How to use and manage the communication infrastructure intuitively". Five steps are proposed to achieve this goal:

1. Definition of business objectives in a business language

2. Translation of the business objectives into technical terms

3. Deriving rules and policies for network systems

4. Automatically breakdown of goals into specific objectives per sub-system and entities

5. Enablement of network elements to interpret, deploy, and comply with these goals

For each of these steps, it is strongly suggested to use standard protocols, languages, and models or develop them if no standard exists.

6.3.1   Define Business Objectives in a Business Language

Let us take a closer look at each of these steps. The first step means that an application developer or a network operator defines high-level objectives, such as "online purchases at my company's website should result in a positive experience for the customer". While this sounds evident, the challenge is to define goals unambiguously, i.e. they cannot be interpreted differently by multiple people. One method to achieve this is Business Process Management (BPM) tool that define, measure, and optimize processes in an organization. It is important to state that optimization can target either cost reductions or increased flexibility, or a combination of both. Starting at the top, the business objectives need to be described in sufficient details, which is usually done through abstraction and decomposition. This ultimately leads to process models and ultimately Enterprise Architectures. An Enterprise Architecture describes the structure of an organization in all aspects, including the organizations core objectives and strategic goals, business processes, information and communication systems, as well as people. Enterprise Architecture frameworks illustrate and document the architecture in a standardized form with graphical representation. The goal of the models is to get insights of the processes in order to validate, optimize, or redesign them. Commonly used frameworks are the Zachman Framework for Enterprise Architectures, the Open Group Architecture Framework (TOGAF), the ARIS Framework, and others.

In the Service Provider world, the enhanced Telecom Operations Map (eTOM), defined by the TeleManagement Forum[3] is commonly used. The eTOM model serves as a reference framework for categorizing all business activities of a service provider and analyses them into different levels of detail according to their significance and priority for the business. The eTOM structure establishes the business language and foundation for the development and integration of Business and Operations Support Systems (BSS and OSS). The Autonomic Communications Forum (ACF)[4] is organization with a tight focus on defining autonomic frameworks and standards, see [8].

When discussing architectures, it is important to distinguish between the Business Architecture (describing the business processes), the Information Architecture (how and where information are collected, processed, stored), the Application Architecture (describing the IT applications and their relationships), and the Infrastructure Architecture (describing the information and communications technologies (ICT) components).

When applying business objectives to an organization, the concept of services becomes relevant. A service is a function or an operation that creates a defined outcome. A nucleus service is self-contained and independent of the status or context of other services, whereas a composed service consists of multiple nucleus services.

6.3.2   Translate the Business Objectives into Technical Terms

At this step, the business objectives are translated into technical terms that can be performed and measured by IT applications. Services can be considered as the interface between business objectives and technology implementations. Business people define services in abstract terms, and the models and tools should help to translate these into measurable operations and results related to technology domains. Using well-defined business objectives is essential to the successful implementation of IT solutions, as it they enable validation and measurement how IT solutions solve business problems.

---

[3]TMF, http://www.tmforum.com, document GB921: The Business Process Framework
[4]http://www.autonomic-communication-forum.org

A simple example: the business objective "deliver outstanding customer maintenance service" could be translated into multiple technical objectives, such as "spare parts are shipped within 24 hours", "call center waiting time is below 15 seconds in 95% of all customer calls", and "customer inquiry emails are answered within 4 hours during business hours". Note that while each of these technical objectives are delivered by a function or service and can be measured and evaluated, they are derived from the overall objective and only provide indirect metrics to the overall goal.

Services have a lifecycle, starting with the definition phase; followed by the design and modeling phase; test phase with simulation tools; execution phase; constant monitoring and validation phase; potential optimization phase; and ultimately the end-of-life phase.

A useful framework for the service life cycle is the Information Technology Infrastructure Library (ITIL)[5], which provides best practices related to IT Infrastructure and IT Services. It comprises a set of concepts and techniques for IT development, management, and operations.

While ITIL describes the Service Management Processes, a Service-Oriented Architecture (SOA) helps to translate the business objects into technical functions that can be executed and measured. The World Wide Web Consortium (W3C) defines SOA as "A set of components which can be invoked, and whose interface descriptions can be published and discovered" [6]. SOA models functions as distinct units, which can be combined and reused to build business applications. These distinct units have standardized interfaces, which hide the implementation complexity from application developers, allowing them to construct new applications by using and connecting a set of predefined services. The service composition concept was introduced with the Common Object Request Broker Architecture (CORBA), a middleware developed by the Object Management Group (OMG). In a service-oriented architecture, service requests are initiated by a service consumer, resulting in service responses from the service provider. Figure 6.2 illustrates a service consumer sending a service request get content) to a service provider. In this example, the service provider offers three types of content: music, video, and text. Each service is composed of sub-services.

Web services are an instantiation of a service-oriented architecture, where network services (also referred to as endpoints) can be described by using the Web Services Description Language (WSDL, defined by W3C) and using SOAP to interconnect them. A Service Registry is an (optional) central instance that offers information about available services, as published by the Service Provider, towards the Service Consumer. Note that the service registry is only theoretically optional, because it is a required component for

---

[5]ITIL is a Registered Trade Mark of the Office of Government Commerce in the United Kingdom and other countries.

[6]http://www.w3.org/TR/ws-gloss/#defs

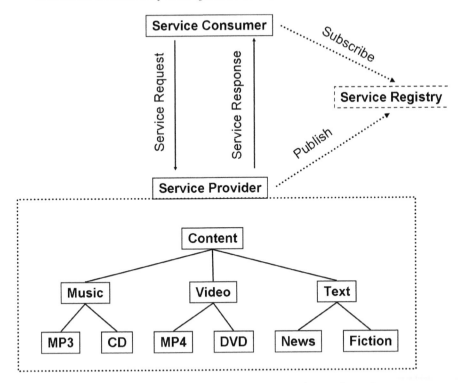

FIGURE 6.2: Relationship between service provider and service consumer

service consumers to locate service providers. UDDI (Universal Description, Discovery, and Integration) is a means to implement a Service registry. The Service registry uses techniques such as Publish/Subscribe to advertise services so that Service Providers can offer services and Service Consumer can use them. The Simple Object Access Protocol (SOAP) provides an envelope for exchanging the messages between the components; alternatives are HTTP, MIME, SMTP, FTP, JMS, etc.

The concepts of network service decomposition relates to the breakdown of objectives: as illustrated above, a service can be composed of one or multiple levels of sub-services. Theoretically, each service can be decomposed into its technical elements down to a level where each sub-service is a single unit. Even for simple network services, applying decomposition until the bottom of the hierarchical structure (i.e., each sub-service is a nucleus service) is reached, could result in rather huge hierarchies. However, in a business environment decomposition to the lowest level is might not always be necessary. Instead, decomposition into reusable and well specifiable service components is the norm. These service components should be available for implementing individual services or act as building blocks for composed services [9], [10], [11], [12].

In summary: we need to separate the technical objectives into a set of reusable service components.

Figure 6.3 applies a "reality check" to figure 6.2; it decomposes a Unified Communication service. This service offers either voice-only communication or alternatively voice and video. The transport offerings are a layer 2 / layer 3 pipe or VPN. Both the voice and video sub-services depend on the Media/Apps service; similarly, IP VPN requires the Transport and Access subservice. Note that the Transport, Access, and Media/Apps sub-service need to be modeled in more details.

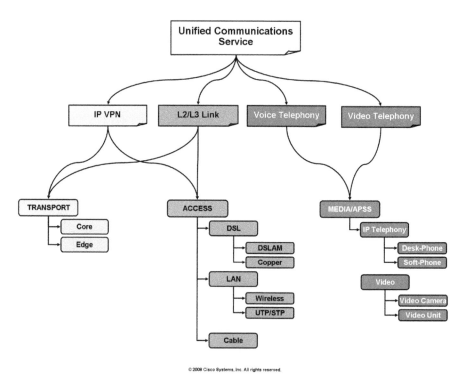

FIGURE 6.3: Service decomposition

6.3.3   Derive Rules and Policies for Systems

Now that we have translated the business objectives into technical objectives, and instantiated service elements and interactions between the elements, metrics can be defined for measuring the performance of service delivery.

While Step I (Define business objectives) offers a very generic definition of a service ("A service is a function or an operation that creates a defined

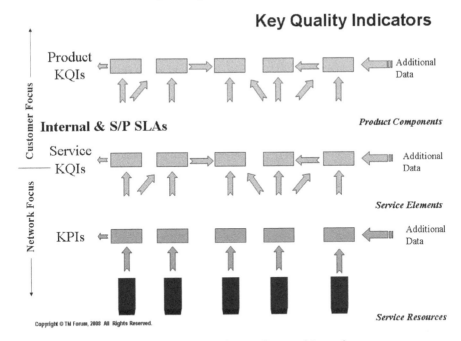

FIGURE 6.4: TMF key indicator hierarchy

outcome"), specific definitions are required for the technical specifications. Starting at the highest level, a policy is a deliberate plan of action to guide decisions and achieve rational outcome(s) [7].

Kephart and Walsh describe three types of policies related to autonomic computing systems; these are action, goal, and utility function policies [13]. A network service is a function providing network connectivity or network functionality, such as applications, databases, file systems, network services (NIS, DNS, DHCP, FTP, NTP, ICMP), and so on. Service level defines a certain degree of quality, related to specific metrics, in the network with the objective of making the network more predictable and reliable. Service level management is the continuously running cycle of measuring traffic metrics, comparing those metrics to stated goals (i.e., performance, response time, packet loss, etc.) and ensuring that the service level meets or exceeds the agreed-upon service levels. Service level agreement (SLA) is a contract between the provider of a (network) service and a customer of this service. It defines service deliverables, service quality, and remedy processes. Another way of expressing is "SLA is the formalization of the quality of the service in a contract between the customer and the service provider."

---

[7]Wikipedia, http://en.wikipedia.org/wiki/Policy

While network service, service level management, and service level agreements are generic terms, the TeleManagement Forum (TMF) has defined two specific service indicators. These are the Key Performance Indicators (KPI) and Key Quality Indicators (KQI), both are defined in the TMF GB923 document[8]

"KPI provide a measurement of a specific aspect of the performance of a service resource (network or non-network) or group of service resources of the same type. A KPI is restricted to a specific resource type. KQI provide a measurement of a specific aspect of the performance of the product, product components (services) or service elements and draw their data from a number of sources including the KPIs." A good example for the definition of service parameters such as KPI and KQI is the "Wireless Service Measurement Handbook GB923A" from the TeleManagement Forum (TMF). Figure 6.4 describes the relationships between the different KQIs and KPIs.

6.3.4 Automatically Breakdown Goals

After deriving rules and policies for systems, we have to breakdown goals into specific objectives for each sub-system and the related entities. A system is a set of functions, which cooperate to achieve a goal, such as delivering a service. Figure 6.5 illustrates from a high-level perspective how to manage the communication infrastructure intuitively.

Unfortunately, the reality is more complex than described by figure 6.5; therefore, figure 6.6 describes the same concept in more details. This leads to multiple questions, such as the following. How can business objectives automatically be translated into policies and processes? How can sub-systems derive sub-goals from the overall goals without human intervention? How can SLAs be defined and verified across multiple transport domains, most like across multiple ISPs? How can we avoid that operators bypass the NMS systems and access network elements directly? Can we introduce rules and policies how operators should manage the network or do the management applications have to adjust to the operator's behavior?

Note that the feedback loop is quite important, because it is the final part that turns a one-way operation into a managed system. The feedback mechanism can be implemented as a response to a request from the upper layer (pull-model) or can be initiated by the sub-system (push-model) without a request. The latter is also described as "Event Driven Architecture" where a state-change at a layer causes a local action or sending a notification to its neighbors or both. The notification initiates actions at the upper layer, such as trigger the invocation of processes or services, potentially resulting in changes to the overall system. The message exchange between components

---

[8]http://www.tmforum.org/page28825.aspx

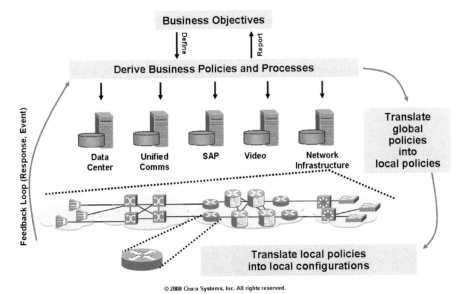

FIGURE 6.5: Manage the communication infrastructure intuitively

could be implemented as point-to-point communication or a publish-subscribe bus.

A relevant question in this context is to what levels of detail autonomous network (sub) systems need to share information with their neighbors, because we want to avoid data overflow as well as a lack of information.

Here is an example from grid computing: let us assume that a specific application simulates a car design, for example the aerodynamics of a spoiler. The spoiler designer (called Designer-S) knows that the simulation application can run on multiple servers at several high-performance data centers; however, he does not know any details how to get access to a specific server because the simulation application does this for him. However, the application needs to request access to these resources, maybe via an agent, negotiator, or resource broker. By sending a request "I need access to servers N1, N2, N3 during non-business hours next Monday to Wednesday" to the resource broker, the application has identified a need and has to wait for the response. The resource broker splits the request into sub-systems that deliver the required resources and define sub-system objectives:

- Objective A: Network transport has to be prioritized

- Objective B: Servers (N1, N2, N3) need to be available and have processing resources available

- Objective C: Monitoring of service level agreements (SLA) need to be

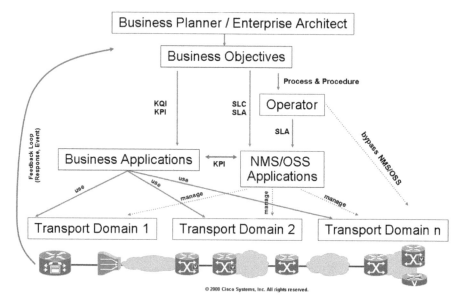

FIGURE 6.6: Top-down approach for business objectives

active for proving service quality and potential trouble shooting procedures have to be prepared.

In today's world, these three subsystems (network transport, data center, SLA monitoring) would be configured with detailed instructions how to achieve the goals:

Objective A": identify user application either by deep packet inspection of application specific criteria or access control list and set QoS DiffServ value = AF11 (low drop precedence). These configuration parameters need to be activated at the start of the simulation and removed at the end.

Objective B": check maintenance window to see if the servers are available during the interval, specify maximum memory and maximum CPU resources, and assign it to the particular user or application.

Objective C": verify that passive monitoring is enabled between the source and destination points in the network and enable active monitoring for the network path quality and server processes. Deactivate these at the end of the request.

By applying network self-management capabilities, we still need the subsystem objectives A, B, C and the detailed objectives A", B", C". However, by defining application requirements and application policies, which are then passed to the sub-systems and components, the detailed configuration can be derived at the local element level. Related to the spoiler design example above, the application requirements can be specified first, e.g. 50 Mbit/s throughput, 100 ms round-trip delay, packet loss $\leq 0.5\%$, server memory at least 10 GB, server uptime = 99.999%. All network components should have access to these requirement definitions. An additional application-specific objective (A') is introduced in-between the objectives (A, A"), in this case:

A': During interval T, application of type **design simulation** from host X needs to be transported according to the network specific parameters of the application-specific objective.

B': During interval T, application of type **design simulation** from user Designer-S needs to be transported according to the server specific parameters of the application-specific objective.

C': During interval T, SLA monitoring between Designer-S and the target servers needs to be enabled, according to the overall SLA parameters defined by the application-specific objective.

The question how to automate the translation from overall system goals into sub-system objectives still remains open.

6.3.5   Enable Network Elements to Interpret, Deploy, and Comply with These Goals

Now we have reached the point where the derived objectives are detailed enough for a component to be interpreted and implemented. This applies the various flavors of network self-management that have been described before, such as self-configuration, self-optimizing, self-diagnosis, self-healing [14], and self-protection. Another relevant area is self-knowledge [15], even though not used consistent in literature, it covers the aspect of a component to "understand" the impact of the goals and global policies on themselves.

So far, we have described a structured approach for building network self-management systems from a top-down perspective: identify business goals $\longrightarrow$ derive technical objectives $\longrightarrow$ break into sub-systems goals $\longrightarrow$ execute autonomously at the component level. Fortunately, a number of building blocks for network self-management already exist; an alternative to the top-down methodology is an ad-hoc approach for a bottom-up design. The next paragraph describes some existing solutions, classified by virtualization first and device manageability instrumentation second.

A hot trend these days is the virtualization concept. Virtualization provides an abstraction layer between applications or users of a service and the physical computing and networking resources they interact with [16]. What started in the network with virtual private networks (VPN) and virtual local area networks (VLAN), where multiple subscribers share the same transport infrastructure without having access to each other's data, has moved to data centers where customers share computing resources. Virtual machines make it possible to shift applications between physical servers with a seamless switchover. Virtualization of storage devices offers large storage capacity, independently of the physical location. Network Management could benefit from virtualization as well; figure 6.7 illustrates the connection between NMS/OSS applications and the network infrastructure via an abstraction layer. The virtual network layer abstracts the implementation details, such as different operating systems and command syntax from the applications. A modified command syntax at a network element does not imply modifications at the application layer any longer, as the virtual network layer translates (standardized) commands into vendor specific configuration commands. Adam and Stadler propose a service middleware for self-management [17].

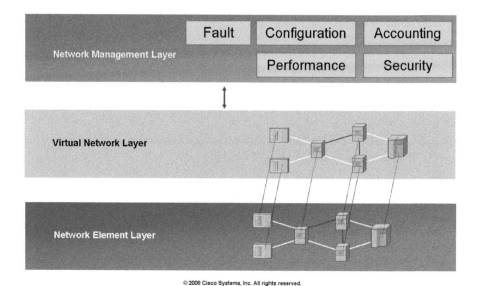

FIGURE 6.7: Virtualization layer for network management

Device Manageability Instrumentation is the second key building block for self-managing systems. Network elements such as routers, switches, security appliances, load balancers, data collection engines, etc. have significantly improved performance capabilities. Routers are capable to do so much more than just building routing tables and forwarding packets. Intelligent device instrumentation, such as network management agents that monitor traffic statistics, application types, CPU and memory usage by processes, etc. is a key component for network self-management. The following table offers an overview of functions that are already implemented at the network element level:

Table 6.1 summarizes the examples and applied technologies and features.

Some of these features should be explained in more details, such as the Cisco Embedded Event Manager (EEM). EEM implements a set of network management concepts directly at the router, as illustrated by figure 6.8. Combining IPFIX (RFC 5101) traffic collection with packet sampling (IETF PSAMP: Sampling and Filtering Techniques for IP Packet Selection) is an example for

Service Objective	Technical Example
Self-Monitoring	RMON-MIB (RFC 1757)
Self-Healing	EVENT-MIB (RFC 2981)
	EXPRESSION-MIB (RFC 2982)
	Cisco Embedded Event Manager (EEM)
	TCL scripts
	IETF IPFIX (RFC 5101) and PSAMP approaches
	Fast Routing Algorithms
Self-Optimizing	Fast Convergence
	High-Availability concepts
	Link, module, device failover
	Virtual Router Redundancy Protocol (VRRP, RFC 2338)
	Cisco Performance Routing (PfR) aka Optimized Edge Routing
Self-Configuration	Cisco IP SLA Enhanced Object Tracking
	Zero-Touch deployment concepts
	Policy based configuration models
Self-Protecting	Intrusion Detection Systems (IDS)
	Network Access Control (NAC)

Table 6.1: Summary of existing self-management solutions building blocks

self-monitoring combined with self-optimization. Monitoring and collection traffic flows with IPFIX has a CPU impact at the network element, which could become a bottleneck during heavy utilization. By defining CPU utilization thresholds, EEM can trigger a change from a full collection to packet sampling to reduce CPU impact. If CPU utilization returns to normal conditions, sampling could be disabled to allow full collection of IPFIX records.

Other standardized embedded manageability features are the Event-MIB (RFC 2981) and Expression-MIB (RFC 2982). The Event-MIB monitors MIB variables constantly and sends a notification if a certain variable exceeds a threshold. The Expression-MIB allows defining new MIB objects. The following example defines a simple "capacity planning" feature at the network element level, sending a trap if the link utilization is above 50% during one hour.

- Step 1: create a new expression with Expression-MIB:
  my-utilization = (ifInOctets + ifOutOctets) * 800 / hour / ifSpeed

- Step 2: define the action with the Event-MIB:
  if my-utilization of interface $Ethernet_1 is$ above 50% of the bandwidth during one hour, generate an event "link abc needs more bandwidth"

Even though this is a simple example, it nicely illustrates the concept of self-monitoring, as the alternative would be the NMS system polling devices

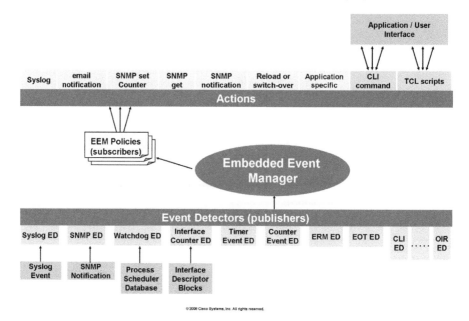

FIGURE 6.8: Cisco embedded event manager

frequently, monitoring the results, and generate an alert if bandwidth is above threshold. Transferring these tasks to the network element reduces SNMP traffic in the network and frees up resources at the NMS system for more relevant tasks.

Service-level-contracts (SLC) typically include thresholds for latency, however just monitoring the link quality is not sufficient if a provider depends on underlying services from other providers. For example, a Content Service Provider guarantees premium access to content within less than 300 msec latency. This threshold is based on the assumption that link latency between the consumer and the content service provider, offered by an Internet service provider, never exceeds 200 msec. Figure 6.9 demonstrates this scenario. Cisco IP SLA Embedded Object Tracking (EOT) can monitor the links latency of three different ISP and modify the routing table to shift traffic to a different ISP if the 200 msec threshold has been breached more than once during a 5-minute interval.

A building block for self-protecting networks is enforcing user authentication or verifying compliance policies, such as installed and updated virus scanner software, application versions, etc. at the clients. Figure 6.10 shows a scenario where a user wants access to a certain web server and the network elements automatically reroutes the request to an authentication server (typically AAA server). Only successful authentication and authorization requests are allowed access to (all) network resources, whereas non-compliant users or

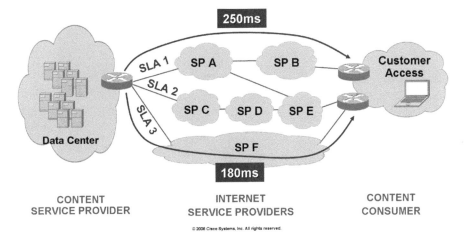

FIGURE 6.9: Self-optimizing with Cisco IP SLA

clients are moved to the quarantine section with limited network access.

6.4   Research Outlook

As mentioned in the abstract, this article wants to trigger questions and challenge the reader to develop innovative answers. The implementation examples described earlier are strongly focused on the network elements, and even though enabling self-management at the network element level is an important building block, it is not sufficient. An orchestrated approach is required at all levels to achieve the vision to "Manage the Communication Infrastructure Intuitively".

The following set of open questions is related to the larger discipline of network self-management. Algorithms are required to help devices to understand the "purpose in life", i.e. the context of their environment. How do we enable this intelligence at the device level?

All network elements need capabilities to sense their environment, including neighbors and their status. Can sensors networks help to build integrated network management functionality at the element level? How shall network self-management deal with out-of-policy behavior (worst-case scenario, abnormal or undefined situations)? How could unknown situations be modeled in advance? A key aspect is to make it possible for a device to detect "deviation from normal" conditions. How can this be applied without keeping large trend record files at the network elements?

Network elements need to identify traffic types to apply self-protection

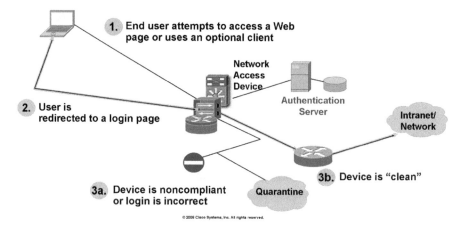

FIGURE 6.10: Self-protection through user authentication

mechanisms. How could they distinguish between "good" user traffic and "bad" hacker traffic? Virtualization requires the development of multi-purpose devices, where today there are multiple special-purpose devices in place. For example, load-balancers are usually not designed for intrusion detection and deep packet inspection. At the same time, multi-purpose devices are expected to offer the same performance levels as highly customized appliances. How can this contradiction be solved?

Develop concepts and models to link self-managed network elements to self-managing network management applications. Which features are best deployed at the network element level and which should reside in an NMS server? Most existing work concentrates on individual self-managing systems, however if we would also consider a network element as an autonomic system it results in an entire network system composed of a large number of self-managing sub-systems. What are the characteristics of such composed complex systems?

Operating a complex system composed of self-managing sub-systems requires the "right" level of delegation as well as the "right" level of visibility to the operator. What rules and criteria can be derived to identify suitable levels and what interaction models (such as human interface, graphical presentations, etc) can be used to operate such a complex system?

Do the existing frameworks, such as TOGAF, Zachman, ARIS, ITIL, eTOM and others define a sufficient level of details to translate business objectives into a set of actions and commands to manage application servers and network elements? How can network self-management help controlling peer-to-peer networks, such as wireless ad-hoc networks, on-the-fly connectivity for defense networks, or mesh networks build by mobile access routers in cars that form networks in motion? These p2p systems require cooperative management skills, such as local storage of events, error messages, accounting records, etc. until information can be shared with neighbors. In such environments, each device is an autonomous element and ideally self-managed, however when connecting towards new neighbors, a process for connecting multiple self-managed systems needs to be developed. How tightly or loosely should multiple network self-management systems be linked?

References

[1] Balasubramaniam, S., Barrett, K., Donnelly, W., van der Meer, S., Strassner, J.: Bio-inspired policy based management (biopbm) for autonomic bio-inspired policy based management (biopbm) for autonomic. Policies for Distributed Systems and Networks, 2006. Policy 2006. Seventh IEEE International Workshop on (0-0 2006) 3–12

[2] Herrmann, K., Mhl, G., Geihs, K.: Self-management: The solution to complexity or just another problem? IEEE Distributed Systems Online **6**(1) (2005)

[3] Wolter, R.: Self-monitoring and -management of cisco network elements. IEEE Distributed Systems Online **28**(4) (2005)

[4] Gellersen, H.W., Schmidt, A., Beigl, M.: Multi-sensor context-awareness in mobile devices and smart artifacts. Mob. Netw. Appl. **7**(5) (2002) 341–351

[5] de Campos, G.A.L., de P. Barros, A.L.B., de Souza, J.T., Celestino Junior, J.: A model for designing autonomic components guided by condition-action policies. Network Operations and Management Symposium Workshops, 2008. NOMS Workshops 2008. IEEE (April 2008) 343–350

[6] Jacob, B., Lanyon-Hogg, R., Nadgir, D., Yassin, A.F.: A practical guide to the ibm autonomic computing toolkit. technical report, ibm international technical support organization. (2004)

[7] Dobson, S., Denazis, S., Fernández, A., Gaïti, D., Gelenbe, E., Massacci, F., Nixon, P., Saffre, F., Schmidt, N., Zambonelli, F.: A survey of autonomic communications. ACM Trans. Auton. Adapt. Syst. **1**(2) (2006) 223–259

[8] Raymer, D., Meer, S.v.d., Strassner, J.: From autonomic computing to autonomic networking: An architectural perspective. Engineering of Autonomic and Autonomous Systems, 2008. EASE 2008. Fifth IEEE Workshop on (31 2008-April 4 2008) 174–183

[9] Vianna, R.L., Polina, E.R., Marquezan, C.C., Bertholdo, L., Tarouco, L.M.R., Almeida, M.J.B., Granville, L.Z.: An evaluation of service composition technologies applied to network management. Integrated Network Management, 2007. IM '07. 10th IFIP/IEEE International Symposium on (21 2007-Yearly 25 2007) 420–428

[10] Bartsch, C., Shwartz, L., Ward, C., Grabarnik, G., Buco, M.J.: Decomposition of it service processes and alternative service identification using ontologies. Network Operations and Management Symposium Workshops, 2008. NOMS Workshops 2008. IEEE (April 2008)

[11] Sifalakis, M., Louca, A., Mauthe, A., Peluso, L., Zseby, T.: A functional composition framework for autonomic network architectures. Network Operations and Management Symposium Workshops, 2008. NOMS Workshops 2008. IEEE (April 2008) 328–334

[12] Momm, C., Hallerbach, I.P., Abeck, S., Rathfelder, C.: Manageability design for an autonomic management of semi-dynamic web service compositions. Network Operations and Management Symposium Workshops, 2008. NOMS Workshops 2008. IEEE (April 2008)

[13] Kephart, J.O., Walsh, W.E.: An artificial intelligence perspective on autonomic computing policies. policy **00** (2004) 3

[14] Perazolo, M.: A self-management method for cross-analysis of network and application problems. Network Operations and Management Symposium Workshops, 2008. NOMS Workshops 2008. IEEE (April 2008) 357–363

[15] Carroll, R., Strassner, J., Cox, G., van der Meer, S.: Policy and profile: Enabling self-knowledge for autonomic systems. In: DSOM. (2006) 239–245

[16] Clemm, A., Granville, L.Z., Stadler, R., eds.: Managing Virtualization of Networks and Services, 18th IFIP/IEEE International Workshop on Distributed Systems: Operations and Management, DSOM 2007, San José, CA, USA, October 29-31, 2007, Proceedings. In Clemm, A., Granville, L.Z., Stadler, R., eds.: DSOM. Volume 4785 of Lecture Notes in Computer Science., Springer (2007)

[17] Adam, C., Stadler, R.: Service middleware for self-managing large-scale systems. Network and Service Management, IEEE Transactions on 4(3) (Dec. 2007) 50–64

Chapter 7

Policy-Based Self-Management in Wireless Networks

Antonis M. Hadjiantonis and George Pavlou

Center for Communication Systems Research, Dept. of Electronic Engineering, University of Surrey

Abstract   Self-management is expected to motivate significant research efforts, both in academia and in industry, because of the apparent benefits it can offer. Future networks and systems will transparently integrate self-management capabilities, relieving users and managers from painstaking tasks. As research progresses, the current separation of self-* properties in configuration, healing, optimization and protection will diminish, gracefully amalgamating all in a Self-maintaining operation. We envision a policy-based system as a future-proof solution, where business objectives and user preferences will be encapsulated in policies. Context-awareness will provide secure and accurate feedback to the system, assisting in fully customized and personalized user experience. Eventually policies and context will vanish inside systems, allowing users to enjoy truly ubiquitous networking.

7.1   Introduction, Background and State-of-the-Art

7.1.1   Self-Management Concepts and Challenges

Self-management refers to the ability of independently achieving seamless operation and maintenance by being aware of the surrounding environment. We propose this definition to assist in the exploration of the newly emerging field of Self-Managing Networks and Systems. This ability is widely embedded in the natural world, allowing living organisms to effortlessly adapt to diverse habitats. Take, for example, our ability to regulate our body temperature. Without planning or thinking about it, our body's functions work in the background to maintain a constant temperature. By attempting to imitate this ability for Systems, we need to provide the logic and directions for their operation and in addition the means to sense their operating environment. Sensing the environment is crucial in order to achieve awareness of surrounding conditions, threats and resources. This collective awareness of a system's operating environment is referred to as 'context-awareness'. Through context-awareness, a System can combine context with its provided logic and directions in order to adapt to changing conditions. Beyond context-awareness, in order to achieve true Self-management, a flexible and reliable way to provide necessary logic and directions is needed. For this purpose, we rely on policies and policy-based management (PBM). Once a System is informed of the policies governing its behavior and achieves context-awareness, it can independently operate and maintain itself, thus becoming a Self-managed System. The work presented in this chapter attempts to investigate Self-management through the interaction of context-awareness and policy-based management, focusing on a management framework for wireless ad hoc networks.

Among pioneering research efforts, IBM had introduced the concept of Autonomic Computing in 2001, which encapsulates the aspects of self-management in an architectural blueprint [2]. The concept was inspired by the ability of the human nervous system to autonomously adapt its operation without our intervention and has appealed to researchers worldwide. IBM's vision [3] has fueled intense research efforts both in industry and academia. In essence, autonomic computing and self-management are considered synonymous. According to IBM, autonomic computing is 'a computing environment with the ability to manage itself and dynamically adapt to change in accordance with business policies and objectives.' This fundamental definition continues to identify the quintessential four properties of a self-management system, frequently referred as self-* or self-CHOP properties: 'Self-managing environments can perform such activities based on situations they observe or sense in the IT environment rather than requiring IT professionals to initiate the task. These environments are self-Configuring, self-Healing, self-Optimizing,

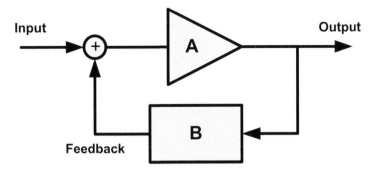

FIGURE 7.1: Closed-loop controller

and self-Protecting.'

Beyond theory, actual realization of self-management properties in computing systems poses significant challenges and remains an open and active research topic. Self-management is closely related with control systems [1] and particularly to closed-loop controllers (Fig. 7.1). By using a system's output as feedback, a feedback loop allows the system to become more stable and adapt its actions to achieve desired output. While such control loops are widely used in electronics (e.g., operational amplifiers), these concepts are increasingly used in computing after the introduction of the autonomic manager (AM) component (Fig. 7.2), as proposed in [2]. This architectural component has become the reference model for autonomic and self-managing systems and will serve as our reference point for the rest of this chapter. The autonomic manager is a component that manages other software or hardware components using a control loop. The closed control loop is a repetitive sequence of tasks including monitoring, analyzing, planning and executing functions. The orchestration of these functions is enabled by accessing a shared Knowledge base. The reference model is frequently referred to as K-MAPE or simply MAPE, from the initials of the critical functions it performs. By analyzing the definition for Self-Management, we identify policies as the basis of such systems, encapsulating high-level business objectives. Policy-Based Management (PBM) is the first building block of the Self-Management framework presented in this chapter and policies are its cornerstone. Equally important and complementary is the system's ability to sense and observe its surrounding environment. To enable these, context-awareness is employed as the second building block of the framework. As a result a policy-based context-aware framework is designed as the foundation for the implementation of self-management properties. A policy-based framework can serve as the Plan and Execute components of a self-management system, as presented in Fig. 7.2. Policy design and specification constitute the Planning phase of autonomic management while policy enforcement constitutes the Execute

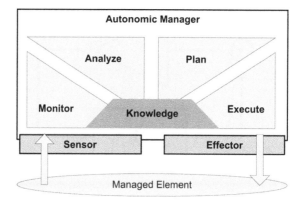

FIGURE 7.2: Functional diagram of IBM's autonomic manager (K-MAPE)

phase. On the other hand, a context-aware framework is assigned the Monitor and Analyze functionality, thus closing the necessary feedback loop. Context sensing and collection constitute the Monitoring phase while context aggregation and inference rules constitute the Analyze phase. The specification of policies and context together with their interaction form the essential Knowledge element. Policy and context repositories are the Knowledge centerpiece of both frameworks gracefully integrating the presented self-management solution. Figure 7.3 illustrates these concepts, in parallel with IBM's autonomic manager [2]. In this chapter a policy-based context-aware framework is presented, aiming to offer a platform for the self-management of wireless ad hoc networks. Wireless networks pose significantly different requirements in their management. As a result, existing solutions for fixed networks are often inapplicable, causing severe traffic overhead and performance degradation. In addition, the emergence of pervasive computing and the proliferation of wireless devices accelerate the spontaneous formation of ad hoc networks without any central administration. The investigation of these issues motivates the presented research efforts of this chapter, aiming to offer a customized solution for wireless ad hoc networks.

7.1.1.1 Self-Management in Autonomic, Pervasive, Ubiquitous Computing

Having introduced the basic concepts of self-management and autonomic computing, we further elaborate and delve into pervasive and ubiquitous computing realms. Often these concepts are used interchangeably, although slight differences exist. Pervasive and ubiquitous computing is mostly targeting user-created networks by transparently integrating appropriate hardware and software within relevant devices and infrastructure. Autonomic computing on

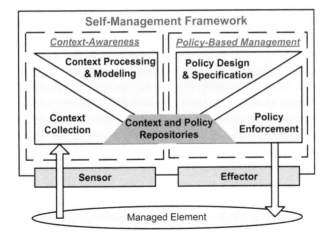

FIGURE 7.3: Mapping of proposed high-level framework to autonomic manager component

the other hand traditionally targets large-scale enterprise networks, aiming to relieve network administrators from painstaking management tasks. The realms of autonomic, pervasive and ubiquitous computing embrace the concept of Self-Management and therefore may be used interchangeably as long as their difference in focus is considered. Pervasive management is receiving intense interest from academia and industry, aiming to simplify and automate ubiquitous network operations. Pervasive management aims to vanish inside devices, relieving users from tedious configuration and troubleshooting procedures. Ideally, autonomic elements exhibit self-configuration, self-optimization, self-protection and self-healing capabilities. When combined, these capabilities can lead to adaptive and ultimately autonomic systems. In reality, the deployment of ubiquitous networks is withheld from several obstacles that need to be overcome in order to realize such a vision. These issues provide motivation for researchers, aiming to realize a system with self-management capabilities and fueling efforts for gradual transition to autonomous self-management.

7.1.2 Open Issues and Motivation

Ubiquitous networking has received both academic and commercial interest. In [4] a detailed description of the challenges for ubiquitous computing is presented from different perspectives. With the proliferation of wireless networks and increasingly networked environments, different approaches have been adopted. In [5], ubiquitous computing is proposed for home networks and in [6, 7] spontaneous approaches to networking are presented, focusing on user's interaction and services. Today, there exists an increasing interest towards wireless and particularly ad hoc networking, as the enabling technologies of ubiquitous environments. In wireless ad hoc networks, users own mobile nodes and can move randomly and unpredictably. Their devices need to organize themselves arbitrarily; thus the wireless network's topology may change rapidly. Conventional wireless networks require some form of fixed network infrastructure and centralized administration for their operation. In contrast, since wireless ad hoc networks are self-creating, individual nodes are responsible for dynamically discovering other nodes they can communicate with. This way of dynamically creating a network often requires an equally dynamic ability to manage the network and supported services, according to higher-level management goals (i.e., policies) and taking into account the surrounding conditions (i.e., context).

As introduced earlier, policies and context are critical building blocks for the self-management of wireless ad hoc networks; therefore we present a thorough review of the applicability of Policy-Based Management (PBM) and Context-Aware Systems (CAS). Research efforts have shown that such highly dynamic environments can benefit from a PBM approach and the emerging context-driven autonomic communications trend. One of the major advantages of adopting a policy-based approach is the relevant 'controlled programmability' that can offer an efficient and balanced solution between strict hard-wired management logic and unrestricted mobile code migration and deployment.

The diverse nature of wireless ad hoc networks calls for differentiation from traditional organizational models. All views tend to adopt a distributed model and several variations exist. In this chapter we present a distributed and hierarchical (hybrid) organizational model, recognizing the emerging trend of peer-to-peer (P2P) computing for network management. The high degree of decentralization and robustness to node disconnection are useful features to be considered. While P2P systems generally scale well, there is a limiting obstacle of provisioning and synchronizing all nodes.

An additional burden in the management of wireless networks is the increasing heterogeneity of participating devices. Therefore interoperability of devices and networks is a critical issue to be addressed. Through literature, we observe that neither policy-based nor context-aware paradigms have been fully standardized, leading to increased fragmentation of research and market value. As the maturity of both paradigms is reached, there is a crucial need for interoperability. Standardization efforts are necessary to promote the usability and penetration of every new protocol or paradigm, hence we base our solution on existing IETF standards.

At the same time, the wide adoption of ad hoc networking in various aspects of our communication needs has diversified the role of managing and managed entities. An individual user of an ad hoc network demands control over his/her device and is not willing to grant permission for reconfiguration of personal settings. Such a reconfiguration might be implied from the enforcement of policies in a node that just joined the network. In [38] the author states 'personal freedoms come into play' and 'no absolute control will be accepted' from future wireless network users. The transient nature of ad hoc networking should be considered and the voluntary entrance or departure of a user from it differentiates the traditional policy enforcement paradigm. Obviously, user control is an important issue which adds special requirements on the design of management frameworks. Policies are affected by this fact, while non-uniform policy enforcement may need to be considered. Additionally context-aware systems are affected; in order to cater for the above requirements, user preferences and input need to be collected and processed.

Context is inevitably connected with the personal information of every user, especially in the case of wireless personal networks. Hence, privacy issues are raised and need to be addressed. When it comes to managing a network where the networked devices belong to individuals rather than organizations, issues like privacy and data protection should be considered. In the European Union for example, strict legislation by the European Data Protection Supervisor (EDPS, Directive 95/46/EC, http://www.edps.europa.eu) mandates the processing and acquisition of personal data and national authorities have been established to monitor their enforcement. Different regulations apply in the US, where a territorial approach is adopted. It is evident that the management of a network consisting of individual's devices should or is legally obliged to respect the directives regarding the collection and processing of personal data. The advancement of wireless devices and peripherals can accurately provide context which can help network management, e.g., a GPS receiver providing location data. However sensitive data like user location are private data and the user should be able to explicitly permit or deny access to them. In general, context-aware management could exploit context information available on a user's device, but a user's permission must be granted.

The issue of privacy is tightly coupled with security. The assumption of secure and trusted environments is made for the majority of presented literature, as well as for the proposed solution. However, once the assumption is lifted, major concerns are raised and need to be addressed. Self-managing systems are vulnerable to intrusion and compromise. Once again, the nature of wireless ad hoc communication adds to the problem's complexity and requires significant effort to ensure secure networking. Security is a continuous issue for every network system. In ad hoc networking, the security issues are more difficult because the wireless interface is used and access to it can not be controlled [46, 64]. Security features should exist in the management system without making it resource demanding and hard to implement. Also the lack of a centralized coordinator makes this task harder, since we cannot rely on a certificate authority for example. Furthermore novel security techniques like the detection of misbehaving wireless nodes (malicious or selfish) could be devised in order to find and block nodes that may compromise system's reliability and safety [33].

7.2   Policies and Context for Self-Management

7.2.1   Policy-Based Management (PBM) Principles

Policy-Based Management (PBM) simplifies the complex management tasks of large scale systems, since high-level policies monitor the network and automatically enforce appropriate actions [8, 9, 10, 11, 12, 13, 14, 15, 16]. In general, policies are defined as Event-Condition-Action (ECA) clauses, where on event(s) E, if condition(s) C is true, then action(s) A is executed. PBM approaches for wireless networks have been proposed in [10, 17, 18] and industry envisions autonomic computing as dynamically managed by business rules and policies [3]. In this section we present the basics of Policy-Based Management and provide a thorough literature review of related research efforts and open issues.

The main advantage which makes a policy-based system attractive is the functionality to add controlled programmability in the management system without compromising its overall security and integrity. Real time adaptability of the system can be mostly automated and simplified by the introduction of the PBM paradigm. Policies can be viewed as the means to extend the functionality of a system dynamically and in real time in combination with its preexisting hard-wired management logic [14, 19]. Policies offer to the management system the unique functionality of being re-programmable and adaptable, based on the supported general policy types. Policies are introduced to the system and parameterized in real time, based on management goals and contextual information. Policy decisions generate appropriate actions on the fly to realize and enforce those goals.

7.2.1.1   PBM Basics

The components of a PBM system are shown in Fig. 7.4 in block fashion and also in simplified UML notation. This framework had been originally proposed by the IETF and has been widely used and accepted in research and industry [12, 20, 21, 22, 23, 24, 25]. The Policy Repository (PR) is an integral part of every policy-based system because it encapsulates the management logic to be enforced on all networked entities. It is the central point where policies are stored by managers using a Policy Management Tool (PMT) and can be subsequently retrieved either by Policy Decision Points (PDP) or by one or more PMT. Once relevant policies have been retrieved by a PDP, they are interpreted and the PDP in turn provisions any decisions or actions to the controlled Policy Enforcement Points (PEP). Although a PR is a centralized concept, various techniques exist to physically distribute its contents. The reasons for distribution are obviously resilience and load balancing [11, 26, 27]. Typical implementations of a PR are based on Lightweight Directory Access

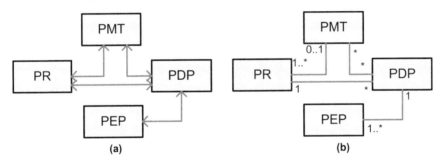

FIGURE 7.4: IETF's framework for PBM (a) block diagram, (b) generic UML notation

Protocol Servers (LDAP v3, RFC 4511 [28]), also known as Directory Servers (DS). We will refer to a DS with its directory content (i.e., policies) as a directory. A single point of failure would make policy-based systems vulnerable; therefore replication features of DS are often exploited. When designing a Policy Repository for the policy-based management of wireless networks, there exist additional requirements that need to be taken into account, e.g., tolerance against connection intermittence and multi-hop communications. These issues are examined in the proposed framework and motivate the design of a Distributed Policy Repository (DPR). In brief, standardization efforts within IETF Policy WG have specified an LDAP schema to represent policies that follow IETF specifications. Originally this representation was targeted towards the representation of QoS policies for IntServ and DiffServ architectures. However the appealing benefits of policy-based management have led different bodies from industry and academia to extend the specification and independently develop new ones, both in terms of application domains and representation. A PBM approach needs to be examined in contrast with the popular mobile code techniques as well as other traditional management schemes [29]. The benefit of policy-based management which makes it applicable to wireless ad hoc networking is the ability to control the re-programmability of the system by allowing the manager to install and remove software modules with the desired functionality on the fly. On the contrary, mobile-code techniques allow full re-programmability of the system but are quite vulnerable to malicious code execution. These security concerns have been the main obstacle in the wider adoption of mobile-code paradigms, although at first sight they appear attractive. PBM overcomes this obstacle, since the programmability of the system depends on the supported generic policy types. Although this may seem restrictive, it provides the assurance that the installed modules have been pre-approved during the system compilation and can be instantiated safely. As detailed later, there are still unresolved issues which may cause system instability and this happens when policy conflicts occur. Research in policy analysis and conflict detection and resolution is intense and may provide solutions in the near future.

7.2.1.2   Policy Representation and Definition

The representation of policies in a system and the policy language used is an important issue. Both depend on the selection of an appropriate information model which will provide the common ground for identifying managed objects and defining policies. Standardization efforts have focused on the development of an Information Model rather than a formal language for policy definition. These efforts are driven by the combined work between IETF [24] and DMTF [25]. The defined Information Models are conceptual models for representing and managing policies across a spectrum of technical domains. Their purpose is to provide a consistent definition and structure of data (including policies), using object-oriented techniques. These models define policy classes and associations sufficiently generic to allow them to represent different policies [48]. The Policy WG of IETF [12] has concluded during 2004 and the output was a series of RFCs defining the Policy Core Information Model (PCIM) [21, 22] and extensions for QoS (RFC 3644, RFC 3670), as well as mapping guidelines for the LDAP [36] model representation [60, 61]. However work under DMTF has continued, producing newer versions of the information model, referred to as CIM (Common Information Model) Policy Model (v2.9 Jan 2005) [60] which slightly differs from IETF's PCIM (which was based on CIM Schema v2.2). We must outline that IETF/DMTF do not define a policy language but implicitly provide a generic definition of policy rules. This definition is in the form of: if <condition >then <action >, where a policy is defined as a set of rules to administer, manage and control access to network resources [21]. IETF's policies can have some additional functionality like policy roles, grouping and prioritization, which are defined in the PCIMe information model RFC [22]. A missing element from IETF's PCIM solution is an explicit triggering mechanism which would make the system event-driven. This is important in a policy-based system, since the generic policy rule event-condition-action is widely accepted [42, 46]. Recent work on the DMTF CIM Policy Model suggested a triggering mechanism and a special query language (CIM Query Language (CQL) [47].

In academia policies have also received intense research interest. Significant work was done at Imperial College which defined a formal policy language named Ponder [42]. Ponder is a declarative, object-oriented language for specifying security and management policies for distributed object systems. Ponder does not rely on an information model for policy definition. A formal grammar is introduced instead and policies must comply with it [42, 49]. Ponder has four basic policy types: authorizations, obligations, refrains and delegations and three composite policy types: roles, relationships and management structures that are used to compose policies [49]. Other efforts for a policy language specification for security and management are presented in [46].

Beyond the selection of a language to define policies, another issue to address is what policies are to be defined for the purpose of managing a network. The issue of policies definition is mostly independent from the policy language and representation. It is more related to what are the management goals and objectives rather than what is to be managed. A standardized information model, e.g., CIM schema, can be used to implement object-oriented design aspects of the network, using managed objects (MO). In the case of wireless ad-hoc networks, no information model describes efficiently their diverse features. However, extension of the CIM model, or any other proper information model, is possible. Literature efforts [31] have proposed an extension to SNMP MIB, called anmpMIB. An interesting effort is presented in [50] and [47], where the concepts of CIM as an extensible information model are used in combination with the Ponder policy language. Therefore the use of both Ponder as the policy language definition and an expanded version of CIM as the managed objects definition are possible and could also apply in ad hoc network environments. Additionally some specific management goals have to be defined in conjunction with a proper case study, in order to extract and refine the low level policies to be implemented. Refinement is the process of deriving a more concrete specification from a higher-level objective [46]. The task of refinement is a complex issue regarding policies, since a fully automated process is not possible.

7.2.1.3   Policy Provisioning and Storage

In a ubiquitous environment like wireless ad hoc networks, a special distribution technique of policies is vital for their effective and reliable dissemination. Since the centralized architectural model is not applicable, a central entity to disseminate policies across all nodes would become a single point of failure. Distributed and collaborative ways are needed to fulfill the special requirements of wireless ad hoc network. In order to share the overhead in the network and avoid bottlenecks, special distribution protocols need to be designed. Those protocols should cater for the varying needs of ad hoc networks and provide fast, low in overhead, reliable, resource-conscious and secure policy distribution. We should note that the policy distribution issue is tightly related to policy storage. Previous efforts on implementing policy distribution protocols are limited and here the most notable are examined. Regarding policy distribution and provisioning, we first review IETF's standardized Common Open Policy Service (COPS) and then we examine the approach of the Ponder toolkit. Finally some promising XML based solutions are reviewed.

Efforts from IETF's Resource Allocation Protocol Working Group (RAP WG) [43] have produced COPS (Common Open Policy Service) Protocol [44] and COPS-PR for Policy Provisioning [45]. COPS is a simple query and response protocol that can be used to exchange policy information between a policy server (Policy Decision Point or PDP) and its clients (Policy Enforcement Points or PEPs). The basic model of interaction between a policy server and its clients is compatible with the framework document for policy based admission control [20]. The focus of IETF's efforts has been mainly to provide a protocol to carry out the task of policy distribution mostly related to QoS parameters and setup. Beyond the initial deployment efforts in industry during 2000 [66], COPS has not found general acceptance and interest. In academia the effort of K.Phanse described in [37, 112] utilize COPS-PR solely for the purpose of QoS configuration. The intermittent nature of ad hoc communications, though, would require that the PEP in IETF's architecture be less dependent on PDP. Therefore the usage of COPS is not a strong candidate for policy provisioning. Furthermore, different architectures introduce a dual node functionality [19], where each managed device acts both as a PDP and as a PEP, thus making the usage of COPS unnecessary. More deficiencies of COPS and COPS-PR are outlined in [65], while researchers are looking into emerging technologies to substitute COPS completely [69]. Looking into policy distribution and provisioning techniques of Ponder toolkit [42] such a protocol does not exist. Instead, remote procedure techniques are used to propagate policy decisions towards the enforcement points, using Java RMI. Researchers are looking into alternative policy provisioning techniques, mainly using XML-based architectures [63, 64] to exchange XML fragments wrapped in an HTTP message. The use of web based protocols (e.g., SOAP over HTTP) for the dissemination of policies and the usage of Web-Services has also been considered [51]. Authors of [69] have investigated substitution of COPS with NETCONF or SOAP and their evaluation results look promising.

Concurrently with policy distribution, issues of policy storage need to be considered. The existence of a policy repository (PR) in most architectures requires an efficient policy storage implementation. The prevailing solution for policy storage is the Lightweight Directory Access Protocol (LDAP) but not the only one. As mentioned before, XML based solutions exist and should be considered as an alternative. Independently of the storage technology used, other equally important issues arise and their solution is vital for the survivability of ad hoc networks. These issues include the distributed storage of the repository through replication and/or fragmentation. Similarly, issues like how to compose a distributed repository for policy updates and policy lookup, or how to check if all copies exist and are updated need to be addressed. These special issues are important in a distributed environment, but in a wireless ad hoc network they should be considered as mandatory in order to realize a robust and efficient management system.

The LDAP protocol [36, 53] is designed to provide access to the X.500 Directory while not incurring the resource requirements of the Directory Access Protocol (DAP) [52]. This protocol is specifically targeted at simple management applications and browser applications that provide simple read/write interactive access to Directories. The reasons for the dominance of LDAP as a policy repository are some of the useful features it has to offer. The object-oriented design and implementation of a Directory using LDAP makes storage of policy objects very convenient and easy to access. The operations/services it offers like search, modify, add, etc. as well as filtering and authentication capabilities can be used in a natural way for policy retrievals, modifications and look-ups. Furthermore, the capabilities to distribute and/or replicate the directory among network nodes make it very attractive to wireless ad hoc networks management. The LDAP directory can be distributed on several physical nodes by utilizing the inherent LDAP's replication capabilities. Finally, LDAP's built-in security mechanisms can provide various levels of access control over the contents and the access to the policy repository. On the other hand, we should note that LDAP technology is more optimized towards frequent search and look-up operation rather than updates and modifications. These limitations should be considered in combination with the frequency of policy modifications in ad hoc networks [53, 54, 55, 56, 57, 58].

On another perspective, for the purpose of storing policies, XML based solutions are considered as an alternative to LDAP. The reasons are the significant penetration of XML in several devices and systems and the wide support it receives as a uniform and interoperable technology for sharing and representing data [59, 62, 63, 64]. Previous attempts in policy-based management systems for ad hoc networks have adopted different solutions for their policy repositories. In [32] a relational database is used as a Policy Repository. Specifically, MySQL database server stores policies in a proprietary way not described. This database, named CMDB, is also used for the storage of configuration and monitoring data on every node. The approach in [37, 112] is based on the COPS-PR protocol and considers a Policy Information Base (PIB) as the policy repository.

7.2.1.4 Policy Enforcement and Conflicts

Some further policy-related issues need to be addressed in order to achieve a complete and efficient PBM system. These issues relate to the decision making process and to the enforcement of policies in the network. Furthermore one has to consider whether the enforcement of policies will be uniform or choice will be given to nodes. Policy decisions are made at a PDP according to the IETF's architecture. However different architectural approaches require readdressing the decision making process and solutions are expected to be highly distributed. Both intelligence and management logic could be shared between nodes according to capabilities and roles. Local, remote or delegated decisions tactics should be considered according to the examined scenario. Traditionally, policy enforcement is expected to be uniform, i.e., all nodes conforming to same policies. However in a user-created wireless ad hoc network this is not necessary, since the purpose and formation of such networks is different from fixed ones. An important issue emerges, regarding whether the policies should apply to all users and how their preferences are respected.

A revolutionary realization of policy enforcement would be to allow network nodes to partly conform to a global policy set. These concepts are introduced in [38] and motivate solutions that consider an enforcement mechanism which would respect user preferences and special requirements [10]. In [10, 92] cases are examined where no absolute control from an authority is accepted, discussing whether all policies should apply to all users and how their preferences should be respected. In [106] a 'promise theory' attempts to provide 'political autonomy' to entities and decentralize policy management. Such requirements significantly increase system's complexity, but yet need to be addressed in combination with the user's need to control their devices and respect their privacy.

Moreover, in an environment where a number of policies need to coexist, there is always the likelihood that several policies will be in conflict, either because of a specification error or because of application-specific constraints. It is therefore important to provide the means of detecting conflicts in the policy specification [102, 103]. Considering different conflict types, it is possible to define rules that can be used to recognize conflicting situations in the policy specification. These rules usually come in the form of logic predicates and encapsulate application-specific data and/or policy information as constraints. Examples on how these rules can be used as part of a detection process can be found in [104, 105].

7.2.2   Context and Context-Awareness

Having discussed the properties of self-management systems and their first pillar, i.e., policy-based management (PBM), we turn our attention to the second one: context-awareness. A context-aware framework is assigned the Monitor and Analyze functionality of a Self-Management system. Context sensing and collection constitute the Monitoring phase while context aggregation and inference rules constitute the Analysis phase of management. In this section we present an in-depth analysis and literature review of context and context-awareness for self-management.

In the literature, context is defined by Dey [74] as any information that can be used to characterize the situation of an entity. An entity is defined as the person, place or object that is considered relevant to the interaction between a user and an application, including the user and applications themselves. Another definition from Malatras [72] defines the context of a system as the set of information of every nature that describes the system, influences system aspects and that is being affected by the system's operation, the ownership of which is not necessarily solely held by the system. According to [72], context awareness refers to the ability of a system to adapt dynamically and continuously its status and operation according to context information. In essence, context is synonymous to information and it is this information that needs to be collected, modeled and processed to become useful Knowledge. Self-management systems can exploit Knowledge, combined with policies.

Regarding self-management of networked devices, context refers to their computational and physical environment and is strongly coupled with the employed management framework. Therefore we first present a taxonomy of context, to assist the reader in understanding the diversity and different forms it can take. Using the taxonomy, a system designer can characterize available context and decide on what context is needed and when. We then present an overview of context models in the literature. Context modeling is necessary to achieve true context-awareness since various context sources produce different data that have to be structured and organized under a unified representation scheme. In other words, a context model acts as a communication protocol among context aware entities, allowing interoperable and efficient processing. Finally, context storage mechanisms are discussed, addressing the need for efficient and reliable storage and retrieval of Knowledge.

7.2.2.1 Taxonomy of Context Information

A classification is presented in [71], where context is distinguished by its persistence, medium and nature. We further extend and elaborate based on Fig. 7.5 that represents a sample context information taxonomy:

- By its persistence:

  - Persistent: No updating is needed as context does not evolve or change in time (e.g., name, ID card)

  - Temporary: Updating is needed for context information that doesn't remain constant over time (e.g. position, health, interface load). Distinguished by its temporal situation:
    * Past: This category is for that context which took place in the past. The implied context history contains all previous user contexts.
    * Present: This category is for the current context, valid at the invocation moment, e.g., where am I at this moment, etc.
    * Future: Context that can be scheduled and stored a priori for future actions, i.e., the venue where a meeting will be held tomorrow morning. Prediction of future context would be very useful.

- By its nature:

  - Physical: Measurable context information that is tangible, e.g., geographical position, network resources, temperature.

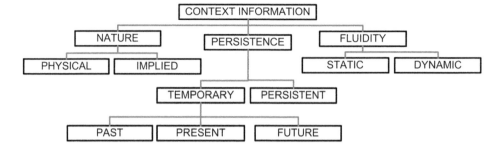

FIGURE 7.5: Taxonomy of context information

> * Necessary: Context information that must be retrieved for a specific task to run properly.
>
> * Optional: Additional context information which could be useful for better performance or completeness.
>
> − Implied: Non-measurable by means of physical magnitudes, e.g., name, hobbies (it is likely that this kind of information will be introduced by the users themselves).

- By its fluidity:

  − Static: Context that does not change very quickly, e.g., the temperature along the day.

  − Dynamic: Context that changes quickly, i.e., the position of a person who is driving.

7.2.2.2   Context Modeling

Context can be exploited if it can be represented in a notion comprehensible by the entities that want to use it for decision making or monitoring. Hence a context model is required. Important requirements are simplicity, ease of deployment, performance, scalability, applicability and usefulness. From literature, two prominent context models are presented, namely based on entity-relationship model and based on Unified Modeling Language (UML). Other approaches also exist and are mentioned below.

7.2.2.2.1   Context model based on Entity-Relationship model
This context model is based on the concepts of entity and relationship and is derived from previous definition of context as well as the one given in [75]. Based on object oriented modeling, an entity can be understood as anything

that could have any kind of influence or relevance at any time in the performance of the activity of an application or service addressed to a user or a managed system [71]. An entity is composed by a set of intrinsic characteristics or attributes that define the entity itself, plus a set of relationships with other entities. The relationships belong to a specific type, among a set of available types of relationships. The concept of local context of an entity can be understood as the information that characterizes the status of the entity. This status is composed by its attributes and its relationships. Moreover, the relations that can exist between the different entities inside the model, as well as entities, can represent many different types of influence, dependence, association and so on, depending mainly on the type of entities that these relationships connect. Using this model, a network of entities and relationships can be constructed to represent the world that surrounds the activity of a context-aware service and that can influence its development. According to [71], the steps to realize a context model based on entity-relationship concept are:

1. Classification of contextual information

2. Mapping context information into an entity-relationship model

   (a) definition of generic entities
   (b) definition of generic relationships

3. Representation and implementation tools

7.2.2.2.2 Context model based on UML principles

Based on work presented in [76, 18, 96, 77, 78] another context model is presented that is better suited to resource-constrained devices. This model exploits design principles of Unified Modeling Language (UML). The popularity and usability of UML makes it appropriate for representing context. In brief, using this model, complex information is derived from a collection and combination of simpler pieces of information. The context of a device is constituted of higher level contexts that have been deduced from simpler ones. Each context can be split into atomic attributes that fully describe the initial context and are not composed of any simpler attributes. Hence, context is composed of self-explanatory attributes and perhaps other contexts, leading to more complex context structures. In addition, semantic-based relationships infer high-level context and can span from simple inference rules such as mathematical functions to semantic or user-defined operations. Semantic information is stored as context metadata, that describe its functionality, operation or meaning accordingly. The potential use of ontologies to identify semantic proximity and pattern matching is a promising research direction. UML has also been used in [79, 80] to model context information, though these solution incur a significant level of complexity on basic UML options.

7.2.2.2.3   Other approaches on Context Models

Depending on the data structures used to exchange contextual information in the respective system [81], relevant approaches are mentioned here:

- Key-Value Models: Schilit et al. [82]

- Mark-up Models: Comprehensive Structured Context Profiles (CSCP) [83]

- Graphical Models: Henricksen et al. [79], Object-Role Modelling (ORM) [80]

- Object-Oriented Models: TEA project [84], Active Object Model [85]

- Logic-Based Models: McCarthy [86]

- Ontology-Based Models: Otzturk and Aamodt [87]and CoBrA [88]

7.2.2.3   Context Storage Mechanisms

Beyond modeling of context, the actual representation of these data is critical, in order for the system to be able to process and handle them. Extensible Markup Language or XML has been widely used due to its inherent advantages:

- it is a architecture independent mark-up language suitable for structured information representation

- it can be validated and checked for errors using mature tools like XML Schemas or DTDs (Document Type Definition)

- due to its architecture independence, it is extremely interoperable and can be used as a mechanism to exchange and store data

- the context represented is searchable and can be manipulated using XQuery, a powerful XML search engine

However XML has some drawbacks:

- it is a hierarchical language that restricts database oriented architecture

- due to its verbosity, it increases the amount of information to be stored

To overcome these drawbacks and comply with the major requirement of minimizing the amount of context information transferred throughout the network, common characteristics of context can be used to reduce overheads and to avoid traffic bottlenecks and bandwidth consumption:

- Context aggregation: Context information is periodically aggregated and average values sampled over time are actually transmitted

- Normalization of context values: Context values are normalized in certain ranges, allowing for less data to be transmitted.

- Threshold criteria: Criteria associated with specific contexts may result in context transmission only when certain thresholds have been exceeded.

7.2.3 Management of Wireless Ad Hoc Networks and Self-Management Capabilities

Among various wireless technologies, Mobile Ad hoc Networks (MANET) [30] have received intense interest, especially from the research community. This interest however has not led to significant industrial exploitation or widespread adoption. According to [89], the major reason for the negligible market impact of the 'pure general-purpose MANET' paradigm is the lack of realism in the research approach. As a result, MANET are normally deployed in labs or by a few experienced users. To avoid such pitfalls, frameworks should be based on realistic assumptions and tightly coupled design with implementation and deployment on wireless testbeds. The notion of 'hybrid mobile ad hoc networks' [89] was introduced by relaxing the main constraints of pure general-purpose MANET, i.e., to consider the deployment of a network that consists of user devices with limited infrastructure support and connectivity. This assumption allows the MANET paradigm and its research results to be applied to several interesting paradigms and cases studies, e.g., mesh networks [99]. We refer to this paradigm as wireless ad hoc for the rest of this chapter, generalizing the deployment of MANET.

The deployment of wireless ad hoc networks suffers from limitations in wireless link connectivity and capacity, due to the design of Physical (PHY)/ Data Link (MAC) layers and the wide use of TCP/IP which is optimized for fixed networks. The capacity and throughput are limited and severely degrade as the user population and number of hops grow [90]. Intermittence and interference amplify the problem, since enabling wireless technologies need to share the same spectrum and ISM (industrial, scientific and medical) frequency bands are by definition subject to interference. In spite of these drawbacks, the percentage of ad hoc networks in cities worldwide accounts for an average 10% of total WLAN deployments [91], reaching 13% in Paris. In addition, the results of field measurements during CeBIT in 2006 (trade show for Telco and IT), counted 291 wireless connections of which 42% were in ad hoc mode [91]. These facts confirm that there is an increased demand for self-management of wireless ad hoc networks. By facilitating easy and efficient deployment of ad hoc networks, one can take advantage of MANET routing protocols and mesh principles to deploy wireless ad hoc networks, on top of which services can be provided. Mobile Ad Hoc Networks (MANET) offer fast and cheap deployment without the need of an existing infrastructure while emerging Mesh technologies attempt to combine the benefits of MANET with the support of wired access points. Managing MANET and wireless ad hoc networks in general is an extremely challenging task. If we depart from cases of special-purpose deployments such as emergency scenarios and military operations, these networks typically consist of heterogeneous devices deployed by their users spontaneously in order to serve a relatively short-term purpose, e.g., file-sharing, online gaming or Internet connection sharing. These devices cannot be fully controlled from a network manager and this fact provides a fruitful ground for self-management solutions. Traditionally, network managers have authority over managed devices (routers, switches), but in ubiquitous wireless networks, users own the managed devices (laptops, PDAs). The increased heterogeneity of devices enlarges relevant problems.

Another issue that needs to be addressed in wireless ad hoc networks is the assured forwarding of packets among the participating nodes. This is one of the basic requirements for any networked application to be deployed over multi-hop ad hoc networks. The duties of fixed routers are carried out by participating wireless nodes and network operation relies on their good intentions to forward the received traffic. This not always the case and often selfish or malicious nodes refuse to forward packets, leading to congestion or, even worse, bringing the network down. Incentives mechanisms have received a lot of research interest; their deployment, though, is limited. Detection mechanisms are also investigated, aiming to determine which nodes are misbehaving and taking appropriate measures against them.

One of the crucial problems of wireless ad hoc networks is the establishment of connectivity without central administration. The basic connectivity settings for devices joining existing WLANs, e.g., public hotspots or home networks, are automatically provisioned by the controlling wireless access point (WAP). Lower levels (PHY/MAC) are automatically configured by the wireless hardware drivers, based on the WAP control packets (beacons). For ad hoc networks, the apparent obstacle is how to establish communication in the absence of a WAP. In general, one of the ad hoc devices assumes the role of a master, acting as a WAP for the rest of the devices. In most cases, initial MAC/PHY configuration is arbitrarily set at the master device by adopting default software driver and/or hardware dependent parameters. Because of the variety of software drivers and hardware-specific implementations, many wireless configuration problems arise during the initial MAC/PHY setup. The use of 'default' settings may work for isolated networks, but in cases of simultaneous network deployments can lead to interference and performance degradation. Imagine a conference venue, where different groups attempt to form ad hoc networks for file exchange, using the default settings. Most likely they will use the same channel (frequency), causing severe interference to each other and throughput reduction.

Beyond framework design, realistic implementations are needed to verify and benchmark frameworks. Simulations have been widely used in academia, e.g., using the ns2 simulator [68], mainly to overcome the obstacle mentioned above and to enable a large scale deployment of networks. Unreservedly, simulations can provide insightful results and indications of problems and bottlenecks. They need however to be used with caution and with careful parameter setup. Research [108] has shown how careless simulation can lead to major errors and false results, as well as how the simulated results differ from reality. They have even demonstrated how different simulator tools provide different results for exactly the same simulation setup. All of these lessons have identified the need for realistic network deployments of frameworks and protocols. In real life, in order to deploy wireless network testbeds, the family of IEEE 802.11 standards [93] is usually considered, since it is the most widely deployed technology. Devices based on 802.11(a,b,g,n) are operating in ISM radio bands and can arbitrarily use any of the defined channels for deployment. The design of appropriate MAC layer algorithms makes these technologies fairly tolerant against interference and noise, but this comes at a price. Speed and performance are sacrificed in order to allow multiple stations to share the same wireless medium, i.e., the available spectrum. CSMA/CA (Carrier Sense Multiple Access with Collision Avoidance) protocols attempt to reduce the collision probability by sensing the wireless channel and backing off if it is sensed busy. The classic problem of hidden terminal is quite common. An additional measure to prevent collisions is used, the RTS/CTS handshake (Request To Send / Clear To Send), but this introduced the exposed terminal problem [94]. The use of Spread Spectrum modulation techniques can cause increased collisions due to interference between different channels (co-channel interference). This happens because channel spacing is overlapping for maximum frequency reuse. Depending on the enabling technology and modulation, different channels are likely to interfere with each other and interference increases the nearer the channels are. For example, 802.11bg technology defines 14 channels in the 2.4GHz ISM band, with center frequency separation of only 5 MHz and overall channel frequency occupation of 22 MHz. Recommended deployments in the Federal Communications Commission (FCC) region use three non-overlapping channels (1,6,11) [95]. All of the above problems provide a challenging application domain for the design of self-management capabilities. In Section 7.4 we provide a thorough solution to these issues.

In order to address the specific problems of wireless ad hoc networks, we need to review existing solutions. We present here the work on the management of mobile ad hoc networks (MANET), which are a special case of wireless ad hoc networks. Related literature in mobile ad hoc networks management is limited and proposed solutions attempt to only partly solve relevant issues. These solutions, however, take into account the requirements of ad hoc communications and are an excellent starting point towards realizing a framework for the self-management of wireless ad hoc networks in general. Existing approaches vary regarding the adopted organizational model. Recently there has been a shift towards PBM systems through a hierarchical approach. Also the proposed deployment of mobile agents amplifies their inherent security implications. The policy-based paradigm [109, 110] offers a promising so-

Table 7.1: Taxonomy of related work on MANET management

	Tiers	Hierarchical	Distributed	Policy-based	Agent-based	Modules	Storage	Managers
A.Hadjiantonis et al. [18]	2	+	+	+	-	2	LDAP	≥1
R.Chadha et al. [17]	2	+	-	+	+	1	mySQL	1
C.Shen et al. [34]	2	+	+	-	+	4	MIB	1
K.Phanse et al. [37]	2	+	-	+	-	2	PIB	1
R.Badonnel et al. [70]	2	+	+	-	-	1	anmpMIB	1
W.Chen et al. [111]	2	+	-	-	-	1	anmpMIB	1

lution since it allows dynamic alteration and controlled programmability of management logic based on the supported policy types.

The first efforts to tackle MANET management were presented in [111]. The suggested Ad hoc network management protocol (ANMP) was based on hierarchical clustering of nodes in a three level architecture. The two proposed clustering algorithms limit severely its applicability due to their centralization. The 'Guerilla' architecture [34] adopts an agent-based two-tier distributed approach where at the higher level 'nomadic managers' make decisions and launch active probes to fulfill management objectives. Mobile agents exploit a utility function to decide their migration and probe deployment. In [17] a PBNM system using intelligent agents is proposed. Policy agents are deployed and manage the network through a two tier hierarchical architecture. Policy definitions follow the principles of IETF but the use of several proprietary protocols (YAP, AMPS, DRCP/DCDP) restricts its wider adoption. Another PBM approach is presented in [112] in order to provide QoS in MANET. The proposed k-hop clustering scheme and extensions to COPS for policy provisioning (COPS-PR) protocol add policy server delegation and redirection capabilities. Although in RFC status, COPS and COPS-PR have found little acceptance and their relatively heavyweight nature may limit their applicability to MANET. Recently, a modified version of ANMP [70] has proposed the probabilistic management of MANET, where a percentage of nodes is guaranteed to be efficiently managed, depending on their connectivity properties. Finally, a context-aware solution for MANET management was introduced in [18], integrating many interesting features. This work introduces a novel organizational model specifically targeted to the needs of MANET by incorporating context awareness to dynamically adapt to the continuously changing conditions. Context information can be used to trigger cross-layer changes (network and application configurations) according to policies, leading to a degree of autonomic decision-making. Table 7.1 presents a taxonomy and summarizes related work in the area.

The context-aware PBM framework introduced in [18] was the first to consider exploiting context information in conjunction with policies for self-management of MANET, hence it has been selected for detailed presentation in this chapter. Further extensions will be introduced later, implementing realistic self-management capabilities for wireless ad hoc networks

7.3   A Framework for the Self-Management of Wireless Networks

Wireless networks pose major research challenges because of their diverse nature and their ubiquity. Their significant differentiation from fixed networks, in terms of requirements and applicability, makes traditional management approaches infeasible and expensive, thus motivating the research efforts presented here.

As networks become more and more complex, it is evident that frameworks with self-management capabilities can significantly expedite and simplify management tasks. Towards this direction we design a self-management framework for wireless networks based on policies and context to realize an adaptive closed feedback loop. The combination of these two concepts, namely policy-based management (PBM) and context-awareness, has made possible the implementation of realistic case studies on wireless testbeds. The proposed architecture combines policy design, specification and distribution with context gathering, processing and dissemination. As it will be explained later, policies and context interact by exchanging information to proactively achieve management tasks. Policies express high-level objectives, guiding the self-management of the wireless networks and providing guidelines as to what action should be executed when certain conditions are met. At the same time, context monitoring achieves a real-time understanding of the network conditions and of the surrounding environment and is used for policy conditions evaluation.

In order to achieve self-management according to higher-level objectives specified as policies, the described process is repetitive, leading to an adaptive closed loop of control. The adaptation loop is initiated with the deployment of uniform high-level policies, which are dynamically translated into management logic and distributed to capable wireless nodes. Policies can drive context gathering, i.e., the monitored context depends on the types of policies deployed, and in turn the gathered context drives policy activation and execution, leading thus to autonomic decision making.

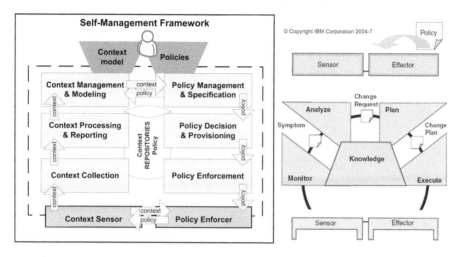

FIGURE 7.6: General diagram of closed-loop management with context and policies

7.3.1 High Level Framework Overview and Design

A policy-based framework can serve as the Plan and Execute components of a self-management system, as described by IBM's K-MAPE blueprint of an autonomic manager [2]. Policy design and specification constitutes the Planning phase of autonomic management while policy enforcement constitute the Execute phase. On the other hand, a context-aware framework is assigned the Monitor and Analyze functionality. Context sensing and collection constitute the Monitoring phase while context aggregation and inference rules constitute the Analyze phase of management. The specification of policies and context together with their interaction form the essential Knowledge element. Policy and context repositories are the Knowledge centerpiece of both frameworks gracefully integrating the presented self-management solution. Figure 7.6 illustrates these concepts, in parallel with IBM's autonomic manager. The large-scale deployment of wireless networks suggests that centralization is not an option and motivates alternative paradigms. The organizational aspects of such networks have been investigated, aiming to provide scalable and robust management. The presence of intermittence, limited capacity and other characteristics inherent to wireless ad hoc networks need to be taken into account. Furthermore, the employment of a policy-based context-aware design affects this design, adding requirements like reduced traffic overhead or anticipation of frequent disconnection and data loss.

Looking at the two extreme cases of organizational models, we have on one hand strictly hierarchical ones and on the other fully distributed ones. Each is better suited to different networks, but for the needs of wireless ad hoc net-

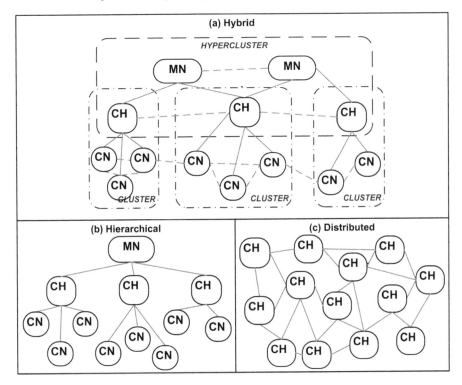

FIGURE 7.7: Organizational models: (a) hybrid, (b) hierarchical, (c) distributed

works, we adopt a hybrid approach. We aim to offer a balance between the strictness of hierarchical models and the fully-fledged freedom of distributed ones. As it will be explained, beyond hybrid deployment, the proposed model embraces both paradigms and can also be deployed as either distributed or hierarchical (Fig. 7.7). The hybrid design is based on a loose two-tier hierarchy by employing node clustering to achieve locality and restrict dissemination of traffic overhead. On top of clusters, a distributed management federation forms the 'hypercluster,' including one or more privileged nodes (managers) with extended capabilities. The multi-manager paradigm and the hyper-cluster formation are two of the distinctive elements of the introduced organizational model.

7.3.2 Policy-Based and Context-Aware Organizational Model

The interaction of policies with context benefits highly from a hybrid organizational model. Traditionally, policy-based management is used on hierarchical models for fixed networks. In such networks, over-provisioning of bandwidth and physical resources eliminates any single points of failure and traffic bottlenecks. Obviously, this solution can not be applied to wireless networks because resources are quite limited. Battery power and bandwidth can be optimized by employing the proposed model. Specifically, wireless networks have an amplified element of locality, which is evident in a wide range of applicability scenarios. For example, ad hoc networks can be formed for a corporate meeting or can be formed from an emergency response unit, reporting to a confined disaster area. Bearing in mind the characteristics of wireless links, unacceptable delays and traffic flooding can be restricted if decision making is performed locally. To achieve that, we need a local control loop, capable of provisioning the network with fast and reliable responses. Ideally, self-management capabilities should accelerate the decision making process. By enabling clustering for management purposes, the element of locality is preserved and the requirements mentioned above are achieved. Hence, a role-based policy design is integrated to the model, e.g., cluster heads are employed in a policy-based hierarchy as local self-managing elements. A cluster head (CH) is responsible for aggregating context from its cluster nodes (CN) and provisioning them with appropriate policy enforcement decisions. Such local clusters can operate autonomously but remain fully aware of network-wide conditions and management decisions. Network-wide awareness is achieved through the collaboration of cluster heads and managers within the hyper-cluster. Based on policies, every CH reports only critical events (context) to other CHs, drastically reducing context dissemination overhead. The aggregated cluster-wide collected context can provide a collaborative network wide view of network conditions. The presence of privileged nodes as manager nodes (MN) allows the review of overall management objectives and the specification of appropriate policies. Policies in turn benefit from the clustered organization model, as they can be selectively applied, e.g., only to cluster heads or only to a specific cluster when needed, further reducing management overhead. Presented case studies will confirm these claims by providing tangible benefits. The introduction of roles and a policy hierarchy motivate the definition of different 'policy enforcement scopes.' Paired with layered context aggregation and dissemination, both architectural elements fully exploit the benefits of a hybrid organizational model. Subsequent sections will further elaborate on policy enforcement scope and layered context aggregation (Section 7.3.6).

Before delving into the policy-based and context-aware details, we first introduce some concepts on role-based management and cluster formation. Clustering is widely used in ad hoc networks for the reasons already explained. Roles are naturally introduced, to cope among others with the complexities of cluster creation and maintenance. For the purpose of our design, we rely on three roles, namely Manager Nodes (MN), Cluster Heads (CH) and Cluster Nodes (CN). These roles have been traditionally used in network layer clustering schemes for MANET routing. However, in the presented work, clustering is used at the application layer for management purposes and each role is associated and guided by special policies. In addition, the introduced hypercluster is formed to distribute and load-balance management tasks among resource-constraint wireless nodes. The hypercluster consists of CHs and MNs, emerging as an overlay above clusters. It can execute distributed algorithms for its own maintenance (e.g., reformation, reaction to node disconnection, etc) and eliminates the single point of failure of a strict hierarchy. A wide range of algorithms can be used for cluster formation and maintenance [126, 127], depending on the applicability scenario and the network composition. For example, ad hoc deployment for tactical operations has quite different requirements than user-initiated social networks. The flexibility of a policy-based design allows the integration of different clustering schemes and in addition the real-time parameterization of their operation. Based on previous work we refer the reader to [18, 96], for a complete solution based on the Dominating Set algorithm. In brief, each node executes a distributed algorithm to assign a role to each device and to select the most capable ones to form the hypercluster. These nodes create the dominating set (DS) of the graph of capable nodes, thus ensuring one-hop accessibility for the remaining nodes. The idea is borrowed from backbone overlay networks used for routing in MANETs where the use of DS is prominent [116, 117, 118]. The novelty proposed is the exploitation of a context-aware capability function as an optimization heuristic. In [18, 96] full details and implementation specifics of such a context-aware clustering scheme are provided.

Returning to node roles, we look into the internal components necessary for each role. A highly modular design architecture is used that takes into account the heterogeneity of wireless networks and their wide applicability range. Each role has an increasingly more complex structure and added functionalities as shown in Fig. 7.8. Their components are separated in policy-based and context aware ones. The typical PBM components are used, i.e., the Policy Management Tool (PMT), the Policy Decision Point (PDP), the Policy Enforcement Point (PEP) as well as a special version of the Policy Repository (PR), the Distributed PR (DPR). Aiming to form a closed feedback loop, we complement the above components with their context aware ones. Hence we introduce respectively the Context Management Tool (CMT), the Context Decision Point (CDP), the Context Collection Point (CCP) and the Context Repository (CR). The major design difference is that the flow of information is reversed to the one in PBM systems, where a top-down approach is adopted. Here, context is collected and processed at the lower layers of the architecture and is passed to the higher layers for management decisions to be taken. At each level, respective components interact, aiming to achieve self-management and autonomic decision making. How this is achieved is detailed in subsequent sections and case studies.

Looking at a bottom up network composition, we present roles from the simplest one to the most complex. It should be noted that simpler nodes employ a component subset of the more complex ones, aiming for increased modularity and functionality reuse. This decision is mainly motivated by the

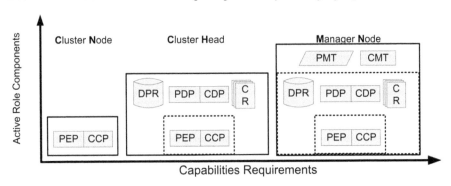

FIGURE 7.8: Block diagram of each role and internal components

need for simplified management interfaces and uniform software design. A device in Cluster Node (CN) role only uses a PEP and a CCP. Being at the lower tier of the hierarchy, such nodes participate in a single cluster and are responsible for reporting to their CH. In addition to its own local PEP and CCP, a device in Cluster Head (CH) role additionally employs a CDP and a PDP. The latter components are responsible for managing the devices belonging to their cluster, on one hand by collecting relevant context information and on the other by provisioning them with appropriate enforcement decisions. To assist decision making, a CH may locally use DPR and CR components. However, the activation of these components is also policy-based to preserve resources. Effectively a CH controls a limited number of CN that form its cluster. Finally, to enable uniform management, a role with fully-fledged policy-based and context-aware capabilities is required, hence the need for a Manager Node (MN). The extra components employed are the PMT and CMT. They offer a management interface to the human network manager, where high-level policies can be defined and altered to achieve business objectives. Depending on the applicability scenario, the assignment and responsibilities of MN can vary significantly. Figure 7.9 presents a high-level view of an example deployment using the proposed hybrid organizational model. The interaction of roles and internal components can be seen as well as the complementary policies and context information flows. Traditionally, one logical MN node is employed for network management, strictly specifying through policies the behavior of managed devices, e.g., routers, firewalls, etc. But in the examined case of wireless networks, we allow for more than one logical manager to coexist (Fig. 7.9) to cater for multi-manager scenarios presented later. Having more than one manager gives the flexibility to form networks between distinct trusted administrative authorities. This is performed without any of these being forced to forfeit its management privileges. Instead managers cooperatively introduce policies which guide the overall network's behavior. For example, an ad hoc network can be setup for a corporate meeting be-

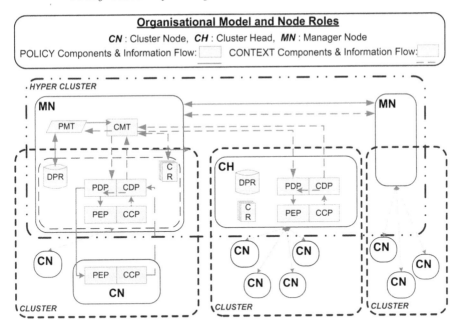

FIGURE 7.9: Hybrid organizational model with internal components and information flow

tween two companies' representatives. The multi-manager paradigm treats the companies' managers as equals and allows both to affect network behavior by introducing policies. In addition, from a functional point of view, in large scale ad hoc networks scalability issues demand more than one manager in order to control and administer effectively the numerous cluster heads and cluster nodes. The managed devices in such scenarios are user-owned devices, like PDAs, media players, etc. These devices are not strictly managed and can benefit from a multi-manager paradigm, e.g., for service provisioning.

Before elaborating on the specifics of the Policy-based and context-aware aspects of the introduced organizational model, we present a detailed schematic representation of internal components in Fig. 7.10. Components are presented regardless of mentioned roles, to better illustrate their interactions.

7.3.3 Policy-Based Design for Autonomic Decision Making

Policies can serve as the Plan and Execute functionality of a self-management system, as described by IBM's well-known MAPE blueprint of an autonomic manager [2]. Policy design and specification constitute the Planning phase of autonomic management while policy enforcement constitutes the Execute phase.

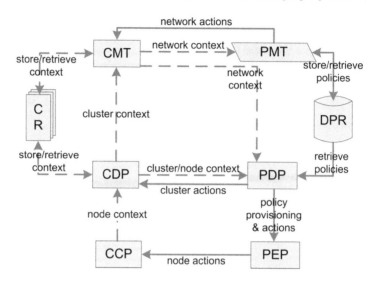

FIGURE 7.10: Policy-based and context-aware components and interactions

Wireless networks and particularly ad hoc ones are difficult to accurately plan, mainly because the majority of the participating devices are mobile users. The unpredictable nature of human users hinders the efforts to provision the network and provide reliable services to them. This is an important point that the presented design attempts to exploit for the benefit of management performance. The main concept is to Plan using policies to dynamically adapt the management capacity of the network according to the users' population and contextual information. Management capacity refers to the distributed use of processing power and resources from user devices for collaboratively serving the wireless network. For example, consider the case of planning the deployment of a temporary wireless network for the needs of a large conference. Instead of over-provisioning the area with several wireless access points, a dynamic number of user devices can be employed as cluster heads and use multi-hop routing protocols to achieve connectivity. In practice, the over-provisioning solution is preferred because of its predictable configuration and administrative simplicity, in spite of its higher cost. In order to entrust the management of a network to user devices, there is a need to infuse predictability and simplicity. To this purpose, policy-based management is employed to provide the means for controlled programmability and self-management capabilities.

The designed policy-based framework provides a highly distributed management environment that can cater for the self-management of wireless ad hoc networks. Management logic is encapsulated in policies that are transparently enforced to devices. For example, Network Operators and Service

Providers use the multi-manager PBM system to introduce the appropriate policies aiming to set guidelines for the management of numerous user devices. Contrary to traditional management systems, the designed system may not require the mandatory enforcement of policies and tight control of managed devices. Instead, the system physically and logically distributes the policies among devices, making them available to vast numbers of users that choose to enforce the relevant policies eventually relieving them from manual configuration. Policy enforcement is the Execute engine of a self-managing system.

In subsequent sections we introduce a sophisticated context-aware model to Monitor and selectively Analyze critical contextual information, complementing the PBM design and providing a real-time understanding of network conditions without adding significant overhead (Section 7.3.4). Context and policies are the essential Knowledge, necessary to every self-management system. Knowledge management depends on accurate context modeling and a Distributed Policy Repository (DPR). The DPR is designed as an extension to the traditional Policy Repository and responsible for the distribution of policies among the network (Section 7.3.5).

Policy Design, Specification and Enforcement
In order to apply the PBM paradigm to our system we adopt the standardized by IETF/DMTF information model for policy representation. Defined policies are represented according to PCIM/PCIMe [121, 122]. To overcome the lack of an event notation in PCIM, we extend the abstract Policy element as PolicyEvent, without loss of interoperability. Based on this representation, mapping policies to the standardized LDAP data model [123, 124, 125] is straightforward. The focus here is placed on the definition of the necessary policies for wireless network management, rather than the formal definition of a policy language. We believe this representation is both effective and lightweight so as to cater for the policy needs in the resource-poor wireless ad hoc environment:

```
{Roles} [Event] if {Conditions} then {Actions}
```

The Event-Condition-Action notion (ECA) is widely used in the literature due to its simplicity and effectiveness. It allows complex policy structures to be formed (e.g., policy groups), as well as increased reuse of both policy Conditions and Actions. Roles element defines which devices will need to apply the specific policy. It also helps grouping policies and easily retrieving them. Event element triggers the evaluation of policy conditions. It can be a periodic, time-based or scheduled event, as well as dynamic real-time event or event correlation. Depending on system's capabilities and complexity, a sophisticated event bus and correlation engine can be implemented. For the purpose of the presented design, a context-aware event service is used. The Conditions element is a Boolean expression containing one or more conditions to be evaluated. If the condition is true, that would trigger the execution of specified actions. The Actions element contains one or more actions needed to be enforced once a specific event has occurred and policy conditions are true.

As already mentioned, to cater for the needs of policy-based design, the four components proposed by IETF are employed and modified. We further elaborate on the internal structure and implementation guidelines of these components, aiming to provide system engineers with an insightful presentation of how they can be employed for the management of wireless networks. In the following section, policy examples will further contribute to this goal. Taking a bottom-up approach we begin with the PEP, followed by PDP and finally PMT. We recommend consulting Fig. 7.10 (pp. 230) for better understanding of each component. The DPR is presented later in a special section (Section 7.3.5):

7.3.3.1 Policy Enforcement Point (PEP)

PEP is the simplest component of a PBM system, responsible for the enforcement of policy decisions on the Managed Objects (MO) it carries, as these decisions are provisioned by the controlling PDP. Depending on the management interface, a PEP may need to be tightly integrated with its host device as it may directly communicate with hardware elements. Contrary to fixed network managed devices, where an MIB agent is normally available, managed devices in wireless networks rarely integrate one; hence a device-dependent middleware layer may be needed to set required configuration parameters and monitor available objects. There is an apparent need for accurate description and interfacing with available MO and, to this end, various information models can be used and/or extended. The presented design extends the PCIM/PCIMe information model to define the actions PEPs need to enforce. Providing PEPs with instructions to what actions need to be enforced is referred to as policy provisioning and it is the responsibility of PDPs.

7.3.3.2 Policy Decision Point (PDP)

PDP is the component responsible for ensuring that policies are applied to all the PEPs it has under its control. It is responsible for evaluating the condition policies and provisioning the required actions to PEPs (policy provisioning). IETF has defined COPS/COPS-PR protocols for this purpose, although in practice they are rarely used. More lightweight approaches are adopted in wireless networks, like XML-RPC, RMI, etc. PDPs also communicate with a Policy Repository to retrieve current policies and instantiate them as Policy Objects (PO). The communication protocol between PDP and PR is primarily LDAP, although recently approaches based on Web Services are explored.

7.3.3.3   Policy Management Tool (PMT)

PMT is the interface between the PBM system and a human manager. Using the PMT a manager can specify high-level directives and management objectives, usually in a simplified graphical user interface (GUI). PMT translates the high-level directives to the internal policy language specification to validate their consistency and feasibility. In addition, it may perform policy analysis, aiming to identify possible conflicts. Conflict detection and resolution (CDR) is a critical task, receiving intense research interest. We elaborate on such issues in the next paragraph. If conflicts exist, a resolution process is initiated that can be either automated or may require human intervention. Once policies are checked, PMT communicates with the Policy Repository to store the new policies, normally over LDAP. A similar procedure is followed on policy modification.

Policy-based management simplifies the complex management tasks of large scale networks, by using policies to automatically enforce appropriate actions in the system. But in an environment where a number of policies need to coexist, there is always the likelihood that several policies will be in conflict, either because of a specification error or because of application-specific constraints. It is therefore important to provide the means of detecting conflicts in the policy specification. Considering the different conflict types, it is possible to define rules that can be used to recognize conflicting situations in the policy specification. These rules usually come in the form of logic predicates and encapsulate application-specific data and/or policy information as constraints. Examples on how these rules can be used as part of a detection process can be found in [104, 105]. By adopting a multi-manager scheme we allow more entities to offer different services to the users, without violating their privacy concerns and preferences. This creates an increased need for an automated conflict detection and resolution mechanism that will prevent policy inconsistencies among different managers and allow for truly self-managed systems to be realized.

7.3.4   Context-Aware Platform for Information Collection and Modeling

Realizing a truly self-managing system requires an integrated Monitor and Analyze functionality to complement Planning and Execute. It has been shown that policy design and specification constitute the Planning phase of autonomic management while policy enforcement constitute the Execute phase. In this section, we introduce a context-aware framework assigned to the Monitor and Analyze functionality of the designed autonomic management system. Context sensing and collection constitutes the Monitoring phase while context aggregation and inference rules constitute the Analyze phase of management.

Based on the above, the IETF policy framework is extended by complementing the policy-related components with a novel group of entities related to context collection and processing. These are necessary for a system to become capable of sensing, communicating with its surrounding environment and adapting to changing conditions. Incorporating context awareness into the policy-based management framework makes it flexible and dynamic in response to the inherently unstable domain wireless ad hoc networks, allowing a degree of autonomy to be reached.

Context information collected from all the nodes forming the MANET refers to their computational and physical environment and is tightly coupled with the policy-based management system since it is this information being monitored that may trigger a certain policy. Every node collects its own context information based on its available sensors. The term sensor is generic since it can refer to a battery monitor, a GPS receiver, etc. To increase performance and scalability, a context model is needed to represent the collected information efficiently and accurately. Based on these requirements we employ a context model based on Unified Modeling Language (UML) design principles, full details of which can be found in [96]. For clarity, we briefly introduce its main features. The model incorporates the notion of semantics to describe context information, sensors and their relationships. The general context of a node consists of higher level contexts that have been deduced from simpler ones, i.e., mobility prediction of nodes deduced from device capabilities, GPS readings, personal diaries, etc. For this purpose, relationships are defined in terms of simple inference rules or mathematical functions. The UML model is inherently associated with the data representation of the collected context. It allows for expressiveness and can be easily mapped to an XML document for interoperable storage. In addition it allows for integrity validation using well-established XML techniques (e.g., XML Schema).

The presented system involves four novel components that deal with context awareness: the Context Collection Point (CCP), the Context Decision Point (CDP), the Context Management Tool (CMT) and the Context Repository (CR). There is an obvious matching of these components to the four elemental components of a PBM system. The major design difference is that the flow of information is reverse of the one in PBM systems, where a top-down approach is adopted. Here, context is collected and processed at the lower layers of the architecture and is passed to the higher layers for management decisions to be taken. For further details on the used context model, the reader is referred to [96]. A detailed case study is also described in [96], where the presented components are customized for the management of MANET. In the following paragraphs, a more generic approach is followed to allow the use of these components and the framework in various wireless scenarios and case studies. As in the presentation of policy-based components, we recommend consulting Fig. 7.10 (page 230) for better understanding (CR is presented later in a special section, Section 7.3.5):

7.3.4.1   Context Collection Point (CCP)

The CCP is installed on every node and its responsibilities include monitoring the environment through the available sensors and collecting the relevant information. Sensors may include a power monitor, a GPS receiver, a Bluetooth monitor, etc. CCP's operation is guided by the enforced action of its collocated PEP. These actions include the definition of context objects to report and monitor configuration of context collection/reporting intervals and various parameters related to the local management of context within a node. A local context manager coordinates the operation of CCP by interacting with the context aggregator/optimizer. After some basic pre-processing, context information is stored in context objects using the available context model. Only the requested information is forwarded from every CCP to their respective Cluster Head's CDP.

7.3.4.2   Context Decision Point (CDP)

The CDP is installed on devices in CH or MN roles, i.e., on devices currently consisting of the hypercluster. Its main responsibility is to extract context information from the controlled cluster: (1) for feedback to the collocated PDP and (2) for forwarding aggregated cluster context to other hypercluster nodes. Like with the PEP/CCP interface, the local PDP configures the collocated CDP, which in turn reports the requested context. The Cluster Context Aggregator/Optimizer communicates with all CCPs belonging to the same cluster through the Cluster PEP/CCP Communication Adaptor to extract the aggregated cluster-wide context. It can use specified inference rules of the context model to infer and combine simple node contexts to more complex cluster-wide ones. Context objects maintain an updated view of current cluster conditions. The local PDP evaluates conditions of cluster-wide Policy Objects with received context, to check whether the actions of a certain policy are triggered for that cluster. Specific aggregated context information from each cluster is also forwarded to their respective MNs, to allow them to acquire a network-wide view of context and conditions.

7.3.4.3   Context Management Tool (CMT)

The CMT is available only at nodes in the MN role and interacts with a collocated PMT to provide a graphical interface for a human manager or administrator. Similarly to CDP, it adds another level in the context hierarchy, the network-wide level. Each CMT collects and aggregates cluster-wide context information from its controlled CHs. This context is processed and exchanged among the other CMTs of MNs. This ensures a network-wide common knowledge regarding context information. It is this context information that is returned from the CMT to the hypercluster PDPs and may trigger appropriate network-wide or hypercluster-wide policy actions. The above ensure network-wide concurrent triggering of policies and thus network-wide adaptation when required.

7.3.5   Distributed Policy and Context Repositories — The Importance of Knowledge Management

Policy and context repositories are the Knowledge centerpiece of the presented framework, gracefully integrating the self-management solution. Knowledge is important at every phase of a self-management system and needs to accurately depict managed resources, management objectives, network conditions, contextual information and devices' status. All of these elements need to be acquired through the network in a consistent, efficient and scalable manner. The specification of policies and context, together with their interaction, form the essential Knowledge element. Policies encapsulate high level directives as well as low level actions to achieve management objectives. Context modeling on the other hand provides a layered view of network conditions by collecting and combining simpler context to complex ones. Both policies and context need to be available among wireless nodes when needed, hence the critical need for Policy and Context Repositories.

The policy repository (PR) is a critical component in every PBM system and we cannot rely on a single node to store it. The idea of storage replication is not new and is widely used in fixed networks as a backup in case of failures. In wireless networks however due to the intermittent nature of links, it is expected that nodes will become disconnected frequently. Thus access to a central repository cannot be guaranteed depending on the networks' volatility and mobility. In order to tackle this deficiency the DPR (Distributed Policy Repository) has been proposed. The key point is to ensure uniform network management. This can be achieved by the presence of a synchronized Distributed Policy Repository among network nodes, which in turn guarantees that all PDPs behave in the same way and enforce the same actions in a network-wide fashion. Because of the unified manner aggregated context is presented to the MNs, the conditions evaluation is the same, ensuring robustness and consistency. The DPR concept anticipates the need for provisioning large-scale wireless ad hoc networks, without the need for over-provisioning management resources, e.g access points, bandwidth or human effort. Because the deployment of such networks varies significantly in terms of spatial and temporal parameters, accurate planning and pre-provisioning is extremely difficult. Hence the proposed distribution of management tasks among Policy Decision Points (PDP) hosted on user devices, based on the policy guidelines stored in the DPR.

The innovation in designing a novel DPR lies in the adoption and customization of features from existing fixed network repositories for use in a wireless environment. When designing a Policy Repository for the policy-based management of wireless networks, there are additional requirements to be taken into account:

- PDPs may be intermittently connected to the ad hoc network and occasionally may not have a route to a Policy Repository instance.

- The nearest PR instance may be several hops away from PDPs, thus introducing significant traffic and latency overhead to the propagation of new or updated policies.

- Multi-hop networks suffer from severe bandwidth degradation as the number of hops increases.

- Additional PDPs may need to be dynamically assigned to anticipate fluctuation in PEPs population. Special conditions may lead to spatiotemporal increase of PEPs density. Current repositories do not consider such conditions.

- Wireless networks increasingly consist of heterogeneous end-user devices that cannot be fully controlled by a network manager.

These requirements prevent the unmodified adoption and deployment of a Policy Repository (PR) using the various techniques targeting fixed networks and have motivated research efforts for an enhanced PR, the Distributed Policy Repository (DPR).

The presented DPR idea was introduced in [18] where different replication states were enforced depending on network's mobility. Details of this proposal are mentioned as a policy example in Section 7.3.6. According to [18], the DPR can have different degrees of replication (e.g., number of replicas) according to how volatile the network is. In [18], the DPR concept is extended to combine a priori knowledge of localized events (e.g., scheduled sport event) with dynamic real-time context information (e.g., processing load or free memory of each PDP). A series of algorithms can be included in the implementation of policy actions, resulting in a highly customizable deployment of the DPR overlay. In effect policies and context guide the DPR behavior and replicas' distribution, ensuring on one hand maximum repository availability (distributed copies) and on the other hand a single logical view of the stored policies (replicated content). Thus, efficient management of clusters can be achieved even when temporarily disconnected from the network manager. The immense importance of Knowledge management has further motivated the integration of a Context Repository (CR) to the presented framework. It can gracefully integrate with the Distributed Policy Repository to effectively create a complete Knowledge management solution for autonomic management of wireless ad hoc networks.

There are a number of significant differences in the design of a Context Repository compared to the design of a Policy Repository:

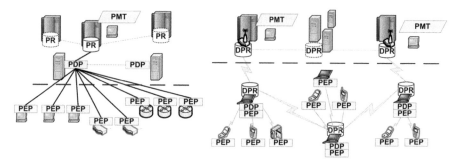

FIGURE 7.11: Traditional (left) and proposed (right) policy repository deployment

- Context has a localized and temporal importance and as such it is not necessary to widely distribute and rigorously require consistency as in the case of policy storage and distribution.

- Context is archived and outdated far more quickly than policies; hence its freshness and accuracy pose different requirements.

- Context at different hierarchy levels can be stored locally and be used independently by each node or cluster of nodes. Only context needed for the inference of network-wide context is required to be distributed.

- Network-wide context is important and needs to be consistently stored and distributed among hypercluster nodes. This is the main motivation for a Context Repository, beyond the local context storage.

A Context Repository instance can be deployed on the nodes constituting the hypercluster. This allows them to maintain a wider view of network conditions and specifically be notified of context information needed for the evaluation of network-wide and hypercluster-wide policy conditions. Contrary to the standardized IETF/DMTF use of LDAP for policy storage, the implementation of context storage is open. Storage of context in the CR is tightly dependent on the employed Context model. For example, use of the aforementioned UML-based context model would imply the use of XML format for context representation and storage.

7.3.6 Context and Policies Interaction for Closed-Loop Autonomic Management

The designed framework is further enhanced by integrating a closed control loop, as described in this Section. Having in mind the component's structure presented earlier (Fig. 7.10, page 230), we introduce the concept of 'enforcement scope.'

To achieve layered closed-loop autonomic control, we define different decision layers to limit context dissemination and control traffic overheads. We define the 'enforcement scope' of a policy as the set of nodes where actions need to be enforced, when the policy is triggered by the context collected in this set. Figure 7.12 illustrates the realization of layered closed-loop autonomic control through component interaction. Based on the above definition, three enforcement scopes are realized for the needs of our design:

7.3.6.1 Cluster-wide Enforcement Scope

Policies can be triggered at a hypercluster node by the context aggregated within its cluster. Decisions are enforced only at the cluster nodes belonging to the cluster where the policy was triggered. These policies are identified by their assignment to the CN and CH role.

7.3.6.2 Hypercluster-wide Enforcement Scope

Policies can be triggered at all hypercluster nodes by the context aggregated within the hypercluster. Decisions are enforced only at the hypercluster nodes. These policies are identified by their assignment to MN and CH node roles only.

7.3.6.3 Network-wide Enforcement Scope

Policies can be triggered at the MNs by the context collected and aggregated from all network nodes. The PDPs of MNs decide to enforce the actions network-wide and delegate those actions to CHs to be enforced to all PEPs of wireless nodes. These policies are identified by their assignment to all three roles (MN, CH, CN).

Regarding the actual policy design, in the following section we present realistic examples of policy types in order to illustrate these concepts. These policies provide a first step towards a flexible and adaptable management framework specifically designed for the needs of a wireless network.

7.3.7 Overview of Applicability and Policy Examples

To demonstrate the effectiveness and combined applicability of policies and context, in this subsection we present illustrative examples, based on a case study for the management of Mobile Ad Hoc Networks (MANET). MANET are a representative example of wireless ad hoc networks and we have already presented the motivation for their efficient management. Through these examples we aim to demonstrate the effectiveness of simple policy rules and the applicability of the defined policy enforcement scope. The chosen policies are not overly complex for clarity and serve as an introduction to policy design. In spite of their simplicity, they provide powerful tools for MANET self-management since they can proactively take measures to prevent the depletion

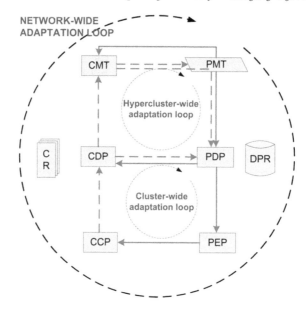

FIGURE 7.12: Diagram of closed-loop adaptation at different levels

of resources (energy conservation example) and degradation of performance (repository replication and routing adaptation examples). In Section 7.4, we present more complicated policy design in the context of various case studies of wireless networks.

7.3.7.1   Energy Conservation Policy with Cluster-wide Enforcement Scope

A major issue in MANET is the conservation of device resources. We tackle this by introducing a policy type that adaptively configures energy consumption according to their current state and environment as well as the overall management objectives:

```
{CN}[T] if {BP=(n..m)} then {TransPow:=k}
```

This policy type is used to manage effectively the device resources by influencing relevant configuration parameters. The Battery Power (BP) context is used here to affect the node's transmission power (TransPow). In implementation k =1,2, where 1=Normal Power and 2=Low Power, so two policies are implemented:

```
{CN}[bp_event] if {BP=(00..33]} then {TransPow:= 2:Low}
{CN}[bp_event] if {BP=(33..99]} then {TransPow:= 1:Normal}
```

The idea is to use a threshold average battery level in order to reduce transmission power and conserve remaining battery power. Policies of this type only need cluster-wide context knowledge since their enforcement is independent among clusters. The PDP of every CH receives context information for the registered variables and enforces the actions to the PEP of the cluster CN. Periodic receipt of individual BP context subsequently generates periodic bp-event, causing the evaluation of the two conditions and triggering of respective actions. In these cases, context information is withheld within the cluster, thus reducing overall traffic load and processing resources.

The effect of this policy is battery power conservation, since one of the main energy consumers of mobile devices is actually their wireless transceiver. This policy is better suited for dense network deployments to avoid node disconnection with their CH. The reduction of transmission power causes a reduction of transmission range that may result in one way link breaks from CN to CH as well as two-way link breaks between CN. To anticipate potential disconnection, additional conditions may be added to the policy, depending on network deployment parameters.

7.3.7.2 Repository Replication Policy with Hypercluster-wide Enforcement Scope

The need for repository replication has already been explained in Section 7.3.5. For this purpose we model a policy type to guide the replication degree of the Distributed Policy Repository. A manager node has the ability to dynamically define the behavior and the replication degree of the DPR by introducing related policies on the fly and without disrupting system's operation or the DPR component:

```
{MN,CH}[T] if {FM=(n..m)} then {ReplDegState:=k}
```

The above policy type is used to guide the replication degree (ReplDegState) of the Distributed PR (DPR) component. The fluidity metric (FM) is a hypercluster-wide aggregated context that represents how volatile the network is. Three states of replication are implemented, namely k=1:Single, k=2:Selective and k=3 Full. These states reflect the need for PR replicas within the hypercluster nodes and adapt according to the volatility of the MANET (Fig. 7.13). As mentioned earlier, the idea is to increase the DPR replication degree when network fluidity increases, hence the three policies below:

```
{MN,CH}[fm_event] if {FM=[00..25)} then {ReplDegState:=1:Single}
{MN,CH}[fm_event] if {FM=[25..70)} then {ReplDegState:=2:Select}
{MN,CH}[fm_event] if {FM=[70..99)} then {ReplDegState:=3:Full}
```

Based on the collected hypercluster-wide information (in this case the FM), the CDP of each CH informs the collocated PDP and policies of this type

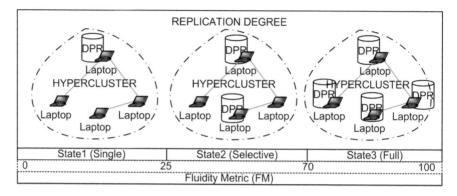

FIGURE 7.13: Replication degrees depending on network fluidity

may be triggered for hypercluster-wide enforcement. Once triggered, their respective actions are enforced only to PEP of hypercluster nodes. The formed adaptation loop ensures the correct replication state is enforced depending on perceived network fluidity among hypercluster nodes. The importance of a reliable and robust DPR has motivated our decision for further analysis of its Self-configuration. Based on the described concepts, an implementation case study is presented in Section 7.4.2, dealing with actual policy provisioning and enforcement. By exploiting LDAP synchronization features, the DPR concept is extended to combine a priori knowledge of localized events (e.g., scheduled sport event) with dynamic real-time context information (e.g., processing load or free memory of each PDP). Thus, a highly customizable deployment of a DPR overlay can be formed.

Based on the above concepts, an implementation case study is presented in Section 7.4. By exploiting LDAP synchronization features, the DPR concept is extended to combine a priori knowledge of localized events (e.g., scheduled sport event) with dynamic real-time context information (e.g., processing load or free memory of each PDP) and a highly customizable deployment of the DPR overlay can be formed.

7.3.7.3 Routing Adaptation Policy with Network-wide Enforcement Scope

A plethora of protocols has been proposed to solve the multihop routing problem in MANETs. A generic classification can distinguish them into proactive and reactive regarding the strategy used to establish routes between nodes. Based on the above, we model a policy type which would enable dynamic on the fly adaptation of the routing protocol. Network conditions/context on one hand and manager defined goals on the other, can both be expressed by this type of policy which effectively alters routing strategy and increases network performance:

```
{MN,CH,TN}[T] if {RM=(n..m)} then {RoutProt:=k}
```

The above policy type is used to adapt network behavior by switching the routing protocol (RoutProt) according to the network's relative mobility (RM). RM is aggregated context information extracted from the network-wide knowledge of node movements, e.g., GPS positioning data, mobility ratio or other context. The simple condition monitors if RM value lies within the range (n..m) in order to enforce the associated action that activates the appropriate routing protocol. For implementation, the idea is to use a proactive routing protocol (OLSR, k=1) when relative mobility is low and a reactive (AODV, k=2) when high. Depending on management goals and on network-wide aggregated context, compound conditions and actions can be introduced in all the proposed policy types, in order to take more parameters into account. Two policies can enforce the described management goals:

```
{MN,CH,TN}[rm_event] if {RM=[00..35)} then {RoutProt:=1:OLSR }
{MN,CH,TN}[rm_event] if {RM=[35..99]} then {RoutProt:=2:AODV}
```

The network-wide enforcement scope of this policy implies that the condition variables used (e.g., RM) should have an aggregated network-wide value. The RM value for example is extracted from the gradual aggregation and processing of simpler node context (e.g., speed) to cluster context and eventually network context. Cluster context is collected at the Context Management Tool (CMT) components of the Manager Nodes and this allows them to compose the network-wide context variables. This higher level context information will drive the triggering of actions that should be enforced globally. Each CMT forwards this value to the local PDP and to the PDPs of all CHs they control. Each PDP enforces the triggered action to all cluster nodes, including itself, and reports successful execution to their MNs. These actions ensure the smooth and controlled execution of network-wide adaptation, in a self-managing manner.

7.4 Implementation and Evaluation of Self-Management Capabilities

Research on autonomic systems has been intense during the past years, aiming to embed highly desirable self-managing properties to existing and future networks. In previous Sections we have thoroughly examined the necessary components to build a scalable framework for managing wireless ad hoc networks, integrating the notion of IBM's autonomic manager to form a

Knowledge-based loop to Monitor, Analyze, Plan and Execute (K-MAPE). In this Section, we elaborate how these components can be used to realize and implement self-* properties:

- Self-Configuration

- Self-Healing

- Self-Optimization

- Self-Protection

A complete self-management solution is not available to the best of our knowledge. Instead, researchers have attempted to partially tackle self-management by implementing some of the desired properties. In the following sections we present attempts to integrate self-* capabilities to the aforementioned management framework. The solutions demonstrate how the framework with the interaction of policies and context forms the basis for implementation and evaluation of self-management solutions. Detailed policy specifications are presented and their deployment on real wireless ad hoc testbeds demonstrates important quantifiable results and experiences.

7.4.1   Self-Configuration and Self-Optimization in Wireless Ad Hoc Networks

The presented case study deals with the dynamic configuration of communication channels in a wireless ad hoc network based on IEEE 802.11. The solution first addresses the Self-Configuration of ad hoc network deployment by initiating communications using the best available wireless channel. The second issue addressed is the Self-Optimization of ad hoc wireless communications by evaluating wireless channel conditions and dynamically switching to a new optimal channel.

Currently, in dense deployments of WLANs (e.g., conferences, stadiums) users manually initiate ad hoc networks without relying on any infrastructure support. This results in poor performance and interference problems among WLANs, even regulatory violations in some cases. The deployment of ad hoc networks and their coexistence with managed WLANs has not received enough research interest, since in most cases the assumption is that an interference free area is available and all ad hoc stations communicate using the same channel. This assumption allowed research to focus on inter-station interference and MAC layer performance, yielding useful conclusions. On the other hand, industrial interest has been limited, mainly due to the lack of a compelling business model.

The described case study of wireless ad hoc networks is suitable to fully exploit the benefits of the aforementioned policy-based context-aware framework. We design the policies and algorithms necessary for the deployment of such networks and evaluate their performance and applicability through testbed implementation. By making appropriate policies available in the Distributed Policy Repository, user devices are assisted by receiving guidelines that transparently configure the ad hoc network, choosing the best available wireless channel to avoid interference and dynamically switching channels if performance degrades. This is an important first step towards the implementation of fully self-managing systems, since the presented solution effectively addresses the Self-Configuration and Self-Optimization need of channel assignment in wireless ad hoc networks.

We argue that by facilitating a predictable and controlled ad hoc network deployment, the performance of both managed WLAN and ad hoc networks can be significantly improved. The solution can be deployed on top of existing and future access networks using a technology-independent policy-based and context-aware management layer. The approach spans among different architecture layers of the protocol stack, exploiting context and cross layer principles but at the same time preserving the layers modularity. This paradigm was deemed necessary, since the applicability domain of ad hoc networks is based on a majority of off-the-shelf end-user devices and only a few special purpose devices, e.g., mesh routers or programmable access points. In addition, standards conformance is important for any solution to be applicable.

Inter-layer communication is used between MAC and Application layers, aiming to make the PBM system aware of the wireless channel conditions. This specialized context collection method provides a feedback mechanism for policies. Based on specified application events (e.g., reduced goodput), the triggered policies can initiate relevant procedures that with the inspection of MAC layer headers provide feedback to the system and possibly trigger further policies to correct the problem or report unresolved issues to the user or the network manager. As already explained in Section 7.3.6, a closed control loop is formed that adds a degree of self-management to the network. There are two important advantages with the adoption of this approach:

1. By using a policy-based design, the system is highly extensible and easily configurable. Policies can change dynamically and independently of the underlying technology.

2. By implementing decision logic at the Application layer, based on policies and inter-layer context extracted from lower layers, modularity is preserved without modifying the MAC protocol.

As mentioned already, today the deployment of ad hoc networks is becoming a popular and convenient solution for quick network setup and spontaneous or opportunistic networking. Unfortunately, user experiences have been disappointing, mostly because of difficulties in setup and poor performance. We have identified two potential obstacles that need to be overcome in order to make the deployment of ad hoc networks easy, efficient and safe:

1. interference between newly created ad hoc networks and existing WLAN

2. regulatory conformance of ad hoc networks deployment.

End-users have no need to be aware of channels and regulations, as long as they are connecting to infrastructure-based WLAN, regardless of their geographic area. In managed WLAN, devices connect to infrastructure-based wireless Access Points and automatically adjust to the correct channel, thus reducing the probability of misconfiguration. The problems described are bound to ad hoc networks, since it is up to the initiating device to select a channel for deployment. In addition, it would be useful to ensure that roaming users are conforming to regional regulations with minimal inconvenience. We attempt to propose solutions to the above problems based on the designed policies for a policy-based management system (Table 7.2).

1. Interference between ad hoc and WLAN networks

Interference between deployed ad hoc networks and existing infrastructure-based WLAN, as well as interference with already deployed ad hoc networks in the same area is the main reason for the disappointing performance of ad hoc networks and it can lead to severe problems in the throughput and coverage of collocated infrastructure-based WLAN. As already mentioned, devices operating in ISM bands can arbitrarily use any of the defined channels and should be able to cope with interference from devices competing to access the same unlicensed bands. The MAC layer can be fairly tolerant against interference and noise at the cost of speed and performance. Choosing a random channel is likely to have a detriment effect for the ad hoc network performance. The above problem has been verified by testbed measurements. To tackle this problem, we design policies P1 to P8 (Table 7.2) that exploit context extracted from MAC layer information, first for the initial configuration and secondly for the dynamic adaptation of the deployed wireless channel

2. Regulatory conformance of ad hoc networks deployment

Although this issue is rarely addressed, it is indirectly affecting the popularity and usability of ad hoc networks. Users attempting to deploy ad hoc networks may be breaking the law, especially if their devices have been configured with the default settings of a different geographic area than their current. For example, the regulatory domain of Japan allows the use of all 14 defined channels of the 802.11bg standards for the deployment of WLAN. For most devices used in this region, the default channel for ad hoc deployment is channel 14. However, the rest of the regulatory domains, e.g., Europe or Americas, explicitly forbid the use of channel 14 by WLAN. In the Americas, channels 12 and 13 are also forbidden, adding to the confusion of ad hoc network users. To prevent such problems, additional policies (Table 7.2:P9,10) are introduced by the regional network managers, which in turn influence the criteria for the policy-based channel selection described above (Table 7.2:P2,3,4,8). For example, for P9 list1 = 1..11 and for P10 list2 = 1..13

To illustrate our proposed solution we investigated wireless networks based on IEEE 802.11 standards, since it is the most widely deployed technology for WLAN and offers support for ad hoc networks. Let us assume that a user

Table 7.2: Wireless ad hoc networks self-management policies

P#	Event	if {Conditions} / then {Actions}
1	Init_new_adhoc	if $\{ready\}$ then $\{scanChannels()\}$, $\{generateScanComplete(results)\}$
2	ScanComplete(results)	if $\{otherWLANdetected = true\}$ ^$\{FC := freeChannels(results), FC = true\}$ ^$\{PC := preffered(FC, ch\_list), PC = true\}$ then $\{optimizeChannel(PC, algorithm_1(criteria_1))\}$
3	ScanComplete(results)	if $\{otherWLANdetected = true\}$ ^$\{FC := freeChannels(results), FC = true\}$ ^$\{PC := preffered(FC, ch_llist), PC = false\}$ then $\{optimizeChannel(FC, algorithm_2(criteria_2))\}$
4	ScanComplete(results)	if $\{otherWLANdetected = true\}$ ^$\{FC := freeChannels(results), FC = false\}$ then $\{optimizeChannel(all, algorithm_3(criteria_3))\}$
5	NewWLANdetected	if $\{dyn\_adapt = true\}$ then $\{generateStartAdapt(newWLANinfo)\}$
6	LinkQualityCheck	if $\{LinkQuality < thr_a\}$ ^$\{dyn\_adapt = true\}$ then $\{generateStartAdapt(cachedWLANinfo)\}$
7	StartAdapt(WLANinfo)	if $\{channel\_distance(WLANinfo, current) < dist\}$ ^$\{app\_specific\_metric < thr_b\}$ then $\{scanChannels()\}$, $\{generateAdaptChannel(results)\}$
8	AdaptChannel(results)	if $\{results\_evaluation() = true\}$ then $\{channel\_switch(all, algorithm_4(criteria_4))\}$, $\{verify\_switch()\}$
9	SystemBoot	if $\{region = FCC\}$ then $\{set\_criteria(approvedChannels[list_1])\}$
10	SystemBoot	if $\{region = EU\}$ then $\{set\_criteria(approvedChannels[list_2])\}$

initiates an ad hoc network using a device supporting 802.11bg. The device is set in IBSS mode (Independent Basic Service Set or ad-hoc/peer-to-peer mode) and device-dependent software and hardware configure the transmission parameters. The device assumes the role of the wireless Access Point and its wireless interface begins to emit beacon messages advertising the existence of an ad hoc network on the statically defined channel. Other parameters are also advertised, like the beaconing interval and any encryption methods used, thus enabling nearby devices to join the ad hoc network in a peer-to-peer manner. If we realistically assume deployment in a populated area and not in an anechoic chamber, such deployment would imply the coexistence of various WLAN (either ad hoc or infrastructure-based) and inevitably their interference. Choosing the default channel or even a random channel is likely to have a detrimental effect for the ad hoc network performance. Unwanted side effects will also be noticed in the operation of nearby infrastructure WLAN or ad hoc networks. The problems arise from the access to the wireless medium and three cases can be identified during the deployment of an ad hoc network on a channel: a) the channel is already in use by other WLAN

b) adjacent or nearby channels are in use by other WLAN

c) no nearby channels in use by other WLAN.

In practice, cases b) and c) are difficult to be separated since co-channel interference depends on unpredictable environmental factors and is also technology dependent.

The above cases were examined on an experimental testbed and measurements were taken. We have deployed a policy-based solution that aims to dynamically assign the best available channel and autonomously adapt to changes in the wireless environment. To prevent the detrimental effects of interference, context information extracted from the headers of Layer 2 frames was used. This can be achieved by two methods and either can be used depending on the scenario and hardware support:

1. The device is using the wireless interface to passively monitor all packets it can hear (also know as sniffing or rf-monitor) and forwards them to the monitoring policies for processing of the 802.11 MAC headers as well as the 802.3 Link Layer headers. Therefore the device can extract useful information about the MAC layer performance for its one hop neighbors and by processing this information can trigger appropriate adaptation policies. The advantage of this method is that it fully exploits management frames and headers of 802.11 without associating to any AP or network. If the device has more than one wireless interface it can also assess its own performance. The drawback of this method is that the monitoring interface cannot be used for communication.

2. The device is using the wireless interface in promiscuous mode and associates to a wireless network as normal. The traffic packets received by the device are examined and information can be extracted from them. In this case not all packets transmitted on the channel are captured, since the device cannot overhear the channel while transmitting. This may be a drawback since the device cannot have a complete view of the neighborhood and may continue to cause interference to other devices without being able to detect that. However, the apparent advantage is that the device can still use the interface for communication, which is important in the case of devices with a single wireless interface.

In order to assess the performance of our policy-based approach we used a wireless testbed to evaluate the implementation's performance. In addition, we used the testbed to measure the effects of interference between devices using the same channel or devices with varying channel distance. Experiments were performed in a confined indoor space, matching the typical conditions of the described case studies.

Our experimental testbed consists of 10 nodes: 2 laptops, 4 PDAs and 4 Internet Tablets. All devices are equipped with internal 802.11b wireless interfaces, while the two laptops have an additional PCMCIA external wireless card. Table 7.3 includes more information on the used equipment. For the

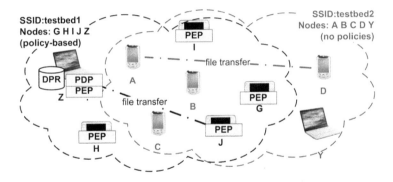

FIGURE 7.14: Wireless ad hoc network testbed deployment

configuration of the wireless interfaces, Linux scripts were used with wireless-tools v28. For monitoring the wireless channel we have modified the source code of airodump-ng, a popular open source 802.11 packet sniffer, part of the aircrack-ng suite [73]. The modifications allowed us to view and dynamically use the captured information within the policy-based interface. Communication between nodes was done either by SSH or by HTTP.

For the purpose of our experiments, the devices were organized in two independent clusters of five nodes as seen in Fig. 7.14. The clusters were set up using different SSID (Service Set Identifiers) in IBSS (ad hoc) mode. The manufacturer's default channel for ad hoc networks creation was found to be Channel 1 (2412Mhz). The network speed (rate) was set to 11Mbps, to allow comparable results among nodes. One of the clusters (testbed1) integrated context-aware policy-based management (PBM) support and the cluster head deployed a PDP for the needs of its cluster. After the PDP had retrieved policies 1 - 8 (Table 7.2) from the nearest DPR, it had accordingly instantiated policy objects (PO) for the monitoring and enforcement of decisions among cluster nodes. For evaluation purposes the PBM support was selectively used to measure its effect on the network performance.

7.4.1.1 Self-configuration for Initial Channel Assignment

In the beginning, we performed static measurements of the channel performance in the presence of multiple ad hoc networks with varying channel distance. According to this scenario, the two clusters would simultaneously attempt to initiate file transfer among peers of the same cluster, as shown in Fig. 7.14. First, the two ad hoc networks were formed on the same default channel (Channel 1). This was possible by using different network names (SSIDs), namely 'testbed1' and 'testbed2.' Afterwards, the same networks were deployed in different channels and file transfers were performed. While 'testbed2' was always deployed on the default Channel 1, 'testbed1' was deployed on Channels 1,2,4 and 6 to vary channel distances and evaluate its effect.

Table 7.3: Wireless testbed specifications

	Operat.System (Linux Kernel)	Processor (MHz -family)	Ram (MB)	Wifi support
Sony Vaio Z1XMP	Debian R4.0 (2.6.18)	1500 - Intel	512	802.11bg
HP iPAQ H5550	Familiar v0.8.4 (2.4.19)	400 - ARM	128	802.11b
Nokia N800	IT OS2007 (2.6.18)	330 - ARM	128	802.11bg

The Cluster Node J of cluster 'testbed1' downloaded a media file from Cluster Head Z and measured the received data download throughput (goodput) and download completion times. The results of the average goodput for each channel combination (T1,T2) are shown in Table 7.4, where T1 the deployment channel of 'testbed1' and T2 that of 'testbed2.' What is worth noticing is that the goodput performance of ad hoc deployment in consecutive channels is even worse than deployment on the same channel by approximately 13%. This can be explained by considering the MAC layer functionality, where on the same channel, all devices hear Request To Sent (RTS) frames and back-off from using the channel and thus can avoid collisions and excessive MAC frames retransmissions. On the contrary, when nearby channels are used, frames from different channels are perceived as interference and increased channel noise, causing the MAC layer to retransmit lost frames and possibly reduce transmission rate to avoid excessive BER. As recorded by our measurements this effect is reduced the furthest apart the channels are, although is still noticeable even when 'non-overlapping' channels are used (e.g., 6,1). This can be explained because of the proximity of most devices which results in the near-far effect.

Additional measurements of missed and sent frames further confirm the detrimental effects of randomly assigning channels to deployed ad hoc networks. All measurements displayed in Fig.7.15 were taken from the cluster head (node Z) of 'testbed1' using its second wireless interface in rf-monitor mode. The purpose was to verify how the device perceives the wireless channel while transmitting using its first interface.

Two sets of measurements are shown, for deployment and monitoring on channel 1 for same channel deployment (T1,T2)=(1,1) and on channel 2 for consecutive channel deployment (T1,T2)=(2,1). Frame measurements provide a good indication of channel utilization and the level of occurred collisions (missed frames). In brief, two points worth noticing are:
(1) Missed frames are in both cases more than sent frames, but in case (2,1) are increased by approx. 15% compared to case (1,1);

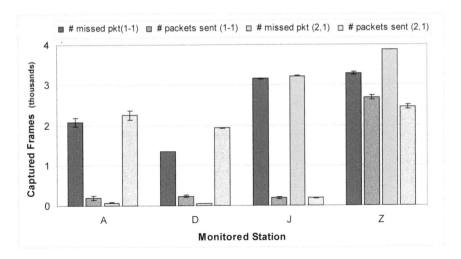

FIGURE 7.15: Packet measurements at node Z for same channel deployment (1,1) and for consecutive channel deployment (2,1)

(2) Node Z can hear a significantly increased number of frames sent from nodes (A,D) of a competing ad hoc network.

By enabling the PBM support for testbed1, the cluster head (node Z) ensures that policies 1-4 are applied during the initial phase of ad hoc deployment. After P1 scanned channels, P2 detects the presence of testbed2 on channel 1 and the scan results indicate channels 2-10 as free (FC=true,PC=true). Since channel 6 of the preferred (non-overlapping) channels list was free, method assignChannel initiates the ad hoc network on the selected channel and the rest of the cluster nodes join using SSID testbed1 on the same frequency. The policy-based initial channel configuration results in the optimum configuration (T1,T2)=(6,1), as confirmed by further measurements. Effectively, the cluster is self-configuring the initial ad hoc deployment and this results in a 20.4% increase of average goodput when compared to using default channels (1,1) and up to 33.3% increase for random channel assignment (2,1). File download completion time is accordingly improved.

7.4.1.2 Self-Optimization for Dynamic Channel Switch

The second implemented scenario investigates the dynamic adaptation of hybrid ad hoc networks to anticipate interference and throughput degradation. Based on the topology of Fig. 7.14, we assume the coexistence of two separate ad hoc networks on the same channel (testbed1 and testbed2 on channel 1). Initially, no traffic transfers are performed between nodes. The

Table 7.4: Initial channel assignment measurements

testbed1,2 (channel)	Goodput testbed1(Mbps)	Goodput decrease (%)	Downl.Time increase (%)
1,1	3.48	-20.38	+20.00
2,1	2.92	-33.27	+46.67
4,1	4.26	-2.68	0.00
6,1	4.38	–	–

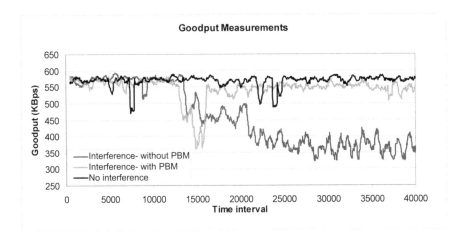

FIGURE 7.16: Policy-based channel assignment measurements

scenario execution has two phases:

Phase 1: ad hoc network testbed1 initiates a file transfer between nodes, with cluster node J downloading a 46MB file from cluster head Z.

Phase 2: ad hoc network testbed2 initiates another file transfer between nodes A and D.

To evaluate the implemented solution, two sets of the described scenario experiments were executed, one set with the PBM solution enabled and enforcing policies 5-8 and another set without any PBM functionality. We have tried to maintain the same execution conditions during all experiments to allow comparison of taken measurements. A representative extract of our measurements is presented (Fig. 7.16) and discussed below.

The measured results demonstrate a significant improvement in network performance when the proposed PBM solution is used (Fig. 7.17). The ad hoc cluster testbed1 is self-optimizing by monitoring events and conditions, resulting in reconfiguration of the transmission channel to avoid interfering WLAN. When the competing ad hoc network (testbed2) initiates a file transfer (phase 2), this results in increased collisions and missed frames for both clusters, which is reflected in reduced Link Quality reported by the wireless interface at node Z. Policy P6, triggered by LinkQualityCheck event, evaluates the moving average of LinkQuality as less than 50% (thr_a) and executes action generateStartAdapt to initiate the adaptation process for channel optimization. In turn, policy P7 is triggered and monitors the specified application metric, in this case the moving average of goodput measurement for the file download between nodes Z and J (app-specific-metric). The measurements of this metric are shown as bold lines in Fig. 7.17 (top), while thin lines show instantaneous goodput measurements in Fig. 7.17 (bottom). Comparing the two graphs of Fig. 7.17, we verify that the use of a moving average smooths goodput fluctuations and prevents false triggering of adaptation policies. Once policy P7 detects the reduction of goodput below 3.67Mbps (thr_b), it acts by scanning the wireless channel, triggering policy P8 and passing scan results (event AdaptChannel). Policy P8 acts by executing channel-switch method using the weighted average algorithm ($algoritm_3$) with specified weights ($criteria_3$). The method indicates that a better channel is available and initiates dynamic switch of ad hoc network testbed1 to channel 6. A channel switch period takes place, causing temporary disconnection of nodes from their cluster head Z. The measurements show that L2 disconnection and connectivity loss occur; however the effect on the ongoing file transfer between J and Z was temporary goodput reduction with a quick recovery to significantly higher goodput. In fact, when compared to the execution without PBM support, the described self-optimization resulted in a 33.5% peak increase of goodput with an average increase of 20.3%. Also, average download time for a 46MB file dropped from 116sec to 50sec.

7.4.2 Self-Configuration of a Distributed Policy Repository

DPR is an enhanced version of the Policy Repository [120] and consists of repository replicas distributed among hypercluster's nodes. Instead of simply replicating the PR among the nodes, we incorporate a sophisticated policy-based replication scheme. By utilizing real-time context information and a priori knowledge, the introduced policies automatically enforce the appropriate replication state among hypercluster nodes.

The Distributed Policy Repository is a set of distributed and/or replicated instances (replicas) of Directory Servers (directories) based on LDAP. Each replica can be either tightly integrated to a master repository instance or loosely coupled to a master or another replica. The design is based on the advanced replication and distribution features of modern LDAP servers. The DPR component is available to nodes consisting of the hypercluster. In order to balance resource consumption and policy accessibility, a selection of the hypercluster nodes activate their DPR component and carry a replica of network policies. The DPR state of each node is imposed by the network policies which define the overall policy replication state. Management objectives reflected

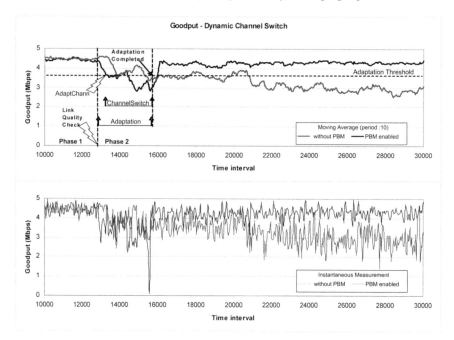

FIGURE 7.17: Testbed measurements of goodput using dynamic channel switch. Top: Moving average, Bottom: Instantaneous

in policies and network conditions influence the DPR replication degree to conserve resources and ensure maximum repository availability.

Specifically, when network mobility is high and links are exceedingly intermittent, reliable access to a remote Policy Repository may be impossible. In this case, policy objects (PO) monitoring network mobility detect the high volatility and proactively increase the replication degree of DPR. Effectively the network will respond with increased decentralization of the policy repository, pushing the storage points (DPRs) closer to the decision points (PDPs). Each manager node (MN) or cluster head (CH) with an active DPR, accommodate a replica of the repository that can serve as an access point for repository requests within their cluster, balancing this way processing load and traffic in the network. A CH with a dormant DPR can access policies from a list of neighboring CHs or MNs with an active DPR. The state of each node is imposed by network policies (Table 7.5).

One of the innovative features of the proposed DPR design is the ability to deploy and maintain special purpose partial replicas of the Policy Repository. These replicas provide a partial view of network policies and can relate to a specific service or location. Accordingly, attached PDPs are responsible only for the enforcement of a policy subset and can be dynamically deployed

Table 7.5: DPR management policies

P	Event	if	{*Conditions*} then {*Actions*}
a	chkDPR	if then	$\{t = t_{Weekday}\}\hat{}\{countPDPs(area_1)/countDPRs(area_1) > thr_1\}$ $\{locatePDPs(area_1)\}$, $\{selectDPRhost(algorithm_1, context_1)\}$ $\{deployDPR(all)\}$
b	chkDPR	if then	$\{t = t_{Weekend}\}\hat{}\{countPDPs(area_1)/countDPRs(area_1) > thr_2\}$ $\{locatePDPs(area_1)\}$, $\{selectDPRhost(algorithm_1, context_1)\}$, $\{deployDPR(all)\}$
c	chkDPR	if then	$\{t = t_{Kickoff-2h}\}$ $\hat{}\{countPDPs(stadium_1)/countDPRs(stadium_1) > thr_3\}$ $\hat{}\{countUsers(stadium_1)/countPDPs(stadium_1) > thr_4\}$ $\{locatePDPs(stadium_1)\}$, $\{selectDPRhost(algorithm_2, context_1)\}$, $\{deployDPR(service_1, service_2)\}$

to provision time-based events or localized conditions. This feature can be employed when there is a need for localized control in areas with dense user population, such as a conference site or a stadium. In such cases, while node population (i.e., users) increases, the management system can deploy special-purpose DPR replicas and accordingly more Policy Decision Points (PDPs) that will be responsible for the distributed enforcement of specific management tasks. Special policies guide the deployment of partial DPR copies, based on a priori knowledge of localized events (e.g., scheduled sport event) with dynamic real-time context information (e.g., processing load on each CH) (Table 7.5).

For the implementation of the Distributed Policy Repository (DPR) we have used OpenLDAP. This selection was made because it is a free and open source implementation of a very fast and reliable LDAP v3 Directory Server for Linux. In addition, the minimum specifications required for running this server allow an extensive range of devices, including low-spec laptops to efficiently host a directory replica. We will refer to a Directory Server with its directory content (i.e., policies) as a directory. The DPR consists of one or more Master read-write directories and several read-only directory replicas (shadow copies). Master directories are hosted and controlled by the managing network entities, i.e., Network Operator and/or Service Providers. These entities are responsible for providing the overall management objectives and guidelines to the wireless network by specifying appropriate policies. The policies are configured and introduced to the systems using the Policy Management Tool and their LDAP representation is stored in a master directory. Based on replication policies (Table 7.5), selected user devices that serve as Policy Decision Points (PDPs) are also chosen to host a directory replica, i.e., a part of the Distributed Policy Repository. To achieve the above, we exploit OpenLDAP's replication engine to enable the policy-based distribution of replicated read-only directories (shadow copies) among the user devices, as well as partial copies for specific purposes (e.g., policies for multimedia services). OpenLDAP implements a Sync replication engine (syncrepl), based on the Content Synchronization Operation (RFC 4533). Syncrepl engine offers client-side (consumer) initiation for replication of all policies or for a custom policy selection, relieving the providing directory (provider) from tracking and updating replicas. This functionality is very useful since the operation of a directory provider is not disrupted by the presence of consumers and can operate even when they are temporarily disconnected because of wireless link intermittence. Upon reconnection, the directory consumers compare their current content with their provider's and retrieve any updates. For the defined policies in Table 7.5, the generic method selectDPRhost($(algorithm_1,criteria)$) can use different algorithms for the best possible placement of replicated directories. The optimal placement solution is a computationally intensive task, hindered by the distributed nature of wireless systems and is out of the scope of this paper. In [18, 96] we have described and evaluated a distributed algorithm based on context-aware heuristics to form a dominating set of nodes that share management responsibilities. The same approach is adopted for the implementation of algorithm1 for policies in Table 7.5. The criteria parameter affects the heuristics used, by modifying the weights of metrics used in the algorithm. Method deployDPR() is used to set up and initiate a replicated directory, part of the Distributed Policy Repository. First the directory configuration file (slapd.conf) is modified and once the replica is initiated, it connects and retrieves policies from its defined master directory. Method parameters (all, service1, service2) define policy groups that will be replicated.

7.4.3 Self-Protection of User Privacy and Preferences

Another interesting extension of the aforementioned framework touches on the sensitive issue of privacy and preference respect of users. The presented case study introduces a self-protection mechanism for networked devices based on the user's input and current regulations [10].

The multi-manager policy-based framework is employed to establish the high-level management objectives of Network Operators and Service Providers. These objectives are encapsulated in policies and distributed among the wireless network. In order to provide a rich and customizable experience to users, these policies often need to collect user information. The collected context can be anything from their location, movement patterns or frequency of service access. The particular type of context differs significantly from network collected context (e.g., throughput, lost packets, etc.) because it is tightly related to the privacy of the individual user. This privacy needs to be protected. To facilitate a protection mechanism for user's privacy, this case study introduced a twofold scheme to modify the presented framework. As explained here briefly, policies and context are both affected by this scheme. For more details, we refer the reader to [10].

Self-management is a compelling functionality and a highly desired attribute of large-scale complex networks. But as already pointed out, wireless

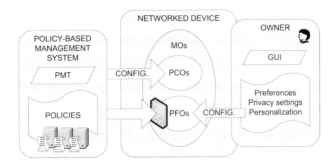

FIGURE 7.18: Policy free and policy conforming objects

networks formed by user devices are not directly managed by an administrative authority and, in addition, often involve processing and collection of user generated personal context. As a first step, the described case study offers the capability to a user to define which high-level managed object (MO) he/she wants to control on his/her own and which he/she entrusts to the PBM system. In addition, users can explicitly specify their own access control rules as preferences and limit the exploitation of specific contextual information. Effectively, this leads to the separation of policy/context objects into Policy-Free Objects (PFO) and Policy-Conforming Objects (PCO). The concept is graphically presented in Fig. 7.18. For example, different policy/context objects can be managed objects controlling access to location data, battery consumption profile, sharing resources, etc. While Services may need to alter those objects, e.g., a Location-based Service (LBS) needs to enable a GPS receiver and access location data, a user may not be willing to allow access to personal context. Hence, using a GUI can set the desired preferences and privacy restrictions to the system.

The second protection mechanism introduced relies on a policy-based regulation scheme. In addition to the explicit user defined preferences, the PBM system has the ability to control unfair exploitation of user data by deploying a regulation scheme with appropriate policies. In a multi-manager case study, we consider a data protection agency (Regulator) that has the control of one Manager Node. Using the PMT interface, the Regulator has the ability to manage the lifecycle of policies and introduce appropriate policies to the managed system according to current regulations. In addition, it can review, edit or disable existing policies so as to ensure users' personal data are not collected or exploited by other entities, e.g., by the Network Operator or a Service Provider.

For example, users who are willing to reveal their location data should be protected from services that can continually track their position. Tracking is possible by frequently polling the user location and comparing consecutive measurements, depending on the accuracy of the available positioning method and users' speed. With the increased penetration in the consumer market of high accuracy GPS-enabled devices and improvement of indoor positioning methods, this issue is becoming quite important. Further than configuration policies, a regulatory body can use the policy-based system to monitor the collection of user data and gather information for offline processing. Simple policies can periodically log information about the services that retrieve user data. The logged details can be reviewed and analyzed statistically to extract information about how Service Providers use the location data of users and investigate their unfair exploitation.

The twofold protection scheme introduced in [10] is an important step to control the exploitation of context information and offer to users a degree of control. By introducing appropriate regulatory policies and user-oriented access control, the PBM system is armed with self-protection capabilities to prevent the unfair exploitation of participating devices. The privacy concerns raised are important requirements in the design of a context-aware solution. Privacy and control of context dissemination are tightly connected with self-protection and self-healing. As discussed in the following Section, although self-protection and self-healing are critical capabilities of autonomous systems there are limited case studies implementing them.

7.5   Conclusions and the Future of Self-management

7.5.1   Summary and Concluding Remarks

Self-management is gradually becoming a reality and it is expected that in the future more research efforts will be looking towards this direction. We have rigorously presented the history and evolution of self-management through extensive literature review. We have seen that a computing environment with the ability to manage itself and dynamically adapt to change in accordance with business policies and objectives defines autonomic computing and encompasses the fundamental Self-Management definition. In future self-management systems, the characteristic four self-CHOP or self-* properties, namely self-configuration, self-heal, self-optimization and self-protection, will be gracefully integrated and become invisible. But in order to reach this point of integration, these properties are currently separated and studied individually. For the realization of Self-Management capabilities, we have adopted closed-loop control as a repetitive sequence of tasks including monitor, analyze, plan and execute functions. A Knowledge base caters for the orchestration of these functions. This reference model by IBM [2] is frequently referred to as K-MAPE or simply MAPE and has been the basis for the presented system analysis, design and evaluation.

The focus of the work presented in this chapter has been in wireless ad hoc networks. This is due to the increasing popularity and penetration of such networks worldwide. In order to apply Self-Management to wireless ad hoc networks, we have analyzed the definition for Self-Management and identified policies as the basis of such systems, encapsulating high-level business objectives. Policy-Based Management (PBM) is the first building block of the presented Self-Management framework and policies are its cornerstone. Equally important and complementary is the system's ability to sense and observe its surrounding environments. To enable these, context-awareness is employed as the second building block of the framework. As a result a policy-based context-aware framework is designed, as the foundation for the implementation of self-management properties. The motivation and applicability of the presented framework is evident considering the closed-loop controller (Fig. 7.1) or IBM's autonomic manager (Fig. 7.2).

The policy-based part of the presented framework is analogous to the initial input and processing unit of a closed-loop control system, i.e., input and unit A in Fig. 7.1. At the same time, the context-aware part is analogous to the returned feedback that is processed and closes the adaptation loop. Similarly, the designed framework can be mapped to the widespread K-MAPE reference model of IBM [2]. Policy-based management can serve as the Plan and Execute components of a self-management system, as presented in Fig. 7.2. Policy design and specification constitute the Planning phase of autonomic management while policy enforcement constitute the Execute phase. On the other hand, a context-aware framework is assigned the Monitor and Analyze functionality. Context sensing and collection constitute the Monitoring phase while context aggregation and inference rules constitute the Analyze phase of management. The specification of policies and context, together with their interaction, form the essential Knowledge element. Policy and context repositories are the Knowledge centerpiece of both frameworks, gracefully integrating the presented self-management solution. Figure 7.6 (page 224) illustrates these concepts, in parallel with IBM's autonomic manager.

We believe that the highly dynamic environment of wireless ad hoc networks can benefit from a Policy Based Management (PBM) and context-aware approach. One of the major advantages of adopting a policy-based approach is the relevant 'controlled programmability' that can offer an efficient and balanced solution between strict hard-wired management logic and unrestricted mobile code migration and deployment. While there has been previous research on deploying PBM solutions for wireless ad hoc networks, the work presented here introduced a novel organizational model specifically targeted to the needs of such networks by incorporating context awareness to dynamically adapt to the continuously changing conditions. Context information can be used to trigger cross-layer changes (network and application configurations) according to policies, leading to a degree of autonomic decision-making and self-management. Based on the introduced distributed and hierarchical organizational model, a Distributed Policy Repository (DPR) is deployed to efficiently cater for the policy distribution and provisioning needs of the network.

A fully operational self-management solution is not currently available to the best of our knowledge. However, researchers have attempted to partially tackle self-management by implementing some of the desired properties. We have presented some notable efforts to integrate self-* capabilities based on the aforementioned policy-based and context-aware management framework. The presented solutions have demonstrated how the framework with the interaction of policies and context forms the basis for implementation and evaluation of self-management solutions. An important consideration regarding the presented research effort is the realistic deployment of ideas on real wireless ad hoc testbeds. Detailed policy specifications and context modeling were implemented and have demonstrated quantifiable performance improvement, as well as lessons and experiences learned.

7.5.2 Future Trends and Challenges

It should be noted that current research on Self-management has mainly focused on addressing the self-configuration and self-optimization properties, while self-healing and self-protection remain in infancy. This is to be expected since the road to Self-management is gradual and complex. Small steps are made each time leading to integration of all properties. For example, presented solutions in Section 7.4 deal with self-configuration and self-optimization of wireless ad hoc networks and only touch on issues of self-protection. According to IBM's roadmap to autonomic computing and self-management [107], five transition levels for gradual adoption are suggested. Today most systems are in either 'Basic' or 'Managed' level, meaning there is a significant amount of human effort in monitoring and controlling operations, assisted by limited management systems. Current research efforts have elevated management to 'Predictive' and 'Adaptive' levels, by integrating automated context correlation and applying high level objectives through manually refined low-level policies.

The ultimate step to 'Autonomic' level has been the focus of self-management research. Major research challenges need to be addressed and resolved before this transition is made. Notably, crucial issues related to policy-based management expected to receive research interest are mentioned here:

- Automation of policy refinement: Today, policy specification languages require a highly technical person to analyze the system and specify the required management objectives as policies. This results in significant effort for setting up the system and writing detailed low-level policies. In addition, policy updates and notifications require same skills and time. It is envisioned that in the future, an automated policy refinement process will take up this time-consuming and difficult task and will automate the creation of low level policies from business objectives. A manager will be able to directly influence and control the system

behavior and always ensure compliance with high level goals.

- Automation of policy analysis: Another crucial issue that needs to be resolved is the analysis of policy specification. This includes the detection of conflicts in policy specification and their resolution. Currently, both issues are manually handled by experienced technical persons, but active research aims to automate these processes. Conflicts can occur either during specification of new or changed policies (static conflicts) or during system run-time because of dynamically changing conditions (dynamic conflicts). Both types can compromise the integrity of the system and disrupt its operation; hence reliable detection and resolution mechanism should be in place, ideally functioning unsupervised.

- Efficient and distributed policy provisioning: As already described, the large scale of managed systems and their highly distributed nature further complicate policy provisioning, i.e., timely providing appropriate enforcement action as required by active policies. Especially in wireless networks, the task is even harder due to mobility and frequent disconnection. Being able to access all or at least the majority of managed devices is important for offering value-added services and generating revenue. While probabilistic management solutions are investigated, p2p computing also offers promising solutions. A highly distributed policy storage facility can ease provisioning, provided it can remain synchronized and up to date. It is expected that this issue will receive intense research efforts in the future.

Paired with policy-based management, Context Awareness will remain an invaluable component of Self-managing systems. More crucial challenges need to be addressed and resolved to achieve truly autonomic computing. Some important open research issues are mentioned here:

- Efficient and secure context modeling: As emphasized earlier, context modeling is important for any context-aware system. It affects the way context is collected and processed and can severely affect the operation of a network. More expressive and efficient models are required to cater for the resource constrained nature of wireless systems, since bandwidth is limited. Future models need to integrate security features to protect and respect the privacy of users.

- Automated extraction of context from heterogeneous devices: With the increasing heterogeneity of devices and networks in general, context collection becomes complicated and fragmented. Overcoming these problems in an automated manner is a challenging research topic. Device and equipment standardization is important for uniform context collection.

- Interoperable context collection and exchange: The current absence of standards further inhibits context extraction and hinders desired interoperability of systems. Software engineering methods and practices can be utilized to create necessary interfaces between context-aware elements, while ontologies and semantics are important tools to be exploited. In addition, protocols for secure and efficient exchange of contextual information should be engineered, departing from the proprietary solutions used today.

Self-management is expected to motivate significant research efforts, both in academia and in industry, because of the apparent benefits it can offer. Future networks and systems will transparently integrate self-management capabilities, relieving users and managers from painstaking tasks. As research progresses, the current separation of self-* properties in configuration, healing, optimization and protection will diminish, gracefully amalgamating all in a Self-maintaining operation. We envision a policy-based system as a future-proof solution, where business objectives and user preferences will be encapsulated in policies. Context-awareness will provide secure and accurate feedback to the system, assisting in fully customized and personalized user experience. Eventually policies and context will vanish inside systems, allowing users to enjoy truly ubiquitous networking.

7.6   Acknowledgments

Research work in this chapter was partly supported by the EU EMANICS Network of Excellence on the Management of Next Generation Networks (IST-026854). The authors also wish to thank Dr. Apostolos Malatras for his contribution to the context-aware aspects of the framework.

7.7   Abbreviations

Acronym Explanation
Acronym	Explanation
AM	Autonomic Manager
CCP	Context Collection Point
CDP	Context Decision Point
CMT	Context Management Tool
COPS	Common Open Policy Service
CR	Context Repository
DMTF	Distributed Management Task Force
DPR	Distributed Policy Repository
DS	Directory Server
IETF	Internet Engineering Task Force

LDAP   Lightweight Directory Access Protocol
PBM   Policy-Based Management
PCIM   Policy Core Information Model
PCIMe   Policy Core Information Model Extensions
PDP   Policy Decision Points
PEP   Policy Enforcement Point
PMAC   Policy Management for Autonomic Computing
PMT   Policy Management Tool
PR   Policy Repository
UML   Unified Modeling Language
XML   eXtensible Markup Language

References

[1] G.F. Franklin, J.D. Powell, A. Emami-Naeini, Feedback Control of Dynamic Systems, Prentice Hall, ISBN-13:978-0131499300

[2] IBM Inc., An architectural blueprint for autonomic computing, accessed online Dec.2007 `http://www-03.ibm.com/autonomic/pdfs/ACBP2_2004-10-04.pdf`

[3] J.O. Kephart , D.M. Chess, The vision of autonomic computing, IEEE Computer, Vol.36/Iss.1, Jan. 2003. pp.41-50

[4] D. Chalmers et al., Ubiquitous Computing: Experience, Design and Science, Ver.4, `http://www-dse.doc.ic.ac.uk/Projects/UbiNet/GC/Manifesto/manifesto.pdf`, Jun.2004

[5] H. Schulzrinne et al., Ubiquitous computing in home networks, IEEE Communications Magazine, Vol.41/Iss.11, Nov.2003

[6] L.M. Feeney, B. Ahlgren, A. Westerlund, Spontaneous networking: an application oriented approach to ad hoc networking, IEEE Communications Magazine, Vol.39/Iss. 6, Jun.2001

[7] J. Latvakoski, D. Pakkala, P. Paakkonen, A communication architecture for spontaneous systems, IEEE Wireless Communications, Vol.11/Iss.3,

[8] M. Sloman, E. Lupu, Policy Specification for Programmable Networks , Proceedings of First International Working Conference on Active Networks (IWAN'99), Berlin, June 1999

[9] D.C. Verma, Simplifying network administration using policy-based management, IEEE Network, Vol.16Iss.2, Mar-Apr.2002

[10] A.M. Hadjiantonis, M. Charalambides, G. Pavlou, A policy-based approach for managing ubiquitous networks in urban spaces, IEEE Int. Conf. on Communications 2007, Glasgow (ICC2007)

[11] R. Boutaba, S. Omari, A. Virk, SELFCON: An Architecture for Self-Configuration of Networks, Journal of Communications and Networks, Vol.3/No.4, pp.317-323, 2001

[12] IETF's Policy Framework Working Group (POLICY WG), concluded in 2004, IETF Website http://www.ietf.org/html.charters/OLD/policy-charter.html Accessed September 2005

[13] D. Verma, Policy-Based Networking, Architecture and Algorithms, Pearson Education, ISBN:1578702267, 2000

[14] P. Flegkas, P. Trimintzios, G. Pavlou, A policy-based quality of service management system for IP DiffServ networks, IEEE Network, Vol.16/Iss.2, March-April 2002, pp.50-56

[15] M. Sloman, Policy driven management for distributed systems, Journal of Network and Systems Management, Vol.2/No.4, Dec. 1994, pp. 333-360, accessed Sep.2005, `http://www-dse.doc.ic.ac.uk/dse-papers/management/pdman.ps.Z`

[16] J. Strassner, Policy-Based Network Management, Solutions for the Next Generation, Morgan Kaufmann, ISBN:1558608591, 2003

[17] R. Chadha et al., Policy Based Mobile Ad hoc Network Management, 5th IEEE Int. Work. on Policies for Distributed Systems and Networks (Policy 2004)

[18] A.M. Hadjiantonis, A. Malatras, G. Pavlou, A context-aware, policy-based framework for the management of MANETs, 7th IEEE Int. Work. on Policies for Distributed Systems and Networks (Policy 2006)

[19] P. Flegkas, P. Trimintzios, G. Pavlou, A. Liotta, Design and implementation of a policy-based resource management architecture, IFIP/IEEE 8th International Symposium on Integrated Network Management, March 2003, pp.215-229

[20] R. Yavatkar et al., A Framework for Policy-based Admission Control, RFC 2753, Informational, Jan.2000

[21] B. Moore et al., Policy Core Information Model-Version 1 Specification, RFC 3060, Standards Track, Feb.2001

[22] B. Moore, Policy Core Information Model (PCIM) Extensions, RFC 3460, Standards Track, Jan.2003

[23] D.C. Verma, Simplifying network administration using policy-based management, IEEE Network, Vol.16/Iss.2, March-April 2002, pp.20-26

[24] Internet Engineering Task Force (IETF), `http://www.ietf.org`

[25] Distributed Management Task Force (DMTF), `http://www.dmtf.org`

[26] IBM Corp., Policy Management for Autonomic Computing, V1.2, accessed on Sep.2007, `http://www.alphaworks.ibm.com/tech/pmac`

[27] D.Chadwick et al., Coordination between distributed PDPs, 7th IEEE Int. Work. on Policies for Distributed Systems and Networks (Policy 2006)

[28] J. Sermersheim, Lightweight Directory Access Protocol (LDAP): The Protocol, RFC4511, Standards Track, Jun.2006

[29] R. Montanari, E. Lupu, C. Stefanelli, Policy-based dynamic reconfiguration of mobile-code applications, IEEE Computer, Vol.37/Iss.7, Jul.2004, pp.73-80

[30] IETF's Mobile Ad-hoc Networks Working Group (MANET WG), IETF Website, accessed September 2005. http://www.ietf.org/html.charters/manet-charter.html

[31] W. Chen, N. Jain, S. Singh, ANMP Ad hoc network management protocol, IEEE Journal on Selected Areas in Communications, Vol.17/Iss.8, Aug. 1999, pp.1506-1531

[32] R. Chadha, C. Yuu-Heng, J. Chiang, G. Levin, L. Shih-Wei, A. Poylisher, Policy-based mobile ad hoc network management for drama, IEEE Military Communications Conf. (MILCOM 2004), Vol.3, pp.1317-1323

[33] R. Badonnel, R. State, O. Festor, Management of mobile ad-hoc networks: evaluating the network behaviour, 9th IFIP/IEEE International Symposium on Integrated Network Management (IM2005), May 2005 pp.17-30

[34] C. Shen, C. Srisathapornphat, C. Jaikaeo, An adaptive management architecture for ad hoc networks, IEEE Communication Magazine, Vol.41/ Iss.2, Feb.2003, pp.108-115

[35] A. Westerinen et al., Terminology for Policy-Based Management, RFC 3198, Informational, Nov.2001

[36] M. Wahl et al., Lightweight Directory Access Protocol (v3), RFC 2251, Standards Track, Dec.1997

[37] K. Phanse, Policy-Based Quality of Service Management in Wireless Ad-hoc Networks. PhD thesis, Faculty of the Virginia Polytechnic Institute and State University, August 2003.

[38] M. Burgess, G. Canright, Scalability of peer configuration management in logically ad hoc networks, eTransactions on Network and Service Management, Vol.1 No.1 Second Quarter 2004

[39] C.S. Murthy , B.S. Manoj, Ad Hoc Wireless Networks, Architectures and protocols, Prentice Hall PTR, ISBN:013147023X, 2004

[40] L.M. Feeney, B. Ahlgren, A. Westerlund, Spontaneous networking: an application oriented approach to ad hoc networking, IEEE Communications Magazine, Vol.39/Iss.6, June 2001, pp.176-181

[41] M. Sloman, E. Lupu, Policy Specification for Programmable Networks, Proceedings of First International Working Conference on Active Networks (IWAN'99), Berlin, June 1999

[42] N. Damianou, N. Dulay, E. Lupu, M. Sloman, The Ponder Specification Language, Workshop on Policies for Distributed Systems and Networks (Policy 2001), Bristol, Jan 2001

[43] IETF's Resource Allocation Protocol Charter (RAP WG, concluded), IETF website, accessed Sep.2005,http://www.ietf.org/html.charters/OLD/rap-charter.html

[44] D. Durham et al., The COPS (Common Open Policy Service) Protocol, RFC 2748, Standards Track, Jan.2000

[45] K. Chan et al., COPS Usage for Policy Provisioning (COPS-PR), RFC 3084, Standards Track, Mar.2001

[46] M. Sloman, E. Lupu, Security and management policy specification, IEEE Network, Special Issue on Policy-Based Networking, Vol.16/Iss.2, pp.10-19

[47] A. Westerinen, J.Schott, Implementation of the CIM Policy Model using PONDER, 5th IEEE Int. Work. on Policies for Distributed Systems and Networks, POLICY 2004, 7-9 June 2004 pp.207-210

[48] DMTF website, CIM Policy Model White Paper for CIM v2.7.0 , accessed September 2005 http://www.dmtf.org/standards/documents/CIM/DSP0108.pdf

[49] N. Dulay, E. Lupu, M. Sloman, N. Damianou, A policy deployment model for the Ponder language, IEEE/IFIP Int. Symposium on Integrated Network Management, May 2001 pp.529-543

[50] L. Lymberopoulos, E. Lupu , M. Sloman, Using CIM to realize policy validation within Ponder Framework, DMTF Website, accessed September 2005 http://www.dmtf.org/education/academicalliance/lymberopoulos.pdf

[51] G. Pavlou, P. Flegkas, S. Gouveris, A. Liotta, On management technologies and the potential of Web services, IEEE Communications Magazine, Vol.42/Iss.7, July 2004, pp.58-66

[52] The Directory: Overview of Concepts, Models and Services, Recommendation X500, ISO/IEC 9594-1 , ITU-T

[53] V. Koutsonikola, A. Vakali, LDAP: framework, practices, and trends, IEEE Internet Computing, Vol.8/Iss.5, 2004, pp.66-72

[54] Sun website, A Technical Overview of the Sun ONE Directory Server 5.2, accessed Feb. 2005 http://www.sun.com/software/products/directory_srvr_ee/wp_directorysrvr52_techoverview.pdf

[55] E.J. Thornton, D. Mundy, D.W. Chadwick, A comparative performance analysis of seven LDAP Directories, TERENA website accessed Septem-

ber 2005 `www.terena.nl/conferences/tnc2003/programme/papers/` `p1d1.pdf`

[56] T.A. Howes, M.C. Smith, S.G. Gordon, Understanding and deploying LDAP directory services, 2nd edition, Addison-Wesley Professional ISBN: 0672323168, 2003

[57] W. Dixon, T. Kiehl , B. Smith, M. Callahan, An Analysis of LDAP Performance Characteristics, General Electric Global Research, Technical Information Series, tech. report TR-2002GRC154, Jun. 2002

[58] X. Wang, H. Schulzrinne , D. Kandlur , D. Verma, Measurement and Analysis of LDAP Performance, Int. Conf. on Measurement and Modeling of Computer Systems (SIGMETRICS'2000), Santa Clara, CA, Jun. 2000, pp. 156-165

[59] Organization for the Advancement of Structured Information Standards, Directory Services Markup Language v2.0, OASIS Standard 2002, accessed Sep. 2005 `www.oasis-open.org/committees/dsml/` `docs/DSMLv2.doc`

[60] J. Strassner et al., Policy Core Lightweight Directory Access Protocol (LDAP) Schema, RFC3703, Standards Track, Feb.2004

[61] M. Pana et al., Policy Core Extension Lightweight Directory Access Protocol Schema (PCELS), RFC 4104, Standards Track, Jun.2005

[62] A. Vakali, B. Catania, A. Maddalena, XML Data Stores: Emerging Practices, IEEE Internet Computing, Vol.9/Iss.2, 2005, pp.62-69

[63] A. Matheus, How to Declare Access Control Policies for XML Structured Information Objects using OASIS' eXtensible Access Control Markup Language (XACML), 38th Annual Hawaii Int. Conf. on System Sciences 2005, HICSS'05, Jan.2005

[64] M. Lorch, D. Kafura, S. Shah, An XACML-based policy management and authorization service for globus resources, 4th Int. Work. on Grid Computing, Nov. 2003, pp.208 - 210

[65] J. Schoenwalder, A. Pras, J.-P. Martin-Flatin, On the future of Internet management technologies, IEEE Communications Magazine, Vol.41/Iss.10, 2003, pp.90-97

[66] R. Sahita, COPS Protocol Provides New Way of Delivering Services on the Network, Intel website,accessed Sep.2005 `http://www.intel.com/` `technology/magazine/communications/nc05021.pdf`

[67] The openLDAP Foundation, `http://www.openldap.org`

[68] Network Simulator v2, accessed Sep.2005 `http://www.isi.edu/nsnam/` `ns/ns-build.html`

[69] T. F. Franco et al., Substituting COPS-PR: an evaluation of NETCONF and SOAP for policy provisioning, 7th IEEE Int. Workshop on Policies for Distributed Systems and Networks (Policy 2006)

[70] R. Badonnel, R. State, O. Festor, Probabilistic Management of Ad-Hoc Networks, IEEE/IFIP Network Operations and Management Symposium (NOMS 2006)

[71] G. Pavlou, A. Hadjiantonis, A.Malatras (editors), EMANICS NoE Deliverable D9.1. Frameworks and Approaches for Autonomic Management of Fixed QoS-enabled and Ad Hoc Networks, online, Dec. 2006, `http://emanics.org/component/option,com_remository/Itemid,97/func,fileinfo/id,47/`

[72] A. Malatras, Context-Awareness for the Self-Management of Mobile Ad Hoc Networks, PhD Thesis, Univ. of Surrey, 2007

[73] Aircrack-ng WLAN Tools, `http://www.aircrack-ng.org`

[74] A. K. Dey, , Understanding and using context, Journal of Personal and Ubiquitous Computing, Vol.5/Iss.1, pp.4-7, 2001

[75] A. K. Dey, G. D. Abowd, Towards a better understanding of context and context awareness, ACM Conference on Human Factors in Computer Systems (CHI 2000), April 2000

[76] A. Malatras, G. Pavlou, S. Gouveris, S. Sivavakeesar, V. Karakoidas, Self-Configuring and Optimizing Mobile Ad Hoc Networks, IEEE Int. Conf. on Autonomic Computing (ICAC'05), 2005

[77] A. Malatras, G. Pavlou, Context-driven Self-Configuration of Mobile Ad hoc Networks, IFIP Int. Work. on Autonomic Communications (WAC 2005), October 2005

[78] A. Malatras, S. Gouveris, S. Sivavakeesar, G. Pavlou, Programmable Context-Aware Middleware for the Dynamic Deployment of Services and Protocols in Ad Hoc Networks, 12th Annual HP-OVUA Workshop, 2005

[79] K. Henricksen, J. Indulska, A. Rakotonirainy, Generating Context Management Infrastructure from High-Level Context Models, 4th Int. Conf. Mobile Data Management (MDM2003) Jan. 2003, pp.1-6

[80] K. Henricksen, J. Indulska, A.Rakotomirainy, Modeling context information in pervasive computing systems. LNCS 2414, 1st Int. Conf. on Pervasive Computing (Switzerland), 2002, pp.167-180

[81] J.M. Serrano, J. Serrat, Context Modeling and Handling In Context-Aware Multimedia Applications, IEEE Wireless Communications Magazine 2006, Special Issue on Multimedia in Wireless/Mobile Ad-hoc Networks, October 2006

[82] B. N. Schilit, N. L. Adams, R. Want, Context-aware computing applications. IEEE Work. on Mobile Computing Systems and Applications, 1994

[83] A. Held, S. Buchholz, A. Schill, Modeling of context information for pervasive computing applications, SCI 2002/ISAS 2002

[84] ESPIRIT PROJECT 26900: Technology for enabled awareness (tea), 1998.

[85] GUIDE Project: Understanding Daily Life via Auto-Identification and Statistics http://seattleweb.intel-research.net/projects/guide/

[86] J. McCarthy, Notes on formalizing contexts, 13th Int. Joint Conf. on Artificial Intelligence, pp.555-560

[87] P. Otzurk, A. Aamodt, Towards a model of context for case-based diagnostic problem solving, Interdisciplinary conf. on modeling and using context(Rio de Janeiro, February 1997), pp. 198-208

[88] H. Chen, T. Finin, A. Joshi, Using OWL in a Pervasive Computing Broker, Work. on Ontologies in Open Agent Systems (AAMAS 2003)

[89] M. Conti, S. Giordano, Multihop Ad Hoc Networking: The Reality, IEEE Communications Magazine, Vol.45/Iss.4,2007, pp.88-95

[90] P. Gupta, P. R. Kumar, The capacity of wireless networks, IEEE Transactions on Information Theory, Vol.46,2000, pp. 388-404

[91] A. Gostev, R. Schouwenberg, War-driving in Germany - CeBIT2006 accessed Oct.2007, http://www.viruslist.com/analysis?pubid=182068392

[92] M. Burgess, G. Canright, Scalability of peer configuration management in logically ad hoc networks, eTransactions on Network and Service Management, Vol.1/No.1, 2nd Q. 2004

[93] IEEE 802.11 WLAN Working Group, accessed Sep.2007 http://grouper.ieee.org/groups/802/11/

[94] A. Jayasuriya et al., Hidden vs. Exposed Terminal Problem in Ad hoc Networks, Proc. of Australian Telec. Net. and App. Conf., Dec 2004

[95] Cisco Systems, Channel Deployment Issues for 2.4-GHz 802.11 WLANs, accessed on Sep.2007, http://www.cisco.com

[96] A. Malatras, A.M. Hadjiantonis, G. Pavlou, Exploiting Context-awareness for the Autonomic Management of Mobile Ad Hoc Networks, Journal of Network and System Management, Special Issue on Autonomic Pervasive and Context-aware Systems, Vol.15/Iss.1, Mar.2007

[97] OpenLDAP Foundation, OpenLDAP 2.3 Directory Services, `http://www.openldap.org`

[98] K. Zeilenga, J. H. Choi, The Lightweight Directory Access Protocol Content Synchronization Operation, RFC4533, Experimental, Jun.2006

[99] R. Bruno,M. Conti, E. Gregori, Mesh networks:commodity multihop ad-hoc networks, IEEE Communications Magazine, Vol.43/Iss.3, Mar.2005

[100] M. Sloman, E. Lupu, Policy Specification for Programmable Networks, Proceedings of First International Working Conference on Active Networks (IWAN'99), Berlin, June 1999

[101] D. C. Verma, Simplifying network administration using policy-based management, IEEE Network, Vol.16,Iss.2, Mar-Apr.2002

[102] E. Lupu, M. Sloman, Conflicts in policy-based distributed systems management, IEEE Transactions on Software Engineering, v.25, 1999

[103] J. D. Moffett, M. Sloman, Policy conflict analysis in distributed system management, Journal of Organizational Computing, v.4, 1994.

[104] M. Charalambides et al., Policy conflict analysis for quality of service management, 6th IEEE Workshop on Policies for Networks and Distributed Systems (Policy 2005)

[105] M. Charalambides et al., Dynamic policy analysis and conflict resolution for DiffServ quality of service management, IEEE/IFIP Network Operations and Management Symposium (NOMS 2006)

[106] M. Burgess, An approach to understanding policy-based on autonomy and voluntary cooperation, 16th IFIP/IEEE Distributed Systems: Operations and Management Workshop (DSOM2005), LNCS 3775

[107] IBM Inc., Autonomic Computing Adoption Model, Accessed online Dec 2007 `http://www.ibm.com/autonomic/about_get_model.html`

[108] T. R. Andel, A. Yasinsac, On the credibility of manet simulations, IEEE Computer, Vol.39/Iss.7, July 2006

[109] M. Sloman,Policy Driven Management For Distributed Systems, Journal of Network and Systems Management, Vol 2(4), Dec. 1994

[110] J. Strassner, Policy-Based Network Management, Solutions for the Next Generation, Morgan Kaufmann, ISBN: 1558608591, 2003

[111] W. Chen, N. Jain, S. Singh, ANMP Ad hoc network management protocol, IEEE Journal on Selected Areas in Communications, Vol 17/Iss.8, Aug.1999

[112] K. S. Phanse, L. A. DaSilva, Protocol support for policy-based management of mobile ad hoc networks, IEEE/IFIP Network Operations and Management Symposium (NOMS), 2004

[113] A. Malatras, G. Pavlou, Context-driven Self-Configuration of Mobile Ad hoc Networks, IFIP International Workshop on Autonomic Communications (WAC 2005), Oct.2005

[114] P. Bellavista, A. Corradi, R. Montanari, C. Stefanelli, Context-aware middleware for resource management in the wireless Internet, IEEE Transactions on Software Engineering, Vol.29/Iss.12, Dec.2003

[115] S. S. Yau, F. Karim, Y. Wang, B. Wang, S. K. S. Gupta, Reconfigurable Context-Sensitive Middleware for Pervasive Computing, IEEE Pervasive Computing, Jul.-Sept. 2002

[116] P.-J. Wan, K. M. Alzoubi, O. Frieder, Distributed construction of connected dominating set in wireless ad hoc networks, IEEE Infocom 2002

[117] R. Friedman, M. Gradinariu, G. Simon, Locating cache proxies in MANETs, ACM MobiHoc 2004

[118] U. Kozat, L. Tassiulas, Network layer support for service discovery in mobile ad hoc networks, IEEE Infocom 2003

[119] S. Gouveris, S. Sivavakeesar, G. Pavlou, A. Malatras, Programmable Middleware for the Dynamic Deployment of Services and Protocols in Ad Hoc Networks, IEEE/IFIP Integrated Management Symposium (IM 2005), May 2005

[120] A. Westerinen et al., Terminology for Policy-Based Management, RFC 3198, Informational, Nov.2001

[121] B. Moore, E. Elleson, J. Strassner, A. Westerinen, Policy Core Information Model-Version 1 Specification, RFC 3060, Feb.2001

[122] B. Moore, Policy Core Information Model (PCIM) Extensions, RFC 3460, Jan.2003

[123] J. Strassner, B. Moore, R. Moats, E. Elleson, Policy Core Lightweight Directory Access Protocol (LDAP) Schema, RFC 3703, Feb.2004

[124] M. Pana, A. Reyes, A. Barba, D. Moron, M. Brunner, Policy Core Extension Lightweight Directory Access Protocol Schema (PCELS), RFC 4104, Jun.2005

[125] M. Wahl, T. Howes, S. Kille, Lightweight Directory Access Protocol (v3), RFC 2251, Dec. 1997

[126] C. Zhao , G. Wang, Fuzzy-control-based clustering strategy in MANET, Fifth World Congress on Intelligent Control and Automation, WCICA 2004, Vol.2, pp.1456-1460

[127] J. I. Kim, J-Y. Song, Y-C. Hwang, Location-Based Routing Algorithm Using Clustering in the MANET, Future Generation Communication and Networking, Vol.2, Dec. 2007, pp.527-531

Chapter 8

Autonomous Machine Learning Networks

Lei Liu

Sun Microsystems, Inc. 16 Network Circle, Menlo Park, CA 94025

Abstract

The next generation of internet and wireless networks demand pervasive storage graphs. Storage Intrusion Prevention Systems (SIDSs) are known techniques for identifying attacks. In pervasive un-directional graphs or intensive algebraic operations, victim states in final stages tend to be easily identified within a finite-time horizon by sensing abrupt transitions in system and network state spaces. However, in the early stages of such attacks, these changes are hard to prevent and difficult to distinguish from the usual state fluctuations. Most of the traditional prevention techniques employ parametric model approaches. Recently proposed model-free approximation techniques result in intensive computation. In order to provide autonomy to various large, or high dimensional state spaces, we divide the problem of attack classification into two parts: class prediction problems and cluster prediction problems. We propose a formal model-free prevention technique. Also, we propose using an autonomous utility framework in combination with a programable architecture, a set of algorithms and optimization methods which are grounded from Reinforcement Learning (RL) and Dynamic Programming (DP) techniques to tune prevention networks. We propose an autonomous learning network adjusted by the RL algorithms for stochastic intrusion environments in order to derive optimization procedures. Such procedures optimize policies and reduce time and space complexity within large prevention spaces. Moreover, we use attribute and instance selection for data reduction of pervasive learning samples. In numeric and information evaluation, we present convergency analysis and error estimation. Finally, experimental results with autonomic tuning learning methods for large datasets show the effectiveness of proposed approach.

8.1   Introduction

In principle of information security, intrusion detection is a method of identifying attempts to compromise CIAA (confidentiality, integrity, availability and authenticity) of computational resources. In practice of data center operations, intrusion detection may be performed manually or automatically. Manual intrusion detection examines evidence from system calls, log files, audit trails or flows of network packets. Systems that perform automated intrusion detection are categorized as intrusion detection systems (IDSs). Modern IDSs are combinational applications of the above two classes of techniques.

As probable attacks are discovered by IDSs, relevant information could be logged to files or databases and alerts will be generated. In addition, automatic responsive activities and even preventive rule-based controls can be implemented through access control systems.

Storage-based intrusion detection systems (SIDSs) monitor object or attribute modification characteristics, and probe common intruder actions visible between storage interfaces [1]. SIDSs focus on the threats that compromise host systems and continue to operate even after the attacks succeed [2].

Due to advances in information security techniques, intrusion preventive systems (IPSs) provide a practical mechanism to defend against intrusion attacks on computational graphs or prevent abuse of computational resources [3]. IPSs can be divided into two categories: misuse-prevention techniques [4] and anomaly-prevention methods [5]. Misuse-prevention tools are based on a set of signatures [6], which describe a known attack state and are used to match current state against an attack state. If a true attack match one of attack class listed in the database, then it is successfully identified. A disadvantage of this approach is that it cannot prevent unknown attacks (also called zero-day attacks). Anomaly prevention uses profiles about normal states to prevent intrusions by noting significant deviations between the parameters of the observed traffic and those of normal states of networks. Prevention against anomaly attacks has been an active research topic. However, it suffers from high false-positive error rates because unseen normal behaviors may be misclassified as attacks. In this paper, we focus on the optimization of anomaly prevention.

Machine learning-based prevention networks are not new in IDS and IPS literature [7] [8]. This research focuses on autonomous machine learning methods. Our approach is different from traditional learning based IDSs and even traditional AI machine learning literature in following major areas.

- Detection can be initialized with any arbitrary networks with programmable architecture but will be autonomically tuned based on proposed RL utility framework. In [9] [7], neurons require fixed architecture. In order to train learning networks, numbers and types of input nodes, layers, the nodes to present non-linearity and output nodes are predefined based on problem specifications. However, these specifications may not be obtained ahead or even not feasible for online detection. Hence, it strictly limits performance and convergence of detection. In addition, it is a static detection architecture. In this paper, we propose a programmable architecture without fixed input nodes, hidden layers, and output nodes. It is based on reinforcement learning utility framework to autonomically tune the input nodes, hidden layers and output nodes. Therefore, the architecture of the neurons becomes the parameters of proposed RL tuning framework. This is the fundamental difference between our approach and other learning based IDSs in literature. Evolution computing was proposed improve learning networks [10]. However, it inherits the memory bound problem in traditional AI techniques.

- Approximating architecture is not restricted to multilayered perceptron structure with sigmoidal nonlinear functions. Any arbitrary structure, such as Bayes Networks, can be autonomically constructed and tuned.

- Previous learning-based IDSs require supervised training [9] [11] [7] [12]. It requires training with known attack types to map input and output matching unknown non-linearity. However, in dynamic programming(DP) settings, there is no available training set, which could be used for approximation. The only solution is to evaluate by simulation and improve the decision using simulation results. Tuning autonomically with proposed RL based utility, online detection without training can be optimized by simulations. With simulations for unknown attacks, most frequently occurred states are better approximated than average ones. This is another major difference between traditional machine learning methods in IDSs and our proposed tuning detection framework.

- To optimize previous fixed detection models, a tuning-based intrusion detection system was proposed [12]. However, the method utilizes human interactions to continually improve rule-based binary classifiers.

- There are no specified parameters in the feature vectors. This is one of the major advantages of our approach. It can start with any arbitrary number of parameters as input to prevention networks. Traditionally, feature vectors are summarized in heuristic sense to be critical characteristics of each state [9]. Analysis and intuition need to be applied to derive the feature parameters. In our proposed approach, there is no static architecture, and the number of parameters can be tuned by proposed RL utility framework.

- Existing machine learning methods [9] [8] [3] are diversified and computationally intensive with a considerable amount of training errors for large state spaces. Hence, off-line training are used for approximation mapping. This research proposes RL based tuning utility framework to ensure online learning to improve detection in real time. It has the flexibility to control intrusion detection.

- In complex systems with large state spaces, a lot of training data are required for approximation if traditional machine learning networks are used for detection. However, sufficient volume of data will not be available in real time. We propose an autonomous tuning with training and online learning which is a combination of evolution of DP, RL, and Neuro Networks in which RL is used for tuning.

- In traditional learning networks, tuning is only limited within fixed architecture with optimization of weights and biased nodes. We propose an interactive utility framework to tune entire architecture using the least squared error fitting.

- Different from traditional learning networks, our approach is self-managed learning method such that biased datasets may only impact convergence but not detection results.

The purpose of this research is to provide an autonomic prevention technique and framework for problem solving and reasoning within next generation SMDP systems and associative task environments. To serve large continuous system spaces, RL tuned prevention function approximation and algorithms are proposed to provide state-action and reward/cost factor mapping. To construct features of large state and action spaces, it depicts and reasons about capacity and resource contention of computational graph. SMDP theory is applied. Specifically, RL search approximation and learning factor multilayer perceptron are proposed to simplify search decision organization and prediction. In addition, a set of RL interactive algorithmic operations and augmented learning network operations are suggested.

We propose an autonomous tuning of prevention networks without fixed input parameters and layer architecture. The tuning is conducted by a incremental algorithm rooted from reinforcement learning techniques. In addition, different from traditional optimization techniques requiring static state specification, this research uses only arbitrary state parameters for intrusion prevention optimization. For machine learning in large state space, sample reduction methods are used. Specifically, both attribute selection and random instance selection are applied to Q learning of the prevention networks.

The rest of the paper is organized as follows. Section 8.2 formulates the prevention problem and associated properties. Section 8.3 briefly reviews the solution space in anomaly attack prevention literature. Section 8.4 proposes a set of autonomous prevention optimization and tuning strategy including architecture, algorithmic operations and, data structures for stochastic systems. Section 8.5 presents performance evaluation and convergence analysis of the proposed algorithms and data structures. Section 8.6 presents experiments using the published datasets from MIT Lincoln Lab. We compare test results for various learning methods and the proposed anomaly prevention framework. Finally, Section 8.7 concludes the paper.

8.2 Problem Formulation

Anomaly attack is one of stochastic processes, which generates random state transitions of information structure. Intuitively, attack prevention observes the relative abrupt changes [13], which are stochastic within an infinite time horizon. In the above setting of task environment, attack prevention is a dynamic problem, which could be further formulated as either objective function parametric optimization problem or dynamic action state control optimization problem.

Parametric function optimization can be easy implemented for function evaluation with fixed parameters. However, it may lead to high dimensionality and searching over entire solution space has slow convergence. Also, control optimization needs to incorporated in the middle of trajectories. Furthermore, each problem has it own uniqueness. Prevention actions and associated state changes form a stochastic decision process which is usually probabilistic. With such memoryless properties, intrusion prevention is modeled as a class of control optimization problems with a semi-Markov model. Therefore, an autonomous prevention is a state-action control optimization problem to dynamically locate an optimal action in each state in the system.

Since it is a memoryless process, each state transition probability is independent of previous states but not of current state. The transition time is not a unity time steps. Hence, it is a semi-Markov Decision Problem (SMDP). However, in real systems, state transition probabilities and reward discount factors are unknown or hard to compute. Without transition probability matrices (TPM), transition reward matrices (TRM), and transition time matrices (TTM) as the core stone of theoretical model, autonomous prevention is hard to formalize as a traditional Markov Decision Problem (MDP), which can be solved by DP algorithms.

Objective prevention function may have high dimensional random decision variables in a large state space. It may be hard to obtain the probability mass or distribution function of these parameters. Furthermore, by given a distribution function, it may not have closed form solutions. Even with distribution of system inbound random decision variables, estimated transition probability and reward matrices could generate TPMs and TRMs via simulators. The solution does not scale-up to large state and solution spaces. For large state space, state-aggregation is required to approximate model with a manageable number of states.

The goal of autonomous attack prevention is to sense and evaluate abrupt changes with fixed or dynamic minimal average delay to minimize the false alarm ratio. As rewards are returned as feedback to SMDP task environment, these autonomous methods can be used to predict resource contention and dependability to ensure service survivability. Hence, the above goal formulation is involved with performance metrics: the rate of false alarm, detection rate and, the delay of detection. We use a augmented receiver operating characteristic (ROC) curve [14] [8] with the delay of detection. For this infinite horizon problem, long run average reward is used in performance metric. Intuitively, the relaxed version of the problem could achieve the minimized average delay of detection actions for a given fixed false alarm rate.

A MDP for a single agent can be illustrated by a quadruple (S, A, R, T) consisting of

- A finite set of states $S, s \in S$

- A finite set of actions $A, a \in A$

- A reward function $c : S \times A \times S \rightarrow \Re$

- A state transition function: $T : S \times A \rightarrow PD(S)$ It maps the agent's current state and action into a set of probability distribution over inputs.

States Q:

- x_1: type of machine learning networks

- x_2: input nodes of learning networks

- x_3: output nodes of learning networks

- x_4: hidden layer of learning networks

- x_5: numeric weight matrix of learning networks

- x_6: Q factors of ROC values

This conservative state-space definition satisfies the semi-Markov properties. However it results in too large state space to approximate in practice.

With RL optimization, prevention reward can be treated as state value learned by interactions. Therefore, with RL methods, optimal policies can be reached without explicit search over possible sequence of entire states and actions. Hence, dynamic prevention problems can be resolved with large and even infinite space with function approximation.

Consequently, dynamic prevention problem is formulated as RL and DP based SMDP problem with model free prevention function approximation. The attack classification problems can be divided into two sub-problems below.

8.2.1 Attack Prediction Problem

Attack class prediction is to map a discrete attack feature space to labeled attack classes. Specifically, the proposed attack class prediction problem is to assign data samples to specific class category. Intuitively, class prediction problem can be formulated as a computation problem to derive class predictors from known classes for unknown classes. It is a machine learning and prediction task environment.

8.2.1.1 Task Environment Formulation

Environment Properties

- The class prediction process has a set of training samples.

- Since classification is completely dependent on the current feature extraction and attribute values, it is a deterministic task environment.

- It is a discrete task environment with a set of discrete training and incoming datasets.

8.2.1.2 Goal Formulation

Goal Properties

- state: A classification state specified by s_{class}, specifically, it is the classification assignment of a data sample into a labeled class type.

- performance: The receive operating characteristics (ROC) curve [14] was proposed in literature for machine learning performance measurement. This research proposes an augmented ROC curve with the delay of prediction.

8.2.1.3 Problem Formulation

Problem Properties:

- states: A finite set of feature extracted training samples with attack attribute values. For a k sample dataset, there are $42 \times k$ possible predictors for learning attack classification state space. Since the ratio of dimensionality is $k \div 42$, it is a large samples with mediate dimension dataset. Intuitively, the class prediction problem may divide into two sub-problems: attribute or instance selection problem and predictor learning problem. The attribute selection problem is to discover the attack features, which are strongly related to class distinction. Consequently, the class prediction problem is transformed into instance or attribution selection and learning prediction upon the selected samples.

- initial state: There are $75MB$ attack attribute values for 42 attribute samples. The 4 labeled classes for the samples are itemized in the arff files.

- goal state: Discovered classification predictors and built classification model to categorize the samples into the classes.

- $Goal\_Test()$: The classification should align the performance function with the ROC curve. For convenience, AUC, the area under the ROC curve, is selected for performance measurement. The AUC should be greater than 0.9.

- Classification Cost: It is the delay with algorithmic complexities.

Consequently, class prediction problem is formulated as attribute selection problem and predictor learning problem upon selected attributes of large samples with medium dimensionality. Large samples are used to learn and test the prediction with large samples.

8.2.2   Attack Class Discovery Problem

Attack class discovery by attack attributes is the classification of samples into different groups. The training datasets are partitioned into sub-sets so that each cluster shares common traits, which are represented by extracted feature sets. The attack class discovery problem proposed in this research is to group data samples as specific class categories which represents not only current data clusters but also unknown classes. Intuitively, class discovery problem can be formulated as a computation problem to derive cluster predictors from known classes for unknown classes.

It is formulated as a clustering problem [15]. Hence, it is a machine learning and prediction task environment.

8.2.2.1   Task Environment Formulation

Environment Properties

- The clustering process has a set of percepts and training samples.

- Since clustering is completely dependent on the current feature extraction and attack attribute values, it is a deterministic task environment.

- It is a discrete task environment with a set of discrete training and incoming datasets.

8.2.2.2 Goal Formulation

Goal Properties

- state: a clustering state specified by $s_{cluster}$, specifically, it is the cluster classification of given data samples into groups

- performance: The receive operating characteristics (ROC) curve [14] was proposed in literature for machine learning performance measurement. This research proposes an augmented ROC curve with the delay of prediction.

8.2.2.3 Problem Formulation

Problem Properties:

- states: A finite set of feature extracted training samples with attack attribute values. For a k sample dataset, there are $42 \times k$ possible predictors for learning attack cluster state space. Since the ratio of dimensionality is $k \div 42$, it is a large sample with mediate dimension dataset. Intuitively, the class discovery problem may divide into two sub-problems: instance or attribute selection problems and clustering learning problem. The instance or attribute selection problem is to discover the attack features, which are strongly related to class distinction. Consequently, the class discovery problem is transformed into instance or attribution selection and learning prediction upon the selected samples.

- initial state: There are 42 attribute values of samples. The labeled classes for the samples are listed in the arff files.

- goal state: discovered cluster predictors and built cluster models to categorize the samples into the classes. Possible unknown types needs to be identified with the proposed algorithms and data structures.

- *Goal_Test*(): The clustering should align the performance function with the ROC curve which is greater than 0.9.

- Clustering Cost: It is the delay with algorithmic complexities.

Consequently, class discovery problem is formulated as instance or attribute selection problem and cluster learning problem upon selected attributes of small samples with high dimensionality.

8.3   Related Work

One of the main threads of detection studies is a model building centric approach based on check point theory [13]. CUSUM detection and Shiryaev-Pollak detection [16] algorithms provide optimal solution with probabilistic analysis upon independent observations with identical distribution. Recent advances [17] in check point theory eliminate the identical distribution assumption for detection operations.

In addition, Markov Model was proposed for anomaly detection [18]. Markov-chain model represents a profile of computer-event transitions in a normal and usual operating condition of a computer and network system (a norm profile). The Markov-chain model of the norm profile is generated from historical data of the normal system activities. The observed activities of the system are analyzed to infer the probability that the Markov-chain model for the norm profile supports the observed activities. The lower probability the observed activities receive from the Markov-chain model for the norm profile, the more likely the observed activities are anomalies resulting from attacks. Inference algorithms of this anomaly-detection technique based on the Markov-chain model for a norm profile examined performance using the audit data of host machines.

Machine learning methods [7], such as the rule-learning technique [12], the statistical technique [19] [20], and the hidden Markov model (HMM) [21], were used to profile the behavior of these programs. This was viewed as one of the traditional problems in pattern classification [22]. Neural networks have been actively applied to IPSs. Especially, in the 1999 Defense Advanced Research Projects Agency (DARPA) intrusion detection evaluation (IDEVAL), the detection technique based on neural networks showed superior performance to other techniques in preventing host based attacks [9] [23]. However,the profiling of normal pattern was computationally intensive operations due to large audit data. This was the major drawback of traditional neutral network based prevention. In addition, evolutionary learning [10] was proposed for intrusion prevention. A fuzzy rule based approach [24] was also suggested.

Anomaly-detection techniques based on Bayes parameter approximation for building a Markov model for a norm profile and likelihood ratio tests for detecting anomalies, Bayesian networks, hidden Markov models, or principal components analysis, are also investigated. However, the training of all these techniques, taking into account the ordering of events, is computationally intensive [25]. The inference of attacks is also computationally intensive. These techniques require large amounts of memory to store a large set of event sequences. The anomaly-detection techniques based on the traditional Markov model are computationally intensive due to their use of the Bayes parameter estimation for learning the norm profile and a likelihood ratio test for inferring anomalies. Considering large amounts and high frequency of events in a computer and network system, those techniques are not optimal to prevent attack in real time.

Another thread of prevention studies was specification-based anomaly-detection techniques [26] that describe authorization policies of a well-defined subject such as a privileged user process or a host by formal logic, rules, and/or graphs. However, it is hard to enumerate and specify all possible normal activities or security policies of a subject, especially for a large-scale subject such as a host or a graph with interactions. Moreover, activities of many subjects involve uncertainties and variations which cannot be well addressed by formal logic, rules or graphs.

Statistical anomaly-detection techniques [27] [19] [20] [28] have demonstrated promising performance for attack identification with respect to detection rate and false alarm rate. In these approaches, a statistical profile for a subject is constructed from historical normal activity data. These statistical profiles do not consider the order in which the events occur to the subject. While it is possible to improve attack detection performance by incorporating the ordering of events. There exist anomaly-detection techniques which applied recurrent and perceptron neural networks to predict the next event from a series of the past events.

Social-engineering defense was proposed [29] for attacks to compromise security metrics by exploiting end users through natural language based on common deception. Specifically, the suggested natural-language processing utilizes rule-based detection methods for intrusion detection.

Distributed multi-agent IDS architecture was studied with light weight agents across multi-layers [30]. An adaptive sub-eigenspace modeling technique was proposed to optimize the proposed distributed agent detection method. However, co-operativeness and convergency of multi-agent learning is still an open issue.

Other than software-based techniques, researchers have started to work on reconfigurable hardware implementations of data mining methods [31]. In addition, a FPGA-based detection architecture was proposed [32] for high performance and throughput demands compared with software and application-specific integrated-circuit implementation. This is one of emerging research topics in intrusion detection literature. However, it is not the scope of this research.

For prevention action control optimization, MDP yields optimal solutions on a control optimization problem as additional non-linear-programming methods in stochastic dynamic programming such as value and policy iterations. However, for a complex system, it is hard to construct the theoretical model specifically, required for MDP formulation. Law & Kelton [33] applied neural network for function approximation. Elevator scheduling case study [34] addressed multi-agent scenario using semi-Markov models. Choosing the right scheme for function approximation is time consuming for RL algorithmic operations. In addition, there is no convergency proof for the combination of RL with function approximation methods. Hence, there is no strategy formulated for the approximation currently.

As a random variable, the expected immediate reward of state transition with action a from state s_i to state s_j can be conceptually defined as:

$$\bar{r}(s_i, a) = E[r(s_i, a)] = \sum_{s_i \in S} p(s_i, a, s_j) r(s_i, a, s_j) \qquad (8.1)$$

- $p(s_i, a, s_j)$ is the long run transition probability from state s_i to state s_j when MDP is running with action a selected in state s_i

- S denotes a set of decision states in the sample space

- $r(s_i, a, s_j)$ is the immediate reward earned from state s_i to state s_j

If a deterministic stationary policy $\vec{\pi}$ is followed, $\pi(s_i)$ is the action that will be selected in state s_i. Intuitively, the average reward associated with policy $\vec{\pi}$ is derived as:

$$\bar{r}(\pi(s_i)) = E[r(\pi(s_i))] = \sum_{s_i \in S} p(s_i) \bar{r}(s_i, \pi(s_i)) \qquad (8.2)$$

- $\vec{\pi}$ is a n-tuple associated with n states

- $p(s_i)$ is the infinitely long run transition probability of state i when MDP is running with the policy $\vec{\pi}$ followed

- S denotes a set of decision states in the sample space

- Under policy $\pi(s_i)$, $\bar{r}(s_i, \pi(s_i))$ denotes the expected immediate reward earned in the state s_i when action $\pi(s_i)$ is selected in state i.

With the above equation, optimal policies can be evaluated with the maximum rewards. Therefore, the above performance objective function provides a conceptual model for control optimization. To identify the optimal action value and associated policy, the transition probability matrices and transition reward matrices associated with each policy are computed to evaluate the objective performance metrics resulted from each policy. This limits the solution to small state spaces due to the exponential computation for m^n (n states and m actions in each state) introduced by the above conceptual model. To optimize controls, DP algorithms [35] were evolved to solve the MDP problems. DP algorithms were proposed as the linear system of equations [36] to resolve MDP problems. By solving Bellman equations, the value components of value function vector can be utilized to locate the optimal actions or policies. There are various forms of Bellman equations to resolve MDP problems. The Bellman equation for a given policy in the average reward context requires to conduct k iterations of evaluation on linear equations in which each policy is selected in the iteration to derive the optimal policies. Value iteration is another form of Bellman equations. It does not require solving any equation. It has been heavily evolved as the pillar of RL algorithms.

From Bellman optimality equation for average reward

$$V^*(s_i) = \max_{a \in A(s_i), s_i \in S} [\bar{r}(s_i, a) - \rho^* + \sum_{s_j=1}^{N} p(s_i, a, s_j) V^*(s_j)] \qquad (8.3)$$

Where

- $A(s_i)$ denotes the finite set of actions taken in state s_i followed by policy $\overrightarrow{\pi}$

- $V^*(s_i)$ denotes the element of value function vector $\overrightarrow{V}$ associated with optimal policy for state s_i

- $p(s_i, a, s_j)$ denotes the transition probability from state s_i to state s_j under action a.

- $\bar{r}(s_i, a, s_j)$ denotes the expected immediate reward in state s_i as action a is taken

- ρ^* denotes the average reward associated with the optimal policy.

To resolve the value bound problem, relative value interaction was proposed. In the kth interaction

$$V_{k+1}(s_i) = \max_{a \in A(s_i), s_i, s_j \in S} [\bar{r}(s_i, a) - \rho^*$$

$$+ \sum_{s_j=1}^{N} p(s_i, a, s_j) V_k(s_j)] \qquad (8.4)$$

$$- V_{k+1}^*(s_i)$$

In the setting of non-unity transition times, relative value interaction was proposed. In the kth interaction

$$V_{k+1}(s_i) = \max_{a \in A(s_i), s_i \in S} [\bar{r}(s_i, a) - \rho^* t(s_i, a, s_j)$$

$$+ \sum_{s_j=1}^{N} p(s_i, a, s_j) V_k(s_j)] \qquad (8.5)$$

$$- V_{k+1}^*(s_i)$$

Where $t(s_i, a, s_j)$ denotes the transition time from state s_i to state s_j followed by policy $\overrightarrow{\pi}$ with action a taken place. The equation above indicates that the maximum value V^* selection may run away from regular value iterations with deterministic optimal path location.

The above interactive conceptual model requires intensive computation. In addition, it requires a theoretical model.

From RL literature, a provable convergent approach has been developed for learning optimal policies in continuous state and action spaces under average rewards [37]. This technique has been applied to research [38] with non-trivial configurations. However, there is no existing discussion on RL and optimization within search state and action space. We propose self-organization of index structure for state-action value mapping for search function numeric method approximation.

8.4   Methodology

Optimization of intrusion prevention is to tune an initial prevention network with arbitrary network parameters and architecture. The objective is to obtain the approximate optimal architecture with a certain input vector for a specific attacks. The number of resulted networks depends on the average rewards and long-run cost of the detection optimization. It can be both an online and off-line incremental learning method to optimize the traditional neuron based detection techniques. The tuning is a reinforcement learning problem for optimizing neuron architecture. Hence, "dynamic" and "autonomous" are the major characteristics of research methodology of this research.

- Dynamic:Since intrusion prevention is formulated the detection as stochastic DP problem, it owns dynamic properties.

- Autonomous:To handle large state space and unknown attack type, the DP problem transformed into RL based autonomically tuning problem. Autonomous in terms of tuning the networks interactively with programmable architecture, algorithms and data structures.

- Data Reduction: To tune large data instances with mediate to high dimensional attributes, random subset of sample instances are selected by $10 - fold$ extraction to avoid of biased filtering. In addition, to search the space of attribute subset, greedy hill-climbing augmented with a backtracking feature is utilized for correlation based reduction. The data reduction technique results in speedup of detection convergency.

Dynamic Programming Algorithms require computation of theoretical model of a system on transition probabilities, the transition rewards and transition times. These quantities are unknown or hard to evaluate. Specifically, to obtain transition probabilities, multiple integrals with the probability distribution functions of many random variables are involved in the calculation on a complex stochastic system.

Both DP and RL algorithms have dependency with $\sum_{x \leq t} p_x(x)$, the distributions of the random variables that govern the system behavior for a time interval t. For continues random variables, $\int_{-\infty}^{x} f(x)dt$ is applied. However, RL does not need the above theoretical model quantity evaluation or estimations but simulate the system using the distributions of the governing random variables. For autonomous search, one needs the distribution of random decision variables.

However, Reinforcement learning techniques provide near optimal solutions without evaluation of the above quantities. Similar with DP techniques, one element of value function vector is associated with each state variable. RL search algorithm associates each element of the value mapping vector with a given state-action pair (i, a) as $Q(i, a)$ for search cost or reward evaluation to derive optimal polices.

Value iteration algorithms do not compute the system of linear equations. However, value iteration for MDP poses the difficulties for Bellman optimality equation computation with unknown value of the average reward (ρ^*). FOR SMDP in the average reward context gets in to approximation solution space. However, regular DP value iteration becomes unbounded for average reward MDP. Hence, relative value iteration keeps the bound operation. Due to the discounting factor is hard to measure in the real world, average reward tends to be a more popular performance measure.

Discrete Environmental Signal: After each action, the learning agent receives immediate feedback from the above task environment. The following reinforcement immediate reward signals $r \in \Re$ are listed below.

$$r = \begin{cases} a & \text{A penalty resulted from a wrong action} \\ b & \text{A reward is granted as a target is learned} \\ c & \text{The goal state is reached} \\ -1 & \text{Cost of any action during learning} \\ 0.0 & \text{otherwise for all nonterminal states} \end{cases}$$

Moving to the target is less than the absolute value of moving to wrong direction since keeping agent in a valid task environment is more important than getting to the target $(|a| > c > b)$. Intuitively, agent needs to learn to achieve goal without rewards. In other words, the absolute value of penalty should be significantly larger than the reward from the excepted move. The agent is required to stay in task environment to move to the goal state. However, the sooner the action is rewarded, the sooner the agent should learn. The cost of moving is measured by a negative unit reward from task environment. It is larger than the reward from getting to the correct positions so that the steps in the path are considered into search evaluation to achieve the minimized path $(0 < b < c < 1)$.

To further relax the large stages, infinite horizon problem represents a approximation. For large state space, value-mapping vectors are not stored explicitly but function-approximation to solve the high dimensionality problem during RL value iterations. If the time spent in a transition is not unity, the average reward per unit time over an infinite time horizon is defined as:

$$\rho = \lim_{k \to \infty} \frac{E[\sum_{i=k}^{k} r(x_i, \pi(x_i), x_{i+1}]}{k} \tag{8.6}$$

From Bellman optimality equation for average reward

$$V^*(s_i) = \max_{a \in A(s_i), s_i \in S} \sum_{s_j=1}^{N} p(s_i, a, s_j)[r(s_i, a, s_j) + V^*(s_j)] \tag{8.7}$$

Where

- $V^*(s_i)$ denotes the element of value function vector $\overrightarrow{V}$ associated with optimal policy for state s_i

- $p(s_i, a, s_j)$ denotes the transition probability from state s_i to state s_j under action a.

- $\overline{r}(s_i, a, s_j)$

From Robbins-Monro algorithm [39], value and state-action mapping vectors in Bellman equation is defined without the dependency of transition probability:

$$Q(i, a) \leftarrow (1 - \alpha)Q(i, a) + \alpha[\gamma(i, a, j) + \max_{b \in A(j)} Q(j, b)] \tag{8.8}$$

To resolve the divergent iteration problem of traditional Bellman optimality equations and associated DP algorithms, relative value function proposition can be utilized.

$$Q(i, a) \leftarrow (1 - \alpha)Q(i, a) + \alpha[\gamma(i, a, j) + \max_{b \in A(j)} Q(j, b) - Q(i^*, a^*)] \tag{8.9}$$

Where a state-action pair (i^*, a^*) is selected arbitrarily.

Since discounting factor is usually unknown in the real world problems, taking transition time without unity into considerations, Bellman equation has difficulties to apply into value iterations with unknown average reward of the optimal policy ρ^* without normalization of SMDP. However, by estimation of $\widehat{\rho}$, the above equation is given as:

$$Q(s_i, a) = (1 - \alpha)Q(s_i, a) + \alpha[\gamma(s_i, a, s_j) - \widehat{\rho}t(s_i, a, s_j) \\ + \max_{b \in A(s_j)} Q(s_j, b) - Q(s_i^*, a^*)] \tag{8.10}$$

8.5   Evaluation

The fundamental property of the value function is used throughout entire study of the learning agent for buffer management tuning. This can produce

---

DYNAMIC-DETECT(s,a,limit)

    Description: Dynamic detection with learning network function
approximation
        INPUT: state space,action space and maximum iteration jumps
        OUTPUT: solution to the problem
        $S_{sub}, A_{sub} \leftarrow$ **SUB-SPACE**(S,A)
        INIT-FACTORS($S_{sub}, A_{sub}, limit$)
        for each k **in** k_max **do**
            LOCATE-ACTION(S,A,Q)
            $r_{s_i,a,s_j}, t_{s_i,a,s_j} \leftarrow$ **PROBE**(s_i, a, s_j)
            SIMULATE $-$ **ACTION**($r_{s_i,a,s_j}, t_{s_i,a,s_j}$)
            UPDATE $-$ **ACTION**($Q(s_i, a), A$)
            TUNE $-$ **NEURO**($Q(s_i, limit), A$)
            LOCATE $-$ **POLICY**($Q(s_i, a)$)
        return max $Q(l, b)$

FIGURE 8.1: Dynamic machine learning detection algorithm

accurate estimation. Moreover, this estimation is represented as a tabular
with a set of tuples, which is are states or state-action pairs.

$$Q(s,a) = Q(s_t, a_t) + \alpha_t \delta_t \tag{8.11}$$

$$\delta_t = r(s_t, \pi(s_t), s_{t+k}) - \rho_t + \max_a(Q(s_{t+1}, a) - Q(s_t, a_t)) \tag{8.12}$$

$$\rho_{t+1} = \rho_t + \beta_t(\gamma(s_t, a_t, s_{t+1}) - \rho_t) \tag{8.13}$$

Where ρ_t, β_t are positive step parameters set to $\frac{1}{t}$. The discount rate is
$\gamma \in [0, 1)$. The above iterative learning converges to the optimal state-action
value with discount. However, there is no convergence proof in cases of average
reward. In addition, the above equations do not fit for large or continuous
action and state spaces.

Let $\tilde{Q}(s, a, p)$ approximate to $Q^*(s, a)$ based on a linear combination of basis
functions with a parameter vector p:

$$\tilde{Q}(s, a, p) = \sum_{1 <= i < M} p^i \phi^i(s, a) \tag{8.14}$$

Where a column vector with a fixed number of real valued components,

$$p = (p^1, p^2, ..., p^M)^T \tag{8.15}$$

and

$$\phi(s, a) = (\phi^1(s, a), \phi^2(s, a), ..., \phi^M(s, a))^T \tag{8.16}$$

Hence, $\tilde{Q}(s, a, p)$ is a smooth differentiable function of p for all $s \in S$. The
above Bellman equation error can be estimated by the mean-squared-error
over a distribution P of inputs.

TUNE-LEARNING(s_i,limit)

    Description: Tuning learning network function approximation
    INPUT: state space,action space and maximum iteration jumps
    OUTPUT: solution to the problem
    $w(i) \leftarrow RANDOMIZE()$
    $SSE_{old} \leftarrow 100000000000f$
    $tolerance \leftarrow RANDOMIZE()$
    for each m **in** N **do**
        for each j **in** k **do**
            $o_p \leftarrow o_p + w(j) * x_p(j)$
        $w(i) \leftarrow w(i) + \alpha * o_p$
        $\alpha \leftarrow A/m$
        $m++$
    if$(SSE - SSE_{old} < limit)$**BREAK**

FIGURE 8.2: Machine tuning algorithm

$$MSE(\overrightarrow{p_t}) = \sum_{s \in S} P(s)[Q^*(s,a) - \tilde{Q}(s,a,p)]^2 \qquad (8.17)$$

$$MSE(\overrightarrow{p_t}) = \sum_{p=1}^{n} (y_p - o_p)^2 \qquad (8.18)$$

Hence, unless validation is performed, there is no guarantee of the network's performance. The common approach is to split data into two subsets in order to generate neurons and predict the function rewards.

$$p_{t+1} = p_t + \alpha[Q^*(s,a) - \tilde{Q}(s,a,p)]\nabla_{\overrightarrow{p_t}}\tilde{Q}(s_t,a_t,p_t) \qquad (8.19)$$

Where α is a positive step-size parameter, and $\nabla_{\overrightarrow{p_t}}Q(p)$, for any function Q denotes the vector of partial derivatives.

$$\nabla_{\overrightarrow{p_t}}Q(p) = \left(\frac{\partial Q(\overrightarrow{p_t})}{\partial p_t(1)}, \frac{\partial Q(\overrightarrow{p_t})}{\partial p_t(2)}, ..., \frac{\partial Q(\overrightarrow{p_t})}{\partial p_t(t)}\right)^T \qquad (8.20)$$

Where $\phi(s_t, a_t)$ is a vector of all basis functions.
Since

$$\nabla_{\overrightarrow{p_t}}\tilde{Q}(s_t,a_t,p_t) = \phi^i(s,a) \qquad (8.21)$$

The above equation is updated as:

$$p_{t+1} = p_t + \alpha[Q^*(s,a) - \tilde{Q}(s,a,p)]\phi^i(s,a) \qquad (8.22)$$

Hence, the average reward estimate is updated as follows:

$$\rho_{t+1} = (1 - \alpha)\rho_t + \alpha_t(r(s_t,a_t,s_{t+1}) - \rho_t) \qquad (8.23)$$

INIT-FACTORS(s,a,limit)

    Description: Initialize value and state-action mapping,
Visiting Factors for state space S
        INPUT: state space, action space and maximum iteration jumps
        INIT-FACTORS(S,A,limit)
        count $\leftarrow$ max($length[S], length[A]$)
        for each s_i **in** count **do**
            $Q(s_i, a_i) \leftarrow 0$
            $Visit(s_i, a_i) \leftarrow 0$
      $k \leftarrow 0$
      $\alpha \leftarrow 0.1$
      $total\_reward \leftarrow 0$
      $total\_time \leftarrow 0$
      $\rho \leftarrow 0$
      $\phi \leftarrow 0$
      $k_{max} \leftarrow limit$

FIGURE 8.3: Initialization of all parameters

As noted earlier, this choice results in the sample average method, which is guaranteed to converge to the true action value by the law of large numbers. A well-known result in stochastic approximation theory gives us the conditions about the learning rate ($\sum\limits_{1 \leq k < \infty} \alpha_t = \infty$ and $\sum\limits_{1 \leq k < \infty} \alpha_t^2 < \infty$) which is required to assure convergence with probability 1.

As discussed earlier, real world systems and real time measurement have a large state and action spaces resulting in latency and errors to reach a convergent state. Hence, autonomous search depends on the nature of the parameterized function approximation to generalize from a limited subset of current state space over a much larger subset of state and action space. We apply index data structure mapping techniques to lookup a state-action value function to generalize from them to construct numeric value approximation of the entire search function.

8.6 Experiment

8.6.1 Data Samples

A large published data repository hosting intrusion dataset is selected to evaluate proposed approaches for several reasons: (1) it is a subset of standard dataset and is a revision of the 1998 DARPA intrusion detection evaluation dataset originated from MIT Lincon Labs [9]; (2) it is a proved dataset by researchers [40] [41] [42] [12] [28] [32] and have been used for the third in-

LOCATE – ACTION(s, a, Q)

    Description: locate action with maximum Q factor value starting from state s_i, randomly select action a
    Input:state space,action space and Q factors
    Output:action a which generate maximum Q factor
    $s_i \leftarrow$ **RANDOMIZE**(S)
    $a_i \leftarrow$ **RANDOMIZE**(A)
    for each a_i in Q do
        if $Q(s_i, a_i) == \max(Q(i, A))$
          $\phi \leftarrow 0$
          return a_i
    $\phi \leftarrow 1$
    return NIL

FIGURE 8.4: Randomized algorithm to locate action with maximum state-action value mapping

SIMULATE – ACTION$(r_{s_i, a, s_j}, t_{s_i, a, s_j})$

    Description: Simulate action a
    Input: immediate reward from action a,transition time from state s_i to state s_j
    $Visit(s_i, a) \leftarrow Visit(s_i, a) + 1$
    $k \leftarrow k + 1$
    $\alpha \leftarrow A/V(s_i, a)$

FIGURE 8.5: Action simulation algorithm

ternational knowledge discovery and data mining tools competition and the fifth international conference on knowledge discovery and data mining; (3) it is intrusion prediction specific dataset and was motivated to build a network intrusion predictive model for detect intrusions or attacks; (3)it also includes broad intrusions simulated in a military network environment; (4) this dataset has been preprocessed.

In this dataset, features characterizing network traffic behavior has been extracted to compose each record. The values of each data records has two categories. One is symbolic value and another is real numbers. The data size is $75MB$. It contains 22 attack types that can be classified in four main intrusion classes as shown in Figure 8.

8.6.2   Sample Reduction

There are two prevention datasets published with KDDCup99 samples. The first dataset has distinguished distinctions, which are used for classification training. The second dataset whose appearances are similar to evaluate the

UPDATE – ACTION$(Q(s_i, a), A)$

Description: Update State-Action Value Mapping
Input: State-Action Value Mapping
Output: updated State-Action Value Mapping
$Q(s_i, a) \leftarrow (1 - \alpha)Q(s_i, a) + \alpha[r_{s_i,a,s_j} - \rho t(s_i, a, s_j) + \max(Q(s_j, A))]$
return $Q(s_j, a)$

FIGURE 8.6: Update state-action value mapping algorithm

LOCATE – POLICY$(Q(s_i, a))$

Description: Locate a policy with optimal State-Action Value Mapping
Input: state space
$IF(\phi == 0)$
   THEN
      $total\_reward \leftarrow total\_reward + r(s_i, a, s_j)$
      $total\_time \leftarrow total\_time + t(s_i, a, s_j)$
$\rho \leftarrow [total\_reward/total\_time]$

FIGURE 8.7: Locate policy algorithm

correctness of the proposed class predictors. We use the same two datasets in our evaluation.

Specifically, the first training dataset is composed of $75MB$ samples. Within training dataset, samples are labeled as 4 classes of attacks. The class distinction is used for training of class prediction and clustering. Whereas, the second independent test dataset has $45MB$ samples. Regarding to test dataset, there are classes of samples in the test dataset as the training dataset does. The classes categories in the test dataset can be used to compute the learning performance of ROC values. For finer optimization, attack types are used for class labels in the tests.

With the above data samples, this research designs and develops a data feeder to transform the published datasets into the Attribute-Relation File Format (ARFF) for training and prediction. Each parameter in dataset is

Attack Class	Attack Category	Attack Samples
normal	normal	95278
u2r	buffer_overflow, loadmodule, multihop, perl, rootkit	59
r2l	ftp_write, guess_passwd, imap, phf, spy, warezclient, warezmaster	1119
dos	back, land, neptune, pod, smurf, teardrop	391458
prb	ipsweep, nmap, portsweep, satan	4107

FIGURE 8.8: KDDCup99 attack classes

formulated as data feature and a value is the corresponding percept of the variable. Hence, there are 42 attributes existing in the resulted training.arff and test.arff relations. They are used for training and testing purpose. The two arff files will be used for attribute selection in order to identify the sub-space of feature extraction for further classification learning. The specific attribute selection can be applied to classifier level so that compatible results will be achieved. The results show that there are only 10 attributes selected for the classification test. The selected attributes are protocol_type, service, src_bytes, dst_bytes, wrong_fragment, count, diff_srv_rate, dst_host_srv_count, dst_host_same_src_port_rate, dst_host_rerror_rate.

In this research, to conduct sample reduction, attribute space search and evaluation methods are considered for attribute selection. It searches the space of attribute subsets by greedy hill-climbing augmented with a backtracking facility. Best first may start with the empty set of attributes and search forward, or start with the full set of attributes and search backward, or start at any point and search in both directions. In addition, the correlation-based feature subset selection [43] is used to generate the sub-space of selected training and test datasets. It evaluates the worth of a subset of attributes by the individual predictive ability of each feature along with the degree of redundancy between them. Hence, subsets of attributes that are highly correlated with the class while having low inter-correlation are preferred.

Hence, instance selection fits for the sample reduction of this research. Instance-based sample reduction produces a random subset of a dataset using either sampling with replacement or without replacement. Since original dataset fits in memory, the number of instances in the generated dataset are specified as 1%. The filter can be made to maintain the class distribution in the subs-ample or to bias the class distribution toward a uniform distribution. After instance sampling, randomized 1% training and test datasets are created for further classification prediction and clustering prediction.

In addition to format conversion, there are several modifications to the dataset in this research.

- All string type attributes are defined as nominal data types.

- Both relation and attribute definition are inserted into datasets.

- The attack type and attack class labels are included in the arff files to calculate learning performance as different test cases.

- After attribute and instance selection is completed, the above arff files are filtered with only selected attributes or instances in arff format. Hence, the final classification computation are only imported with these selected arff files.

In general, the selected datasets only have 42 attack attribute name and value pairs for 4940 samples in training dataset and 3110 samples in test dataset. This is a significant cost reduction for learning of class prediction and clustering.

The detailed value of the sample instance is presented in Figure 9. As it illustrated, encoding needs to be done for each symbolic data type in Figure 1 in order to apply to initialize an arbitrary network for tuning. For example, protocol_type, service, flag, land, logged_in, is_host_login, is_guest_login, back are required to be encoded with the following sample instance. It will be utilized for classification networks tuning

8.6.3   Initial Arbitrary Network

Although this research proposes RL based interactive optimization for network tuning, the initial arbitrary learning network can be established with heuristics of problem specification. Intuitively, with 42 extracted features of network communications, 42 input nodes are selected to initialize input layer. A hidden layer is initialized with 3 nodes whereas output nodes are initialized with 5 nodes for each class of the listed attack types.

8.6.4   Tuning

Tuning is done by RL interactive algorithms for optimizing initialized network architecture. The network is monitored and modified during training time. The nodes in this network are all sigmoid.

- Types of machine learning networks

- Learning Rate for the back-propagation algorithm

- Momentum Rate for the back-propagation algorithm

- Number of epochs to train through

- The consequential number of errors allowed for validation testing before the network terminates.

- Numbers for nodes on each layer in multiple layer network

8.6.5   Comparison of Class Prediction

In literature, predictive models can be parametric, non-parametric, or semi-parametric. This research is to study parametric machine learning methods. Experiments are designed to utilize the datasets published with KDDCup99 datasets. The experiments in this research are designed with consistent evaluation techniques. First. training of prediction models with the selected dataset is evaluated by $10 - fold$ cross validation. Then the trained models need to be validate by the selected test datasets. The Q learning based classification results are illustrated in following paragraphs. Please note that Q learning is based on correct classification due to large sample classification. *ROC* values are listed in the detailed test result tables.

Discrete Environmental Signal: After each move of machine learning method at top level of selection, the learning agent receives from the above task environment one of the following reinforcement immediate reward signal $r \in \Re$.

$$r = \begin{cases} -10 \text{ if agent makes less correct classification;} \\ 0.1 \text{ if agent makes more correct classification;} \\ 0.4 \text{ if all samples are classified correctly;} \\ -1 \quad \text{the cost of making any selection;} \\ 0.0 \text{ otherwise for all nonterminal states} \end{cases}$$

The move to the correct classification is less than the absolute value of moving to wrong classification ratios since keeping agent on in grid is more important than getting to the target. Intuitively, agent needs to learn to achieve goal without rewards. The agent is required to move to the correct classification. However, the sooner the action is rewarded, the sooner the agent should learn. The cost of moving is measured by a negative unit reward from task environment. It is larger than the reward from getting to the correct positions so that the steps in the path are considered into search evaluation.

A naive Bayes classifier [44] is a simple probabilistic classifier based on applying Bayes' theorem with strong independence assumptions. Depending on the precise nature of the probability model, naive Bayes classifiers can be trained very efficiently in a supervised learning setting. In many practical applications, parameter estimation for naive Bayes models uses the method of maximum likelihood; in other words, one can work with the naive Bayes model without believing in Bayesian probability or using any Bayesian methods. The naive Bayes classifiers often work much better in many complex real-world situations than one might expect. Recently, careful analysis of the Bayesian classification problem has shown that there are some theoretical reasons for the apparently unreasonable efficacy of Naive Bayes classifiers [45]. An advantage of the Naive Bayes classifier is that it requires a small amount of training data to estimate the parameters necessary for classification. Because independent variables are assumed, only the variances of the variables for each class need to be determined and not the entire covariance matrix. Hence, it is a good method for attack classification. The result in Figure 9 also demonstrate the above analysis.

Support vector machines (SVMs) is another set of supervised learning methods for classification [46]. It maps input vectors to a higher dimensional space where a maximal separating hyperplane is created. Two parallel hyper-planes are constructed on each side of the hyperplane that separates samples. The separating hyperplane is the hyperplane that maximizes the distance between the two parallel hyper-planes. The larger the margin or distance between these parallel hyper-planes is, The better the generalization error of the classifier will be.

Logistic regression [47] is a regression model for binomially distributed dependent variables. It is a technique in which unknown values of a discrete variable are predicted based on known values of one or more continuous and/or discrete variables. The model is "simple" in that each has only one independent, or predictor, variable, and it is "binary" in that the dependent variable can take on only one of two values: attack or normal. Models are not restricted to a single independent variable or to a binary dependent variable.

Neural networks [48] are nonlinear arbitrary modeling techniques that are able to model complex functions. Neural networks are used when the exact nature of the relationship between inputs and output is not known. Neural networks learns the relationship between inputs and output through training. Neural network training techniques are backpropagation, quick propagation, conjugate gradient descent, projection operator, Delta-Bar-Delta, etc. Theses are applied to network architectures such as multi-layer perceptrons which is used in this research for class prediction. To evaluate training result, simulation is done with selected training dataset. There were 99.7368% samples correctly classified during training. After training with the selected training dataset, 91.2862% test samples are classified without errors. It is noted that it is a time consuming learning method for large attack classification. The training performance is plotted as Figure 2. The detailed result shows that more than 1 hour iterations required to complete the training for prediction.

A radial basis function (RBF) is a function which has built into it a distance criterion with respect to a center [49]. Such functions can be used very efficiently for interpolation and for smoothing of data. Radial basis functions have been applied in the area of neural networks where they are used as a replacement for the sigmoidal transfer function. Such networks have 3 layers, the input layer, the hidden layer with the RBF non-linearity and a linear output layer. The most popular choice for the non-linearity is the Gaussian.

Logistic Model Trees (LMT) does attack classification for building logistic model trees, which are classification trees with logistic regression functions at the leaves [50]. The algorithm predicts classification with binary and multi-class target variables, numeric and nominal attributes and missing values. PART classifier generates a PART decision list [51]. Uses separate-and-conquer. Builds a partial C4.5 decision tree in each iteration and makes the best leaf into a rule. J48 [52] classifier generates a pruned or unpruned C4.5 decision tree for class prediction. Other than neuro perceptron network, all learning algorithms and related class prediction results are shown in table below (see Figure 9) for comparison.

As it is shown, SelectAttributeClassifier computes dimensionality of training and test data is reduced by attribute selection before being passed on to a classifier [53]. It adds the attribute selection to the classification for cost reduction in addition to the instance-based sample reduction.

Another significant finding from network Q learning result is that SelectAttributeClassifier, Bayes Network, LMT, PART and Neuro Perceptron Network have the highest values and accuracy. It indicates that the 3 algorithms are

ID	Classifier Category	Algorithm	Results(Training/Test)
(1,1)	Tree	J48	99.0891%/91.0932%
(1,2)	Tree	LMT*	99.7368%/91.2862%
(1,3)	Rule	PART*	99.5951%/91.254%
(1,4)	Function	LibSVM	97.4291%/77.6527%
(1,5)	Function	MultilayerPerceptron*	99.413%/90.5145%
(1,6)	Function	RBFNetwork	99.3725%/80.7074%
(1,7)	Meta	SelectAttributeClassifier*	99.3725%/91.4791%
(1,8)	Bayes	BayesNet*	99.0891%/91.0932%
(1,9)	Bayes	NaiveBayes*	95.6883%/77.8135%

FIGURE 8.9: Classification prediction result for Q learning

better algorithms for attack class prediction.

Please note that $10 - fold$ cross-validation on full training set is selected for all algorithm training. If we use full training set without cross-validation, results seems over-fitting with full ROC value and 100% correct classification during training cycle. However, test validation does show the error ratio of classification. Hence, cross-validation is the proper technique to compare algorithm training performance.

Select an action of MultilayerPerceptron(MP) with a probability of $1/9$. The state is marked as s_1 with $r = -0.9$ due to high correct classification result with additional moving costs. This action value mapping increases the number of visited state-action pairs by $V(1) = V(1) + 1$. This state transition also increase the number of interactions by 1. Hence, the $\alpha = 0.1/V = 0.1$. Hence, the Q factor is updated as follows. Since all $Q(s_1, A(s_1))$ pairs are initialized as -10, equation 11 is evaluated as

$$Q(s_1, MP) = (1 - 0.1) * Q(s_1, MP) + 0.1 * [(-0.9) \atop +maxQ(s_1, A) + 10] = -9.009 \tag{8.24}$$

Since iteration is less than maximum iteration, select action BayesNet as learning network with current state s_1. The state is updated to s_2 with reward of $+0.2$ and moving cost of -1 due to the expected action to move. Hence, $r = -0.8$ is the total reward of computation.

$$Q(s_2, BayesNet) = (1 - 0.1) * Q(s_1, BayesNet) \atop +0.1 * [(-0.8) + maxQ(s_2, A) \atop +10] = -9.008 \tag{8.25}$$

Hence, the state s_1 with action BayesNet is selected as a significant state as $Q(s*, a*) = -9.008$ for further Q factor computation.

Now, current state is set to s_2 to select an action of SelectAttributeClassifier with a probability of $1/9$. The state is changed to s_3 with $r = -0.7$ due to the action to taken by agent. This action value mapping increases the number of visited state-action pairs by $V(3) = V(3) + 1$. This state transition also increase the number of interactions by 1. Hence, the $\alpha = 0.1/V = 0.1$. Hence, the Q factor is updated as

$$
\begin{aligned}
Q(s_3, BayesNet) &= (1 - 0.1) * Q(s_1, BayesNet) \\
&\quad +0.1 * [(-0.7) + maxQ(s_3, A) \\
&\quad +9.008] = -9.0702
\end{aligned} \tag{8.26}
$$

Now, current state is set to s_3 to select an action of LMT with a probability of 1/9. The state is changed to s_4 with $r = -0.6$ due to the action to taken by agent. This action value mapping increases the number of visited state-action pairs by $V(4) = V(4) + 1$. This state transition also increase the number of interactions by 1. Hence, the $\alpha = 0.1/V = 0.1$. Hence, the Q factor is updated as

$$
\begin{aligned}
Q(s_4, LMT) &= (1 - 0.1) * Q(s_1, LMT) + 0.1 * [(-0.6) \\
&\quad +maxQ(s_4, A) + 9.008] = -9.0602
\end{aligned} \tag{8.27}
$$

Consequently, the state s_1 with action listed in table 1 will be simulated with 1/9 probability distribution. It concludes that LMT is the optimal prevention method for the current learning samples.

Even though, with exploration activities, MP may have equal probability for further tuning of network architecture. Now, current state is set to s_4 to select an action of MP with a probability of 1/9. The state is changed to s_5 with $r = -0.9$ due to the action to taken by agent. This action value mapping increases the number of visited state-action pairs by $V(5) = V(5) + 1$. This state transition also increase the number of interactions by 1. Hence, the $\alpha = 0.1/V = 0.1$. Hence, the Q factor is updated as

$$
\begin{aligned}
Q(s_5, MP) &= (1 - 0.1) * Q(s_5, MP) + 0.1 * [(-0.9) \\
&\quad +maxQ(s_5, A) + 9.008] = -9.092
\end{aligned} \tag{8.28}
$$

The subsequent tuning of MP prediction architecture is illustrated below.

ID	Architecture	Results(Training/Test)
(2,1)	42,1,4	98.502%/90.0322%
(2,2)	42,3,4*	98.7854%/90%
(2,3)	42,5,4*	99.1093%/90.9003%
(2,4)	42,5,10*	98.502%/90%
(2,5)	42,10,4*	98.7652%/90%

FIGURE 8.10: MP prediction result for Q learning

To avoid greedy local optimum, iterations should continue until it reaches the maximum interaction number, then set current state back to s_1 to continue to select an action. MP may have an equal probability for further tuning of network architecture (42,1,4). Now, current state is set to s_4 to select an action of MP with a probability of 1/9. The state is changed to s_5 with

$r = -0.9$ due to the action to taken by agent. This action value mapping increases the number of visited state-action pairs by $V(5) = V(5) + 1$. This state transition also increase the number of interactions by 1. Hence, the $\alpha = 0.1/V = 0.1$. Hence, the Q factor is updated as

$$Q(s_5, MP) = (1 - 0.1) * Q(s_5, MP) + 0.1 * [(-0.9) \atop +maxQ(s_5, A) + 9.008] = -9.092 \tag{8.29}$$

Now, the current state is set to s_5 to select an action of MP with a probability of $1/9$. Continue to tune network as $(42,3,4)$. The state remains as s_5 with $r = -1$ due to the action to taken by agent. This action value mapping increases the number of visited state-action pairs by $V(5) = V(5) + 1$. This state transition also increase the number of interactions by 1. Hence, the $\alpha = 0.1/V = 0.05$. Hence, the Q factor is updated as

$$Q(s_5, MP) = (1 - 0.05) * Q(s_5, MP) + 0.05 * [(-1) \atop +maxQ(s_5, A) + 9.008] = -8.5832 \tag{8.30}$$

Now, the current state is set to s_5 to select an action of MP with a probability of $1/9$. Continue to tune network as $(42,5,4)$. The state remains as s_5 with $r = -0.9$ due to the action to taken by agent. This action value mapping increases the number of visited state-action pairs by $V(5) = V(5) + 1$. This state transition also increase the number of interactions by 1. Hence, the $\alpha = 0.1/V = 0.033$. Hence, the Q factor is updated as

$$Q(s_5, MP) = (1 - 0.033) * Q(s_5, MP) + 0.033 * [(-0.9) \atop +maxQ(s_5, A) + 9.008] = -8.31565 \tag{8.31}$$

Hence, $Q(s_5, 42 - 5 - 4)$ is the optimal state action. It concludes that the multi-layer perceptron network with 42 of input nodes, 5 of hidden nodes and 4 of output nodes is an optimal prevention method. The experimental evaluation of the proposed astomous tuning is achieved with the following Figure. It is a simplified version of classification goal discovery without numerous non-optimal path exploration.

FIGURE 8.11: Autonomous tuning scenario

8.6.6 Comparison of Cluster Prediction

The attack class discovery problem proposed in this research is to group data samples as specific individual class categories, which are not only representing current data clusters but also predicting unknown classes.

To optimize the clustering methods proposed in the report, this research utilizes Q learning to compute and compare various cluster methods. First of all, to reduce the dimensionality of learning, only selected training and test data set are used for class discovery. In addition, there are several clustering methods are proposed for class discovery with 100% accuracy.

The k-means [54] algorithm is an algorithm to cluster objects based on attributes into k partitions. It attempts to find the centers of natural clusters in the data. With k-means algorithm, attributes holds a vector space. The result from this research shows that k-means algorithms discover the attack classes with 100% accuracy with both training and test dataset. The cluster sum of squared errors is 0.534.

RL iterative algorithm runs an arbitrary cluster on data that has been passed through an arbitrary filter. Like the cluster, the structure of the filter is based exclusively on the training data and test instances will be processed by the filter without changing their structure. Filtered cluster algorithm reports discovery of the attack classes with 100% accuracy with both training and test dataset. The cluster sum of squared errors is 0.563.

Clustering data uses the Farthest-First algorithm [55] to develop a best possible heuristic to resolve k-center problem. Farthest-First algorithm reports discovery of attack classes with 100% accuracy with both training and test dataset.

Hence, this research proposes k-means algorithm, filtered cluster algorithm and Farthest-First algorithm for attack discovery prediction. These algorithms have demonstrated with the accuracy and performance of prediction for the selected training and test dataset.

8.7 Conclusions

General Conclusions from this research are numerated below:

- Prevention problem is the combinational problem of attack classification and attack clustering. Furthermore, to handle learning with large samples, sample reduction with attribute and instance selection reduces both time and space complexities.

- Training of prediction models with the selected dataset is evaluated by $10 - fold$ cross validation. Then the trained models need to be validate by the selected test datasets.

- MIT Lab has provided repeatable training and test dataset. The pre-processed dataset with labels are easy to use for algorithm evaluation. However, the datasets have large samples with mediate dimensionality. Evermore, the major issue of the published dataset is about the imbalance of the number of samples and the dimensionality of the attributes. Hence, the datasets will result in over-fitting issue. We propose the attribute and instance selection procedure to overcome this problem effectively. After attribute selection, only 10 attributes of 1% samples are elected for prediction computation. This significantly reduce the computational complexity and costs.

- The correlation-based feature subset selection and Best-First Search are the recommended attribute selection and dimension reduction methods for attack classification. In addition, random re-sampling techniques are recommended for instance reduction. Sample reduction techniques are critical for the autonomous tuning of prevention network architecture.

- From the class prediction results of this research, SVMs, SMO and Vote are tuned off by the proposed antimonous classification algorithms for attack class prediction. Whereas, SelectAttributeClassifier, BayesNet, LMT, PART and Neuro Perceptron Network are exploited algorithms for autonomous attack class prediction. Specifically, autonomous Bayes Networks and Neural Networks are exploited optimal prevention networks in terms of time and space complexity.

- With the attack clustering results from this research, K-Means algorithm, Filtered Cluster algorithm and Farthest-First algorithm are recommended attack discovery methods for attack discovery.

- Scale-free networks are more robust than random networks for random failure but more vulnerable to targeted attacks.

- A discretionary architecture adopted with autonomous tuning mitigates prevention problems effectively.

This derives the conclusion that the proposed autonomous tuning utility framework is optimized by attribute and instance selection, class prediction and discovery methods. It reduces time complexity and error prediction existing in static learning methods or evolutionary computational tasks.

References

[1] Pennington, A. G. et al.: Storage-based intrusion detection: watching storage activity for suspicious behavior. In: Proceedings of the 12th USENIX Security Symposium. (2003)

[2] Banikazemi, M. et al.: Storage-based intrusion detection for storage area networks (sans). In: Proceedings of the 22nd IEEE/13th NASA Goddard Conference on Mass Storage Systems and Technologies. (2005)

[3] Peng, T. et al.: Survey of network-based defense mechanisms countering the dos and ddos problems. ACM Computing Surveys **39**(1) (2007)

[4] Subhadrabandhu, D. et al.: A framework for misuse detection in ad hoc networks-part i. IEEE Journal of Selected Areas in Communications **24**(2) (2006)

[5] Han, S.J., Cho, S.B.: Evolutionary neural networks for anomaly detection based on the behavior of a program. IEEE Transactions on Systems, Man, and Cybernetics Part B **36**(3) (2006)

[6] Aldwairi, M. et al.: Configurable string matching hardware for speeding up intrusion detection. ACM SIGARCH Computer Architecture News **33**(1) (2005)

[7] Maloof, M. et al. In: Machine Learning and Data Mining for Computer Security. Springer (2006)

[8] Chan, P.K., Lippman, R.P.: Machine learning for computer security. Journal of Machine Learning Research (2007)

[9] Lippmann, R. P. et al.: Evaluating intrusion detection systems: The 1998 DARPA off-line intrusion detection evaluation. In: Proceedings of DARPA Information Survivability Conference & Exposition. (2000)

[10] Hofmann, A., Sick, B.: Evolutionary optimization of radial basis function networks for intrusion detection. In: Proceedings of Int. Joint Conf. Netural Networks. (2003)

[11] Ye, N. et al.: Computer intrusion detection through ewma for auto correlated and uncorrelated data. IEEE Trans. Real. **52**(1) (2003)

[12] Yu, Z. et al.: An automatically tuning intrusion detection system. IEEE Trans. Syst., Man, Cybern. **37**(2) (2007)

[13] Brasseville, M., Nikiforov, I.V. In: Detection of Abrupt Changes. Prentice-Hall, Englewood Cliffs (1993)

[14] Spackman, K.: Signal detection theory: Valuable tools for evaluating inductive learning. In: Proceedings of the Sixth International Workshop on Machine Learning. (1989) 160–163

[15] Jain, A. K.: Data clustering: A review. ACM Computing Surveys **31**(3) (1999)

[16] Shiryaev, A. In: Optimal Stopping Rules. Springer-Verlag, NewYork (1978)

[17] Lai, T.L.: Sequential changepoint detection in quality control and dynamic systems. J. R. Statist. Soc. B **57**(4) (1995)

[18] Ye, N. et al.: Robustness of the Markov-Chain model for cyber-attack detection. IEEE Trans. Real. **53**(1) (2004)

[19] Ye, N. et al.: Multivariate statistical analysis of audit trails for host-based intrusion detection. IEEE Trans. Comput. **51**(7) (2002)

[20] Ertoz, L. et al.: The MINDS-Minnesota Intrusion Detection System. In: Next Generation Data Mining. MIT Press, Cambridge, MA (2004)

[21] Ye, N. et al.: Robustness of the Markov-Chain model for cyber-attack detection. IEEE Trans. Real. **53**(1) (2004)

[22] Cho, S.B.: Incorporating soft computing techniques into a probabilistic intrusion detection system. IEEE Trans. Syst., Man, Cybern. **31**(4) (2002)

[23] Lee, S.C. et al.: Training a neural-network based intrusion detector to recognize novel attacks. IEEE Trans. Syst., Man, Cybern. **31**(4) (2001)

[24] Tsang, C-H. et al.: Genetic-fuzzy rule mining approach and evaluation of feature selection techniques for anomaly intrusion detection. Pattern Recognition **40**(9) (2007) 2373–2391

[25] Warrender, C. et al.: Detecting intrusions using system calls: Alternative data models. In: Proceedings of 1999 IEEE Symp. Security and Privacy. (1999)

[26] Bowen, T. et al.: Building survivable systems: An integrated approach based on intrusion detection and damage containment. In: Proceedings of 2000 DARPA Information Survivability Conference Exposition. (2000)

[27] Denning, D.E.: An intrusion-detection model. IEEE Trans. Software Eng. **SE-13**(2) (1987)

[28] He, D. et al.: Network intrusion detection using cfar abrupt-change detectors. IEEE Transactions on Instrumentation and Measurement **57**(3) (2007)

[29] Stone, A. et al.: Natural-language processing for intrusion detection. Computer **40**(12) (2007)

[30] Shyu, M.L. et al.: Network intrusion detection through adaptive sub-eigenspace modeling in multiagent systems. ACM Transactions on Autonomous and Adaptive Systems **2**(3) (2007)

[31] Song, H. et al.: Efficient packet classification for network intrusion detection using fpga. In: IEEE Symp. Field-Programmable Gate Arrays. (2005)

[32] Abhishek, D. et al.: An fpga-based network intrusion detection architecture. IEEE Transactions on Information Forensics and Security **3**(1) (2007)

[33] Law, A.M., Kelton, W.D. In: Simulation Modeling and Analysis. McGraw Hill, Inc., New York, NY (1999)

[34] Crites, R., Barto, A.: Improving elevator performance using reinforcement learning. Neural Information Processing Systems (1996)

[35] Howard, R. In: Dynamic Programming and Markov Process. MIT Press, Cambridge, MA (1960)

[36] Bellman, R. In: Dynamic Programming. Princeton University Press, Princeton, NJ (1957)

[37] Konda, V.R., Tsitsiklis, J.N.: Actor-Critic Algorithms. SIAM Journal on Control and Optimization **42**(4) (2003)

[38] Vengerov, D. et al.: A fuzzy reinforcement learning approach to power control in wireless transmitters. IEEE Transactions on Systems, Man, and Cybernetics, Part B (2005)

[39] WardiDOI, Y.: On a proof of a Robbins-Monro algorithm. Journal of Optimization Theory and Applications **64**(217) (1990)

[40] Elkan, C.: Results of the kdd'99 classifier learning. In: ACM SIGKDD. Volume 1. (2000)

[41] Kumar, V.: Data mining for network intrusion detection: Experience with kddcup'99 data set. In: Workshop Netw. Intrusion Detection. (2002)

[42] Sabhnani, M. et al.: Application of machine learning algorithms to kdd intrusion detection dataset within misuse detection context. In: Proc. Int. Conf. Mach. Learn.:Model, Technol. and Appl. (2003)

[43] Hall, S.: Feature subset selection: A correlation based filter approach. Ph.d Thesis (1999)

[44] Kotsiantis, S. B. et al.: Logitboost of simple Bayesian classifier. Computational Intelligence in Data Mining Special Issue of the Informatica Journal **29**(1) (2005) 53–59

[45] Zhang, H.: The optimality of Naive Bayes. American Association for Artificial Intelligence (2004)

[46] Cortes, C., Vapnik, V.: Support-vector networks. Machine Learning **20** (1995)

[47] Agresti, A. In: Categorical Data Analysis. New York: Wiley-Interscience (2002)

[48] Frank, M.J.: Dynamic dopamine modulation in the basal ganglia: A neurocomputational account of cognitive deficits in medicated and non-medicated Parkinsonism. Journal of Cognitive Neuroscience **17**(5172) (2005)

[49] Yee, P.V., Haykin, S. In: Regularized Radial Basis Function Networks: Theory and Applications. John Wiley (2002)

[50] Landwehr, N. et al.: Logistic model trees. Machine Learning (2005)

[51] Frank, E., Witten, I.H.: Generating accurate rule sets without global optimization. In: Fifteenth International Conference on Machine Learning. (1998) 144–151

[52] Quinlan, J.R.: Improved use of continuous attributes in c4.5. Journal of Artificial Intelligence Research **4** (1996) 77–90

[53] Frank, E. et al.: Locally weighted Naive Bayes. In: The 19th Conference in Uncertainty in Artificial Intelligence. (2003)

[54] Kanungo, T. et al.: An efficient k-means clustering algorithm: Analysis and implementation. IEEE Trans. Pattern Analysis and Machine Intelligence (2006)

[55] Hochbaum, H.: A best possible heuristic for the k-center problem. Mathematics of Operations Research **10**(2) (1985) 180–184

Biography

Lei Liu serves as a staff engineer at System-Storage Division for Sun Microsystems, Inc. He has about 12 years of experience on full life cycle R&D, product development, integration, testing and partner engineering engagement. Lei has accomplished various tasks from the latest hardware CMT platform and Grid Rack System to the latest software Solaris™10, Java ES™4 and Grid Infrastructure Software along with seven published papers. In addition, Lei has nine patents filed with US patent office and numerous patents pending at Sun. Lei has been actively contributing on Solaris, Grid and SOA initiatives. Lei is a JCP member and a member of the expert group for JSR 229,235,247, etc. Lei is a member of Enterprise Grid Alliance. At present, Lei is focusing on security R&D for next generation storage.

Lei's current research and development interests: Quantum Computer, Bio-inspired Storage, Autonomic Computing, Algorithm and Information Theory, Operating System and Performance Model, Programming Calculus, Machine Learning, Storage Security, Distributed Data Structure, Automata, Ubiquitous Computing, Mobile Ad Hoc Network, Next Generation Internet, Parallel

& Grid Computing, Virtualization, Buffer Management, Attack & Detection and Autonomic Middleware.

Chapter 9

Probabilistic Fault Management

Jianguo Ding, Pascal Bouvry

Faculty of Science, Technology and Communication (FSTC)
University of Luxembourg, L-1359 Luxembourg, Luxembourg

Bernd J. Krämer

Department of Mathematics and Computer Science
FernUniversität in Hagen, D-58084 Hagen, Germany

Haibing Guan

School of Information Security Engineering
Shanghai Jiao Tong University, Shanghai 200030, P. R. China

Alei Liang

Software School
Shanghai Jiao Tong University, Shanghai 200030, P. R. China

Franco Davoli

Department of Communications, Computer, and Systems Science (DIST)
University of Genoa, Genoa 16145, Italy

9.1   Introduction

With the growth in size, heterogeneity, pervasiveness and complexity of applications and network services, the effective management of networks has become more important and more difficult. In order to meet the diverse demands and challenges confronting the networks and to allow for a scalable and manageable growth, the autonomic network paradigm has been proposed to create self-organizing, self-managing and context-aware autonomous networks.

Self-management is a high degree requirement for an autonomic network system to enable it to be usable and stable. Self-management is defined as using autonomic principles to provide management functionalities. An important corollary is that the management function itself is an autonomic service, and is therefore subject to the same requirements as other autonomic services. A self-management system can be summarized by four objectives, self-configuration, self-healing, self-protection and self-optimization, and four attributes, self-awareness, environment-awareness, self-monitoring and self-adjusting [50].

1) self-configuration involves automatic incorporation of new components and automatic component adjustment to new conditions to maintain desired functionalities or provide new functionality;

2) self-healing can detect, diagnose and repair hardware, software and firmware problems;

3) self-protection can automatically defend against large-scale attacks or cascading internal failures from permanent damaging valuable information and critical system functions. It may act proactively to mitigate reported problems;

4) self-optimization can continually seek ways to improve their behavior and enable the system to be more efficient.

As one of the FCAPS management functions, automatic fault management is a crucial issue in achieving the goal of self-protection and self-healing. Individual hardware defects, software errors or combinations of such defects and errors in different system components may cause the service degradation of other (remote) components in networks or even their complete failure due to functional dependencies between managed objects. Hence, an effective distributed fault detection method is needed to support quick fault detection in network management and allow for automation of fault management.

Although the Open System Interconnect (OSI) management standard provides a framework for managing faults in heterogeneous open systems, it does not address methodological issues to detect and diagnose faults. In order to fill this gap, a great deal of research efforts in the past decade have been focused on improving management systems in automatic fault detection and diagnosis. Rule-based expert systems have so far been the major approach to alarm correlation in fault detection [40] [57]. This approach suits well-defined problems where the environment is not very dynamic, but they do not adapt well to the evolving network environment [10]. Case-based reasoning [31] [45] and Coding-based methods [55] [30] offer potential solutions for fault identification and isolation, but they cannot deal with uncertain or unstable situations in networks. Finite State Machines (FSMs) are used to model fault propagation behaviors and to execute the fault identification [48] [4] [35]. However, this approach has difficulties in scaling up to large and dynamic networks. Kätker and Geihs provide Model Traversing Techniques for fault isolation in networks [21], but this lacks of flexibility, especially when fault propagation is complex and not well structured. Lunze and Lamperti et al. investigate the fault diagnosis in discrete-event systems [28][34]. Badonnel exploits the promise theory framework to model voluntary cooperation among network nodes and make them capable of expressing the trust in their measurements during the fault detection process [2]. Most of these solutions are based on certain mechanisms and improve the automatic scheme in fault management. However, they are sensitive to "noise" (such as loss of management information, delay in information collection and response, misunderstanding alarms). That means they are unable to deal with incomplete and imprecise management information effectively in uncertain and dynamic environments.

In networks, due to losses or delays in data collection, it is difficult to obtain full and precise management information. As the complex dependency relationship between managed objects and the cause-effect relationships among faults and alarms are generally incomplete, it is impossible to get a full and exact understanding of the managed system from the viewpoint of systems management.

In daily management, specialist or expert knowledge is very important and useful. However, quantitative expert knowledge is often expressed in imprecise ways, such as "very high," "normal" or "sometimes." When expert knowledge is incorporated into a management system, probabilistic approaches are needed for the quantitative expression of this kind of expert knowledge.

Due to the complexity of networks, it is not always possible to build precise models for fault management. A well-designed strategy for fault management should therefore operate efficiently in the case of redundant, incomplete and unreliable information.

Thus, probabilistic reasoning is another effective approach for fault detection in fault management [17] [52] [49] [6] [46].

Moreover, in real-life networks, dynamic changes are unavoidable due to the enhanced network complexity and the potential degeneration or improvement in system performance. Hence to understand unavoidable changes and to catch the trend of changes in a network will be very important for automatic fault management.

Most of the current commercial management software, such as IBM Tivoli, HP OpenView, SunNet Manager, Cabletron Spectrum and Cisco Works network management software, support the integration of different management domains, collect information, perform remote monitoring, generate fault alarms and provide statistics on management information. However, they lack facilities for exact faults localization, or the automatic execution of appropriate fault recovery actions. From the experience in network management, a typical measurement for on-line fault identification indicates 95% fault location accuracy while 5% of faults cannot be located and recovered in due time [16]. Hence for large networks with thousands of managed components it may be rather time-consuming and difficult to resolve the problems in a short time by an exhaustive search for the root causes of a failure and the exhaustive detection may interrupt important services in the systems.

In large-scale uncertain and dynamic network environments, the autonomic fault management paradigm is an alternative strategy in assisting to achieve the self-management of the networks. In dealing with autonomic fault management systems, there are three major aspects to be considered: 1) to design an architecture to support autonomic behavior; 2) to represent the uncertain and dynamic information that is necessary to an autonomic object to achieve an autonomic behavior; 3) to communicate and to organize the autonomic objects among themselves in a possible large context, particularly to execute probabilistic reasoning between the depended objects.

In this chapter, application issues of probabilistic models for automatic fault management are investigated in order to resolve the fault detection challenges in uncertain and dynamic network environments.

9.2 Probabilistic Inference in Fault Management

Efficient fault management requires an appropriate level of automation and self-management. A serious problem of using deterministic models is their inability to isolate primary sources of faults from uncoordinated network events. Observing that the cause-and-effect relationship between symptoms and possible causes is inherently non deterministic, probabilistic models can be considered to gain a more accurate representation. Bayesian networks are appropriate models for probabilistic management in fault management. From the view of management, a normal operation is to trace the root causes from detected symptoms (alarms). Hence the backward inference based on the probabilistic model from effects to causes is the basis in distributed fault management.

9.2.1 The Characteristics of the Faults in Distributed Systems

Due to the enlarged topology, distributed application and services, and complex dependency between managed systems, the faults in a network are different with those in a centralized system. In general, the characteristics of faults in networks are identified as follows:

- **The sources of faults are distributed.**

  In networks, the topology and distributed services in distributed systems are often expanded in a distributed environment. The managed objects are distributed geographically or logically. The faults which are generated from a network are also scattered in the whole network. Therefore, an efficient fault management system should consider fault detection within the distributed environments.

- **Fault propagation comes from dependency relationship between managed objects.**

  In networks, to finish certain applications or services, managed objects cooperate with each other for the same goal. Often in the cooperation, one managed object depends on another object in transactions. This kind of dependency is denoted by the cause-effect relationship. For example, the disconnection in a switch may interrupt the connection services with other devices which are connected to the switch and on which they depend, Hence the dependency relationship between managed objects is an important factor in fault detection and recovery.

- **The sources of faults are often hidden.**

  In complex networks, due to the fault propagation, the symptoms of a hardware or software fault are often caused by a remote or hidden factor. That means the root of the faults is sometimes difficult to identify directly from the detected symptoms. Consider the following simple scenario: A physical link failure occurs in one of several interconnected networks. The failure is detected and reported by the management component monitoring the physical resource in question. The failure also has sided againt effects on other resources in the network, e.g., connections on the various layers which use the link will experience timeouts. The management components of resources such as, e.g., protocol stacks, will therefore report failures. The result is that the operator's console is literally flooded with reports indicating the existence of some network abnormal condition, making it extremely difficult to determine the real cause of the problem.

- **A managed system holds enough information for fault management.**

  In current networks, lots of devices and application modules keep the records of the numerous states of the operations and services. Most of the information related to faults is also recorded in the managed systems. Hence, it is possible to mine new knowledge from all the recorded information and the statistics of the historical data. For example, the system's event log consists of a large number of individual events sent by all nodes in the system that have event generation capabilities. On a typical day, the log could reach tens of thousands of events [39]. This number is a function of parameters such as the size of the system, the configuration, the type and the amount of traffic being carried, the number and the type of faults that have occurred during that day. Events are collected at a centralized operation and maintenance center (OMC) where the events log is being assembled. It is known that the manual processing of this mass of data tends to become unfeasible as the number of high speed systems in the network increases. However, the

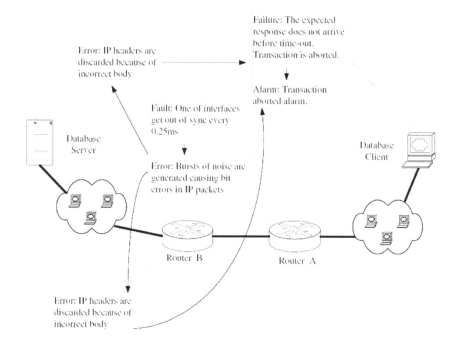

FIGURE 9.1: A model of fault propagation.

statistics in the large volume of history data can be used to retrieve new knowledge which is a potential reference for fault management.

In Figure 9.1, a simple network is presented in which a client accesses a remote database server. An interface of one of the routers between the client and server gets intermittently out of sync causing bursts of bit errors in transmitted IP datagrams. As a result, many IP datagrams passing through the router are rejected by the next router due to header errors, or by the server due to the corrupted datagram body. The client does not receive any response to its query and times out. This example illustrates how a seemingly invisible fault manifests itself through a failure at a location distant from the location of the fault. Since most faults are not directly observable, the management system has to infer their existence from information provided by the received alarms. The information carried within reported alarms may include the following: the identity of the object that generated the alarm, type of failure condition, time stamp, alarm identifier, measure of severity of the failure condition, a textual description of the failure, etc. [15] [51].

Hence, in a network, a single fault may cause a number of alarms to be delivered to the network management center. Multiple alarms may be a result of (1) fault re-occurrence, (2) multiple invocations of a service provided by a faulty component, (3) generating multiple alarms by a device for a single fault, (4) detection of and issuing a notification about the same network fault by many devices simultaneously and (5) error propagation to other devices causing them to fail and, as a result, generate additional alarms [15]. It may be argued that typical networks provide plenty of information necessary to infer existence of faults [55].

9.2.2 Bayesian Networks for Fault Management

Bayesian networks are appropriate for automated diagnosis due to their deep representations and precise calculations. A concise and direct way to represent a system's diagnostic model is as a Bayesian network constructed from relationships between failure symptoms and underlying problems. A Bayesian network represents cause and effect between observable symptoms and the unobserved problems so that when a set of symptoms are observed the problems most likely to be the cause can be determined. In practice, the fault management system is built from descriptions of the likely effects for a chosen fault.

The development of a diagnostic Bayesian network requires a deep understanding of the cause and effect relationships in a domain, provided by domain experts. One advantage of Bayesian networks is that the knowledge can provide clear inner relationship between effects and causes. It is not represented as a black box, as in Artificial Neural Networks. In addition, compared with other logic, Bayesian networks also provide more fine-grained quantitative evaluation in probabilistic models. Thus, humanly understandable explanations of diagnoses can be given.

When a network fault management system is modelled as a Bayesian Network, two important processes need to be resolved:

1. **Ascertain the Dependency Relationship between Managed Objects.**

   A network consists of a number of managed objects. An object is a "part" of the network that has a separate and distinct existence. At the physical level, an object can be a network, a node, a switch, a layer in a protocol stack, a virtual link, a physical element like an optical fiber, a piece of cable, a hardware component, etc. At the logic level, an object can be a software service, such as a process, a piece of code, a URL, a servlet or a service request. Objects in a network consist of other objects down to the level of the smallest objects that are considered indivisible. An indivisible object is defined as a terminal object. The concept of division and appropriate level of division are system-dependent and application-dependent.

   Objects in a network are dependent upon each other in rather complex ways. These dependencies are very important for the alarm correlation and fault identification process. In most cases a failure in one object has sided effects on other objects that depend on it. For example, a link failure has an effect on other resources in the network, e.g., connections on the various layers that use the link will experience timeouts. The knowledge of these dependencies gives us valuable information for alarm correlation and fault localization.

Dependencies: When one object requires a service performed by another object in order to execute its function, this relationship between the two object is called a dependency.

Consider any two objects, say A (such as a service, an application component in software or hardware) and B. A is said to be dependent on B, if B's services are required for A to complete its own service. One weight may also be attached to the directed edge from A to B, which may be interpreted in various ways, such as a quantitative measure for the extent to which A depends on B or how much A may be affected by the nonavailability or poor performance of B, etc. Any dependency between A and B thus arises from an invocation of B from A, which may be synchronous or asynchronous.

Dependency analysis explores causal dependencies among objects and data items, with the goal to trace the fault symptom back to the cause. This is an often used trouble-shooting technique, applicable to any system that is based on collaboration of independent or distributed entities. For instance, deadlocks in databases may be diagnosed by following transactions that are blocked waiting for other transactions.

In computing, there exist many different kinds of dependencies. However, not all references and interactions actually represent causal dependencies that are relevant for diagnosis. Hence the dependencies, which are pertinent to the purpose of the management, are taken into account.

The dependencies among distributed entities can be assigned probabilities to the links in the dependency or causality graph [25] [29]. This dependency graph can be transformed into a Bayesian Network with certain special properties [18].

In networks, the notion of dependency can be applied at various levels of granularities. Sometimes the dependencies that occur between different system components should be defined carefully. For example, the maintenance of an email server obviously affects the service 'email' and thus all the users whose user agents have a client - server relationship with this specific server; however, other services (news, WWW, FTP) are still usable because they do not depend on a functioning email service. Therefore, the inter-system dependencies are always confined to the components of the same service.

Two models are useful to get the dependency between cooperating entities in networks.

- Functional model (from the view of users)

  The functional model defines generic service dependencies and establishes the principle constraints to which the other models are
  bound. A functional dependency is an association between two entities, typically captured first at design time, which says that one
  component requires some services from another.

  The functional dependence between logical objects is determined
  by the implementation and functional support relationships and
  originates in a graph, from which it is possible to correlate a set
  of state changes (which may be considered as a "signature" of a
  problem) to the original cause of the problem.

  The functional model is utilized by a "network state estimator" to
  correlate the changes in the network state. The state changes are
  reported by the received alarms, to which information exogenous to
  the network (such as those related to climatic situations) is added.

- Structural model (from the view of system implementers)

  The structural model contains the detailed descriptions of software
  and hardware components that realize the service. A structural
  dependency contains detailed information and is typically captured
  first at deployment or installation time.

2. **Obtaining the Measurement of the Dependency.**

The faults and anomalies in networks can be identified based on the
statistical behavior of the Management Information Base (MIB) variables and the recordings in log files. When Bayesian networks are used
to model networks, Bayesian networks represent causes and effects between observable symptoms and the unobserved problems, so that when
a set of evidences is observed, the most likely causes can be determined
by inference technologies.

Single-cause (fault) and multi-cause (fault) are two kinds of general
assumptions to consider the dependencies between managed entities in
network management.

In Bayesian networks, a nonroot node may have one or several parents
(causal nodes). Single-cause means any of the causes must lead to the
effect. Therefore, the dependencies between causes and effect for single-
cause are denoted as:

$$P(\overline{e} \mid c_1, \ldots, c_n) = \begin{cases} 100\%, & c_i = F(False), \exists i, i \in [1, n]; \\ 0, & \text{otherwise.} \end{cases} \tag{9.1}$$

e denotes the effect node, c_i denotes the cause of e, $i = 1, \ldots, n$.

The existence of multiple causes means that one effect is generated only
when more than one cause occurs simultaneously. Therefore, the mea-

surement of the dependencies has various possibilities based on the particular problem domain. In the above description, the states of the objects are identified as T(*True*) or F(*False*). In complex systems, it is possible that managed objects hold more than two states.

In networks, the measurement of dependencies between managed objects can be obtained from the following methods:

- Management information statistics are the main sources to get the dependencies between the managed objects in networks.

- The empirical knowledge of experts is another important source to determine the dependency between managed objects.

- For particular dependencies, an experiment provides a way to retrieve the dependencies between the managed objects.

Hasselmeyer [14] argues that the dependencies among distributed cooperating components should be maintained and published by services themselves, and he proposes a schema that allows these dependencies to be obtained.

Some researchers have performed useful work to discover dependencies from the application view in networks [13] [20] [12].

Despite all the methods cited in this section, it has to be observed that obtaining dependency information in an automatic fashion is still an open research problem. In obtaining dependency information, it needs to use available and suitable techniques to deal with every system, layer or type of device separately.

In terms of precision, the behavior of a Bayesian network reflects the quality and the detailing level of its structure, which stems from the object system model. Another factor which affects the Bayesian network model is the precision of the value of the conditional probabilities.

9.2.3 Probabilistic Inference for Distributed Fault Management

The semantics of a Bayesian network determines the conditional probability of any event given any other event. When computing such a conditional probability, the conditioning event is called the evidence, while the event for which we want to determine its conditional probability given the evidence is called the query. The general capability of a Bayesian network to compute conditional probabilities allows it to exhibit many particular patterns of reasoning (inference).

- Causal reasoning is the pattern of reasoning that reasons from a cause to its effects.

- Evidential reasoning is the reasoning from effects to its possible causes.

- Mixed reasoning combines both causal and evidential reasoning.

- Intercausal reasoning involves reasoning between two different causes that have an effect in common.

In case of fault management in networks, we only consider the backward inference (evidential reasoning), which is the basic operation of fault diagnosis.

In our previous research, a Strongest Dependency Route algorithm (SDR) for backward inference in Bayesian networks is presented in [7].

Various types of inference algorithms exist for Bayesian networks, such as exact inference [33] [42] [43] and approximate inference [38]. Each class offers different properties and works better on different classes of problems, but it is very unlikely that a single algorithm can solve all possible problem instances effectively.

In the early 1980's, Pearl published an efficient message propagation inference algorithm for polytrees [24] [41]. The algorithm is exact and works only for singly connected networks. Pearl also presented an exact inference algorithm for multiply connected networks called loop cutset conditioning algorithm [41]. A straightforward application of Pearl's algorithm to an acyclic digraph comprising one or more loops invariably leads to insuperable problems [27] [38].

Another popular exact Bayesian network inference algorithm is Lauritzen and Spiegelhalter's clique-tree propagation algorithm [33]. The clique propagation algorithm works efficiently for sparse networks, but still can be extremely slow for dense networks. Its complexity is exponential in the size of the largest clique of the transformed undirected graph.

In general, the existent exact Bayesian network inference algorithms share the property of run time exponentiality in the size of the largest clique of the triangulated moral graph, which is also called the induced width of the graph [33]. It is also difficult to record the internal nodes and the dependency routes between particular effect nodes and causes. In distributed systems management, the states of internal nodes and the key routes, which connect the effects and causes, are important for management decisions. Moreover, the sequence of localization for potential faults can be a reference for system managers and thus very useful. It is also important for system performance management to identify the relevant key factors. Few algorithms give satisfactory resolution for this case.

Compared with other algorithms, the SDR algorithm belongs into the class of exact inferences and it provides an efficient method to trace the strongest dependency routes from effects to causes and to track the dependency sequences of the causes. The computing complexity of the SDR algorithm is $O(n^2)$. It is useful in fault location, and it is beneficial for performance management. Moreover, it can treat multiple connected networks modelled as DAGs.

9.3 Prediction Strategies for Fault Management in Dynamic Networks

For complex networks, it is important that the fault management be proactive, that is, to detect, diagnose and mitigate problems before they result in severe degradation of system performance. Proactive fault management depends on monitoring networks to obtain the data on which a manager's decisions are based. Fault prediction is to predict a failure in advance based on the current information about the system. It is especially true for large systems that have some components failing all the time, for such a prediction can be done by an analysis of the historical information.

Dynamic changes in networks raise higher barriers for exact fault location. Hence, for large networks with thousands of managed components, it may be rather time-consuming and difficult to locate the unknown causes of faults in due time by the exhaustive search for the root causes of a failure and this process may interrupt or impair important system services. Dynamic updates bring even more challenges in the fault detection.

Systems, whose behavior is not fully understood, are often modeled by Bayesian networks (BNs). However, the BN paradigm does not provide direct mechanisms for modeling temporal dependencies in dynamic systems [1] [56].

We apply Dynamic Bayesian Networks (DBNs) to address temporal factors and model the dynamic changes of managed entities and the dependencies between them.

9.3.1 Dynamic Characteristics in Networks

In real-life networks, dynamic changes in networks are correlated to hardware, software and the dependencies between those components in implementing certain functions. Hence changes in networks demonstrate some particular characteristics.

1. Hard Changes and Soft Changes in Networks

Dynamic updates in networks can be classified into either hard or soft changes [8].

A hard change refers to a change that happens abruptly and most of the time is generated on purpose by the system owner. This kind of change does not depend on the system's history. For example, a router being added or removed from the distributed system may cause an abrupt change in the system topology and behavior. Some intended operations also generate this kind of hard change, such as a change of the configuration of a distributed system. Generally, a hard change does not happen so often, but it depends on the intention of the system manager.

A soft change, in contrast, refers to a change that happens gradually and depends on the system history. A soft change typically results from changes of system properties such as performance degradation, application degeneration, dependency modifications and so on. A soft change may bring some potential problems, such as the network traffic gets slower or certain network services decrease in efficiency. The roots of these kinds of problems are aging devices, conflicted applications or unknown hidden factors in the systems. From the experience of system management, lots of unknown or unlocated causes of faults are triggered by a soft change, which is related to the potential changes and updates of the system. Compared with a hard change, a soft change keeps going on all the time in networks and it is hard to predict using a straightforward approach. In our research we focus on soft changes in networks.

When networks are modelled as graph structures, hard changes can be treated by structure modifications in the models based on the abrupt changes in the systems or based on the intentions of system managers. Soft changes (such as the improvement or degradation in performance of a hardware component or a software component) will not effect in the topology of the network, but will update the weights (dependencies) between the components in the network.

Considering soft changes in networks, one kind of change comes from individual networks entities; another arises from updating dependencies between managed entities. From the viewpoint of management, an entity can be a hardware device, a software component or a certain application.

In networks, real-life dynamic systems are often rife with nonlinearities, many of which are expressed as discrete failure modes that can produce discontinuous jumps in system behavior.

2. Characteristics of Dynamic Changes in Networks

In networks, dynamic changes are often identified as a discrete nonlinear time series.

A time series is a chronological sequence of observations on a particular variable. Time series data are often examined in hope of discovering a historical pattern that can be exploited in the preparation of a forecast. In order to identify this pattern, it is often convenient to think of a time series as consisting of several components: trend, cycle and irregular fluctuations.

Prediction of system faults, anomalies and performance degradation forms an important component of network management. The advent of real-time services on networks creates a need for continuous monitoring and prediction of systems performance and reliability. Although faults are rare events, they have enormous consequences when they do occur. However, the rareness of faults in distributed systems makes their study difficult. Performance problems occur more often and in some cases may be considered as the indicators of an impending fault [37]. Efficient handling of these performance issues may help eliminate the occurrence of severe faults.

9.3.2   Dynamic Bayesian Networks for Fault Management

When a dynamic network is modeled, a time dimension will be considered. Because observations and evidence can be updated over time, a management system should capture the evolution of the system as it changes over time.

DBNs provide a way to model a dynamic system, which describes a system that is dynamically changing or evolving over time [22]. DBNs will enable users to monitor and to update the system as time proceeds, and even to predict further behavior of the system. As there is no standard definition for DBNs, researchers may use different descriptions to accommodate their research requirements. Current literature tends to use the terms "dynamic" and "temporal" interchangeably.

The temporal approaches can be divided into two main categories of time representation models:

- as points or instances or

- as time intervals.

DBNs have various definitions in different application areas. In fault management and dissident systems, DBNs possess a time related function:

$$BN(t) = (V(t), L(t), P(t)) \qquad (9.2)$$

For a soft change in networks, dynamic changes only happen in individual components and on the dependency between components. Under these kinds of changes, the topology of the Bayesian Network keeps stable; hence the time parameter can be omitted in nodes and edges:

$$BN(t) = (V, L, P(t)) \qquad (9.3)$$

DBNs can represent large amounts of interconnected and causally linked data as well as the dynamic properties when they occur in networks. Thus DBNs can model time-related changes in the dependencies between managed objects in networks.

DBN is an extension of BN that models a time series [11]. A DBN is a way to extend Bayesian networks to model the possible distributions over a time series. We only consider the discrete-time stochastic processes, so we increase the index t by one every time a new observation arrives. The observation could indicate that something has changed in networks. Note that the term 'dynamic' means not that the topology of the network changes over time, but that a dynamic system is modeled.

9.3.3   Prediction Strategies for Network Management

Predictive management plays a crucial role in networks. The ability to predict service problems in networks, and to respond to those warnings by

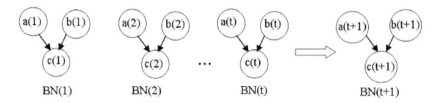

FIGURE 9.2: Model of dynamic Bayesian network.

applying corrective actions, brings multiple benefits. Firstly, the detection of system failures on a few servers can prevent those failures from spreading to the entire distributed system. For example, slow a response time on a server may gradually escalate technical difficulties on all nodes in the attempt to communicate with that server. Secondly, a prediction can be used to ensure the continuous provision of networks services through the automatic implementation of corrective actions.

Correlation serves to diminish the number of alarms presented to the operator in network management, yet ideally, the approach should be able to facilitate the fault prediction, which can predict the faults that have occurred from the alarms and warn the operator before severe faults may happen.

Considering the model of DBN in Fig. 9.2, two possible changes, which are updated over time, are presented in DBNs:

- the possible updates of the nodes (variables)

- the possible updates of the links (dependency between nodes).

When a network is modeled as a DBN, one important task is to capture the trends for the evolution in the network. This amounts to obtain $BN(t+1)$ based on the data set $BN(1), BN(2), \ldots, BN(t)$. Here $BN(t)$ denotes the updated BN at time t.

In DBN, the prediction can be denoted as

$$BN(1), BN(2), \ldots, BN(t) \Rightarrow BN(t+1) \tag{9.4}$$

In DBNs, the following prediction tasks are to be considered as a result of the management requirements.

- **Prediction per individual component.** The state of an individual component in a network can change over time due to the degradation or improvement of the component. The prediction of the individual component's change of states can be denoted as:

$$P(v(1)), P(v(2)), \ldots, P(v(t)) \to P(v(t+1)), v \in V. \tag{9.5}$$

$P(v(t))$ represents the probability of the state of component v at time t.

- **Prediction of the dependency between components.** The modification of dependencies between managed objects derives from the update of the system performance and changes in the correlation between objects. This can be denoted as:

$$P(v(1)|\pi(v(1))), P(v(2)|\pi(v(2))), \ldots, P(v(t)|\pi(v(t)))$$
$$\rightarrow P(v(t+1)|\pi(v(t+1))) \tag{9.6}$$

$v \in V$, $P(v(t)|\pi(v(t)))$ represents the probability of the dependency between node v and its parent $\pi(v)$ at time t.

- **Prediction for potential faults based on backward inference.** When the future state of the effect nodes is estimated, a promising prediction is to trace the causal nodes based on the estimated state of the effect nodes. The prediction from effects to causes is considered as the backward inference:

$$E(t+1) \rightarrow C(t+1) \tag{9.7}$$

$E(t)$ denotes the set of effects at time t, and $C(t)$ denotes the set of causes at time t.

Dynamic Bayesian Networks (DBNs) are applied in fault management in order to address the temporal factors and to model the dynamic changes of managed entities and the dependencies between them. Our related work [DLJ+06] investigated the prediction capabilities by means of the relevant inference techniques when the imprecise and dynamic management information occurs in the distributed system. To identify the dynamic changes in networks as discrete nonlinear time series, Least Square Fit (LSF) is used for the polynomial regression. It is workable in a large scale distributed system with thousands of nodes and links. Based on the given prediction strategies, not only the prediction in an individual entity and the dependency relationship between managed entities are considered, but also the potential reasoning from effects to causes in DBN can be obtained.

To evaluate the approach of probabilistic backward inference and prediction strategy, we design a simulation scheme, which is reported in [5], to construct the simulation in Bayesian networks for probabilistic inference and prediction, so that the simulation in Bayesian networks is close to real life networks and, further, the intelligent decision in management of networks can be obtained.

9.4 Application Investigations for Probabilistic Fault Management

The probabilistic fault management strategies do not intend to replace traditional network management, but to complement the deficiency of traditional management systems in uncertain situations.

9.4.1 Architecture for Network Management

Network management takes place between two major types of systems: those in control, called managing systems, and those observed and controlled, called managed systems. The most common managing system is called a network management system (NMS). Managed systems include hosts, servers and network devices such as routers or intelligent repeaters. Here, we use "online network devices" to represent the term "managed system."

9.4.1.1 Components of Network Management System

As specified in Internet RFCs [47] and other documents, a typical distributed management system comprises:

- **Manager**: A manager generates commands and receives notifications from agents. There usually are only a few managers in a system.

- **Agents**: Agents collect and store management information such as the number of error packets received by a network element. An agent has local knowledge of management information and transforms that information into the form compatible with SNMP. An agent responds to commands from the manager, and sends notification to the manager. There are potentially many agents in a system.

- **Managed object**: A managed object is a vision of a feature of a network, from the point of view of the management system [19]. For example, a list of current active TCP circuits in a particular host computer is a managed object. Managed objects differ from variables, which are particular object instances. Managed objects can be scalar (defining a single object instance) or tabular (defining multiple and related instances). In literature, "managed object" is sometimes used interchangeably with "managed element."

- **Management Information Base (MIB)**: A MIB is a formal description of a set of network objects that can be managed by using the Simple Network Management Protocol (SNMP). The format of the MIB is defined as part of the SNMP.

- **Management protocol**: A management protocol is used to convey management information between agents and network management stations (NMSs). Simple Network Management Protocol (SNMP) is the Internet community's de facto standard management protocol.

Interactions between NMSs and managed devices can be any of four different types of commands: *read*, *write*, *traverse* and *trap*.

9.4.1.2   Protocols for Network Management

1. Simple Network Management Protocol (SNMP)

SNMP (Simple Network Management Protocol) is a communication protocol that has gained widespread acceptance since 1993 as a method of managing TCP/IP networks, including individual network devices, and devices in aggregate. SNMP was developed by the IETF (Internet Engineering Task Force), and is applicable to any TCP/IP network, as well as other types of networks. It defines management entities, typically the NMS, and agent entities, typically the network devices, or more accurately, the processes that run on the NMS and network devices. The information available through SNMP is organized into a Management Information Base (MIB). The structure of this information is defined in the Structure of Management Information (SMI). One or more MIB files define the MIB supported by a given SNMP agent. The bottom line is that SNMP provides management information in a structured manner that is well suited to retrieval and modification via applications.

2. Common Management Information Protocol (CMIP)

Common Management Information Protocol (CMIP) is an OSI-based network management protocol that supports information exchange between network management applications and management agents.

CMIP is a well-designed protocol that defines how network management information is exchanged between network management applications and management agents. It adopts an ISO reliable connection-oriented transport mechanism and has built up security that supports access control, authorization and security logs. The management information is exchanged between the network management application and management agents through managed objects.

CMIP is widely used in the telecommunication domain and telecommunication devices that typically support CMIP.

By far the largest advantage of SNMP over CMIP is its simple design, so it is easy to use on a small network as well as on a large one, with ease of setup, and lack of stress on system resources. Also the simple design makes it simple for the user to program system variables that they would like to monitor. Another major advantage of SNMP is its wide use today around the world. Because of its development during a time when no other protocol of this type existed, it became very popular, and has built in protocol supported by most major vendors of networking hardware, such as hubs, bridges and routers, as well as major operating systems. SNMP is by no means a perfect network manager, but it can fix these flaws due to its simple design.

3. Other Related Protocols for Network Management

Besides the standard protocols SNMP and CMIP, there are still some other miscellaneous protocols which can help in network management.

- **Ping:** Ping is commonly used to check connectivity between devices. It is the most common protocol used for availability polling. It can also be used for troubleshooting more complex problems in the network. Ping uses the Internet Control Message Protocol (ICMP) Echo and Echo Reply packets to determine whether one IP device can talk to another. Most implementations of ping allow users to vary the size of the packet.

- **Traceroute:** Traceroute is most commonly used to troubleshoot connectivity issues. If we know that we cannot reach a host from another host, traceroute will show whether the connectivity loss exists at one of the intermediate routers. Traceroute determines that the destination has been reached when it receives an ICMP destination port unreachable message. Note that we are actually discovering the path that the ICMP timeout messages are taking when they come back.

- **Terminal Emulators:** Terminal emulators are used for many purposes in network management, including users' accesses to network devices. Obviously, access is useful for configuring and troubleshooting devices. There are also times when information or operations on network devices is not available through SNMP and scripts must be written to access this information or capability through terminal access. Telnet is the traditional way of obtaining terminal emulation access to network devices.

- **Syslog:** The syslog protocol was first defined as part of the UNIX operating system to log messages within the OS. Syslogs allow a computer or device to deliver messages to another computer. Syslog messages have a particular format that associates a facility and a severity or priority with a message. The facility code allows syslog to group messages from different sources and take actions based on this facility or group.

9.4.1.3 Extended Architecture for Probabilistic Fault Management

In an extended architecture, a module of Fault Diagnosis Agent (FDA) is added to execute the management tasks for probabilistic fault diagnosis. A detailed model is denoted in Figure 9.3.

The FDA is designed to operate in parallel with a SNMP manager or a SNMP agent. It can communicate with a SNMP manager and get the needed information (such as the states of the interfaces or ports of devices) from the SNMP manager, in emergent situations or special requirements, and it can

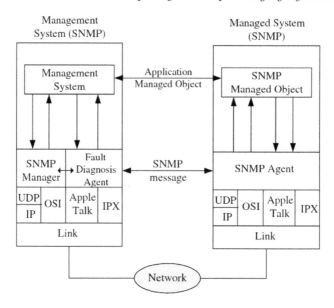

FIGURE 9.3: Detailed model of network management with FDA.

also communicate with agents which reside on the managed devices to retrieve information by SNMP.

In ordinary operation, FDA gets information or fault events from the SNMP manager, so that it saves the traffic of network communication. FDA has interfaces with the logfile analyzer and system managers so that additional events can be integrated into the fault management systems.

9.4.2   The Structure and Function of Fault Diagnosis Agent

Figure 9.4 depicts the inner structure of FDA and the process for fault management. Some important components and their functions in FDA are described in the next subsections.

9.4.2.1   Data Collection and Analysis for Fault Management

The following components are related to data collection and analysis:

1. **Event Configurator**

   Event configurator can process event configuration based on certain management tasks or managers' purpose. Thus we need to determine when and which events to be triggered by taking the following steps:

   - Select the objects or variables.
   - Select the devices and interfaces.

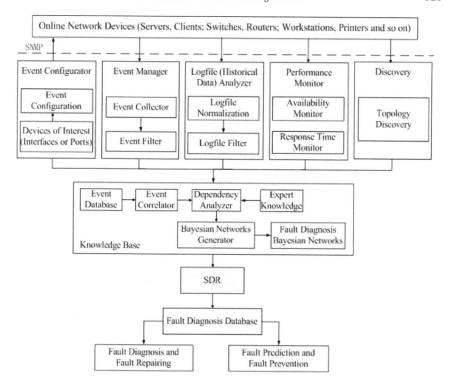

FIGURE 9.4: The structure of fault diagnosis agent.

- Determine the trigger values for each object or object and interface type.
- Determine the severity for the event.

Event configuration can:

- Determine what information is of interesting and what states or levels are normal for the network.
- Evaluate the events. The devices will automatically generate and make sure that the desired events are turned on and the undesired ones are turned off or filtered out.

Configuring Events

There are two ways to configure events:

(a) The easiest and the least costly way is to have the network device check the trigger points and generate the event. The data we want to query already exist in the device on which we are configuring the event.

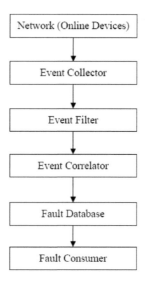

FIGURE 9.5: The procedure of event management.

(b) The other way is to collect data at a management station and analyze that data against thresholds there. Thresholds on devices, also known as agent-based thresholds, allow the network devices to directly generate events when something interesting happens in the network.

2. **Event Manager**

The event manager component can execute event collecting, event filtering and event correlating. Figure 9.5 denotes the procedure of event management in the fault management system. Some components (on-line devices, event collector, fault consumer) in Figure 9.4 are standard or defined by the network device vendors and they can be configured to meet the management requirements. The other components are own developments (such as event filter, event correlator, fault database) which are organized for certain functions in dealing with the event management. Most of the components related to event management are actually software components which are embedded in the management system.

Event Producers

Any device on the network can produce an event for a variety of reasons. In addition, events can be produced by the network management systems or call-tracking systems. Network engineers (expert) can identify issues and directly enter them as events.

Some examples of components that produce events and the types of events they might generate are as follows:

- Devices that provide network connectivity such as hubs, switches, and routers can generate events such as the loss or re-establishment of connectivity between two devices.

- Computers providing services can generate events such as a database or print server becoming available or unavailable.

- Network management systems can generate events such as an event that is triggered if, after calculating the number of collisions versus the network load, a sample falls outside the expected curve.

- An event could be produced by an end user of the network calling the help desk and reporting a problem such as the inability to reach a particular printer. The help desk technician may generate an incident report that would trigger an event describing the incident reported.

Continuous and Discrete Data Sources of Events

There are two kinds of data sources for events: continuous data source and discrete data sources.

(a) **Continuous data sources** present a continuously changing curve of values. Examples of continuous data sources include:

- The number of packets received on a network interface,
- The CPU utilization on a device,
- The number of calls completed on an ISDN interface.

These data sources are sampled at some rate to produce a data stream. The type of trigger applicable to continuous data sources is a continuous threshold. This threshold type generates an event when the most recent value in a continuous data stream becomes interesting.

(b) **Discrete data sources** present data that can have several discrete states. Discrete event is a main character in computer systems. In networks, most of the events comes from discrete data sources. Examples of discrete data sources include:

- The state of a network interface (*up, down, testing*),
- The operational state of a device (*operational, faulty, reloading*),
- The environmental status (*normal, warning, critical, shutdown, notPresent*).

Discrete data can be represented by data types, including enumerated data, Boolean data or text data fields with a fixed range of text values or states.

Types of Events

Events can be divided into two broad categories: state change events and performance events. State change events are triggered when something in the network changes state. Performance events are generated when a possible performance issue is noticed by some component of the network. Most performance events are generated by thresholds that are defined by the network manager on the network devices or NMS stations.

Event Monitoring

Event monitoring can be divided into two types: time-driven monitoring and event-driven monitoring. The former involves periodically obtaining snapshots of the network state, e.g., by polling for the values of MIB variables. The later involves asynchronous receipts of notifications about "interesting" events.

Consider trap messages generated by agents in response to state changes in managed objects. Traps may be the result of completely unpredictable events, such as the loss of an underground line to a backbone, a hardware failure in a router or a software error in a user's application.

Event Collection, Normalization and Filtering

The goal of event management is to collect all event information and determine what, if any, actions need to be taken as a result of each event. There are several steps that must be taken on receipt of an event:

(a) Events must be collected. Usually, events will come in from a variety of sources through several different methods or protocols.

There are two methods for collecting event data: active polling and event reporting.

    i. **Active polling** involves a management station actively obtaining specific management data from network devices. The collected data are then stored in a database and used later for reporting.

    ii. **Event reporting** (polling by exception) denotes that the managed device or agent generates a trap or event which is received and logged by the manager. For event reporting, events are generated when preconfigured thresholds are exceeded, or a change of state or an unusual event such as a fault occurs.

(b) Upon receipt, the events should be normalized to facilitate their processing. Normalization means to format the events in a consistent way, regardless of the delivery mechanism.

(c) Next, this event must be determined if it can be filtered or deleted. Because the volume of events to be processed can be very high, it is important to eliminate undesired events as early as possible. Filtering means to eliminate undesirable events by comparing them to some pattern and eliminating those events that match the pattern. T. Koch provided efficient approaches in automated event filtering and event correlation [23].

(d) Next, the management system should correlate events and determine the faults that exist in the network. Correlation in this context means to examine events to determine the preceding cause of an event or to determine the root cause of a fault. Event correlation will provide help in investigating the dependency between events or between managed components based on certain service.

3. **Logfile Analyzer**

   Logfiles are used by many event producers to record the events they generate. Sometimes, the only way to determine the occurrence of these events is to parse the logfiles and to extract the event information. Some logfiles only have event information. Others have a variety of data mixed with event information.

   In networks, logfiles keep a record of historical events of a system. Efficient analysis and mining of logfiles will help to retrieve the events and to evaluate the system performance.

   Some examples of log files that might have event information include the following:

   - syslog logfiles
   - system console logfiles
   - application message files

   Many systems log messages to files. Windows NT has an event log that records system events. Many UNIX systems use the syslog protocol to log system events. Network devices such as routers and switches can record system events on a syslog server. Applications may also log messages to files. Some of these messages are the result of applications finding that a threshold has been crossed, such as a database management system noticing that a database partition is within 90 percent of capacity.

   Logfiles are often recorded to be less-structured or half-structured. Before the information can be processed further, the logfile should be normalized and filtered in terms of the management requirements.

   For example, consider a syslog message that indicates that an interface went down:

   Sep 19 16:51:07 10.29.2.1 79: Sep 19 16:50:24: %LINEPROTO-5-UPDOWN: Line protocol on Interface Ethernet0, changed state to down

The example syslog message would be normalized as follows:

Source = 10.29.2.1
Time = Sep 19 16:50:24 (may want to convert to UTC)
Priority = 5
Type = linkDown
Variables = Interface Ethernet0

Then the event can be appended into a structured relational database which might contain the elements such as Event ID, Source (IP addresses), Time Stamp, Priority, Event Type, Variables (States of Interfaces or Ports), Destination (IP address) and so on. Event analysis can be executed by the operation on the database.

4. **Performance Monitor**

Availability is the measure of time for which a network or application is available for a user. From a network perspective, availability represents the reliability of the individual component in a network. The availability can be an important parameter in measuring the dependency between network events or between network devices.

Measuring availability requires coordinating real-life measures with the statistic collected from the managed devices.

Measuring Availability

According to Stallings [53], availability is expressed by the Mean Time Between Failures (MTBF) with the following formula:

$$Availability = \frac{MTBF}{MTBF + MTTR} \tag{9.8}$$

where MTTR is Mean Time To Repair.

ICMP pings are the easiest to use and report on measuring availability. The following equation shows the relevant formula:

$$Avail = \frac{(Total\# \ of \ PINGs \ received)}{(Total\# \ of \ PINGs \ sent)} \tag{9.9}$$

Availability Statistics

Availability can be defined as the probability that a product or service will operate when needed. In networks, this can be defined as the average fraction of connection time that the product or service is expected to be in operating condition. For a network that can have partial as well as total system outages, availability is typically expressed as network availability:

$$Availability = 1 - \frac{total\ connection\ time_{outage}}{total\ connection\ time_{in-service}} \qquad (9.10)$$

The availability of objects or services can be an important source to evaluate the reliability of objects.

9.4.2.2 Dependency Analysis for Events

Dependency analysis for events includes the integration work of the topology analysis and event correlation analysis and eventually to ascertain the precedence events for each event.

1. **Topology Discovery**

   Physical network topology refers to the characterization of the physical connectivity relationships that exist among entities in a network. Discovering the physical layout and interconnections of network elements is a prerequisite to many network management tasks, including reactive and proactive resource management, server siting, event correlation and root-cause analysis [3].

   SNMP-based algorithms for automatically discovering network layer (i.e., layer-3) topology are featured in many common network management tools, such as HP's OpenView, IBM's Tivoli, Actualit's Optimal Surveyor and the Dartmouth Intermapper. Recognizing the importance of layer-2 topology, a number of vendors have recently developed proprietary tools and protocols for discovering physical network connectivity. Examples of such systems include Cisco's Discovery Protocol, Bay Networks' Optivity Enterprise and Loran Technologies' Kinnetics network manager.

   The topology of a network can help in investigating the dependency relationship between managed objects.

2. **Event database**: is an integrated database to record all received events and knowledge related to fault management. The events come from a variety of sources such as SNMP notifications, syslog messages, entries in log files, NMS events or event expert knowledge. Refined expert knowledge can be manually appended into the database. Historical data are another possible source to enlarge the database. Some historical data (such as logfiles, historical records of fault management) are available in a network, but most of them are not structured, large-scale and intermingled. Hence data mining technologies are useful tools to help the data classification, refining and integration [44] [54]. Temporal events and period data collection and statistics can be organized as temporal databases, so that the prediction operation can be implemented in the dynamic environments of networks.

The event database should be a formatted and structured database. A typical example structure of event database is depicted as Table 9.1. The data in the event database can be used for event-correlation analysis, dependency calculation and statistics analysis.

Table 9.1:   The Structure of Event Database.

Event ID	$0x003831f49700000055$	...
Event Type	*LinkDown*	...
Priority	*Critical*	...
Time Stamp	1125302050	...
Source IP	10.29.2.1	...
Variables	*InterfaceEthernet0*	...
Precedence Event	$0x003831f49700000000$	...

Event ID is the only identification of an event. It is generated by the event collection system automatically.

Event Type is to identify the type of the event, such as *Link Down*, *system error* or *notification of link overload*, etc.

Priority identifies the level of the events. In event database, ordinal number is used to denote the priority of an event, such as *Information* (4), *Major Warning* (3), *Serious* (2) and *Critical* (1).

Time Stamp records the time when the event is generated. In Event Database, UTC (Coordinated Universal Time) is used to replace GMT (Greenwich Mean Time). For example, GMT: *7:54:10, Aug. 29*, is equal to UTC: *1125302050*.

Source IP denotes the location (IP address) where the event is generated.

Variables depict the characteristics of a concrete event. MIB defines some standard variables for distributed systems management. The values of variables or their collaboration analysis may help one to determine the real value of a fault event. For example, in order to detect a *down* in an interface, there are three primary methods, all using RFC 2233 variables:

(a) Watch for *linkDown* traps from a device.

(b) Watch for interface status change messages in syslog.

(c) Poll the *ifOperStatus* and *ifAdminStatus*. If *ifAdminStatus* is up and *ifOperStatus* is down, the interface is intended to be up.

Precedence Events identify the preceding events (cause events) of a current event. The value of precedent events is the result of event correlation and can be used further for dependency analysis between events. An event may have one or more precedence events, or else the event tends to be a root event if the precedence event is *null*.

3. **Event Correlation and Dependency Analyzer**

Event correlation is an automated process that enables administrators to find, among many events, those revealing critical problems that cannot be ascribed to other issues (root cause analysis). Of particular importance is the detection of a few problems that have an adverse effect on the stability and performance of critical services (e.g., an e-business application server for an online retail shop).

Event correlation is the "smart" part of a management application. It relies on a number of techniques [32]: state transition graphs (finite state machines), rule-based reasoning, binary coding (code-books), case-based reasoning, probabilistic dependency graphs (Bayesian networks), model-based reasoning, neural networks, etc. In today's management platforms, several techniques are often used in conjunction.

A number of researches have already investigated the distribution of event correlation. Some proposals focus on correlating network events with the network topology [4] or intrusion detection [26]. Others propose general-purpose languages [36].

The basic premise underlying dependency models is to model a system as a DAG in which nodes represent system events or system components and weighted directed edges represent dependencies between them. A dependency edge is drawn between two nodes only if a failure or problem with the node at the head of the edge can affect the node at the tail of the edge; if the dependency edge presents, the weight of the edge represents the impact of the failure's effects on the tail node. Heavier edges represent more significant dependencies and therefore more likely main causes.

9.4.2.3 Bayesian Networks for Fault Management

The following components depict the practice details how Bayesian networks are applied to probabilistic fault management.

1. **Bayesian Networks Generator (BNG)** is designed to generate a BN model from the event database by event correlation and dependency analysis. In order to generate a BN for network management, the following steps should be taken:

   (a) Identify the **event family**. Here an event family is defined as the set of one event and all its parents events. Based on the event

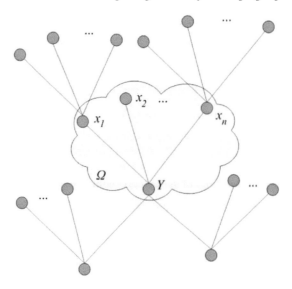

FIGURE 9.6: Dependency analysis in events.

database and correlation techniques, the direct precedence of an event will be identified and recorded. Then an event family will be set up. Figure 9.6 depicts the event family (sub-BN) Ω and the dependency between nodes.

(b) In BNs, the dependency is denoted by a JPD, that means the effect node (son) is effected by all its joint causal nodes (parents), even if some are strong or weak. See Figure 9.6, for an event family Ω; suppose node Y has n parents $(x_1, x_2, \ldots, x_n)$. Table 9.2 depicts how the JPD is obtained between every pair of parent nodes and their son node.

Table 9.2: The JPD obtaining between every pair of parent nodes (X) and their son node (Y).

i	$\overline{Y}$	$x_1 x_2 \ldots x_n$	τ_i	$P(\overline{Y} \mid X_i)$
1	1	$0\ 0\ \ldots\ 0$	τ_1	P_1
2	1	$0\ 0\ \ldots\ 1$	τ_2	P_2
$\vdots$	$\vdots$	$\vdots\ \ \vdots\ \ \ \ \vdots$	$\vdots$	$\vdots$
2^n	1	$1\ 1\ \ldots\ 1$	τ_{2^n}	P_{2^n}

The JPD should have 2^n data items: $P(\overline{Y} \mid X_i), i \in [1, 2, \ldots, 2^n]$, X_i denotes the status of parent nodes of Y at $No.i$. Here 1 denotes the state is in order, while 0 denotes the state is out of order. For example,

$$X_1 = x_1 x_2 \ldots x_{n-1} x_n = 0\ 0\ \ldots\ 0\ 0$$
$$X_2 = x_1 x_2 \ldots x_{n-1} \overline{x_n} = 0\ 0\ \ldots\ 0\ 1$$
$$\vdots$$
$$X_{2^n} = \overline{x_1 x_2} \ldots \overline{x_{n-1} x_n} = 1\ 1\ \ldots\ 1\ 1$$

For each accepted event, it is possible to get the value of precedence events by polling or trap record. All the events in the event family at the time will become a record for the JPD.

In Table 9.2, $\overline{Y} = 1$ denotes Y is out of order. τ_i depicts the number of the statistics of the events where $\overline{Y} = 1$ and X_i occur. τ_i can be also denoted as the statistics of time duration for fault events. Then

$$P_i = P(\overline{Y} \mid X_i) = \frac{\tau_i}{\sum_{j=1}^{2^n} \tau_j} \tag{9.11}$$

Thus we can determine the JPD of dependencies between parent nodes and their son. All these data are the prerequisites for further SDR inference.

(c) Each event family can be bridged by the correlation and the dependency between conjunction events, and as a result a global BN for a distributed system can be created. Both the structure of the BN and the values of dependencies between nodes are recorded as fault diagnosis Bayesian Networks.

The BNG can generate dynamic Bayesian networks based on the temporal data in event database for a given period. Then a temporal JPD will be obtained.

2. **Strongest Dependency Route (SDR) Calculator** is a key component to execute the calculation for pruning and backward inference in BNs. The results after the SDR are:

(a) a spanning tree with strongest dependency routes, which can be acquired by the depth-first search in spanning tree, between effect nodes and cause nodes;

(b) a strongest dependency sequence of candidate causes for specified effects. Normally, a SDR module is started by the manager when the management system cannot locate the root cause of faults, and it can act as an assistant in evaluating the system performance.

The SDR module can work on a dynamic Bayesian network model, and it can result in a temporal spanning tree and temporal strongest dependence sequences.

3. **Fault diagnosis database (FDD)** is a refined database to record the results (both the spanning tree and the strongest dependency sequence of causes) of the SDR operation. A FDD is actually a knowledge base in guiding the probabilistic fault diagnosis in networks.

   In a static model, FDD can be the basis to execute fault diagnosis and fault repairing. In a dynamic model, FDD can provide fault prediction and help for fault prevention.

   Normally a repairing action will be executed on the basis of the knowledge of FDD in an even-driven style. The action module can be controlled by the system manager. Moreover, a human manager can do some repair work to assist system maintenance.

   Another important function for an action module is to send feedback to the FDD and fault diagnosis BN, and further to update FDD and diagnosis BN. The results of both success and failure results will be accepted by the FDA itself, so that it is possible to improve the accuracy and performance for future inference and prediction in fault management.

All in all, the FDA will involve the fault management processes in networks, such as collecting alarms (events), filtering correlated alarms (events), dependency analysis, diagnosing faults, verifying and eliminating faults, taking action to ensure customer satisfaction and developing and implementing corrective plans.

9.4.3   Discussion of Application Issues

For probabilistic fault management of networks, there are still some related topics that should be investigated carefully.

- Efficiency and accuracy are important factors in evaluating a management system. For probabilistic fault management, the scale of the data set, the accuracy of the BN models, the filtering and refining data and iterative correction are main stages to improve the efficiency of a management system. BN models should be defined carefully in appropriate levels to resolve the core problems. The granularity definition of managed objects should be consistent with the management tasks. Hence a managed object may denote a subset, a server or a function module in different levels based on different management goals and tasks.

- The complexity of computation and operation in fault management is related to the complexity of the model, data collection, analysis and the complexity of the algorithm. Building a Bayesian network requires a careful tradeoff between the desire for a large and rich model on the one hand and the costs of construction, maintenance and inference on the other hand. Actually, building a Bayesian network is a creative and iterative process. With the advance of iterative procedures and associated graphical tools for supporting the overall construction process, the quantification task will be addressed within its proper context, which hopefully will contribute to reducing its burden.

- The growth of applications and network services extends a network and suggests network management to migrate to service-oriented application management. Application services are generally composed of a set of distributed components, each of which contributes a specific function to the total services provided by the application. Thus, a fault or malfunction occurring in any of the software components can lead to a problem in the end-to-end service that the customer perceives. In addition, application components function in close cooperation with the elements that comprise the environments like the communication network, the operating system, various middle-ware such as databases, message functions and their services. Thus, determining the source of a problem involves the hunt for the root cause which may lie in any of the components and services that contribute to the end-to-end service to the customer. In service-oriented networks, probabilistic fault management is still an important research topic in maintaining service efficiency. The strategies of fault management which is discussed in this chapter can be also used in application services of the network. Although the sources and the process of data collection are different, the core components and functions of the Fault Diagnosis Agent (FDA) presented in this chapter are still workable in the scenario of application services. Some related research is necessary to discover dependencies from the application view in distributed services [13] [12].

- Security is another important topic in network management. Security not only focuses on protecting the systems itself, but also relates to the protection of management information. In our approaches to fault management, the FDA is designed on the basis of traditional management frameworks. Hence it will not bring new risks to the security of management information and communication data. The key computing and operation of the components are executed locally in network management stations. Management information is more sensitive than other data in networks; thus most management systems are running in reliable and private environments.

9.5 Conclusions

Some unavoidable uncertain factors and dynamic changes in networks raise higher barriers for fault management. Bayesian network is an appropriate tool in modeling the probabilistic environment and is efficient in executing the backward inference, which is an important task in tracing the root causes when faults are detected. In modeling dynamic networks, the time factor should be taken into account, since even the standard Bayesian Networks paradigm does not provide direct mechanisms for modeling the temporal dependencies in dynamic systems. For this reason, Dynamic Bayesian Networks (DBNs) are applied in network management in order to address the temporal factors and to model the dynamic changes of managed entities and the dependencies between them. Furthermore, the prediction capabilities are investigated by means of the relevant inference techniques when the imprecise and dynamic management information occurs in the network.

Application issues of probabilistic fault management are investigated to demonstrate how the probabilistic fault management can be brought into practice. The software architecture of the fault diagnosis agent (FDA) is designed to implement the main tasks of probabilistic fault management, such as data collection, data filter and refinement, Bayesian network generation, SDR algorithm application, inference of causes and fault prediction operations.

Acknowledgement

This work was carried out during the tenure of an ERCIM "Alain Bensoussan" Fellowship Program. This research was also partly supported by the Chinese national program of high-tech research and development 863 project: 2006AA01Z169 and Chinese national 973 project: 2007CB316506. The authors would like to thank the anonymous reviewers for their valuable comments.

References

[1] C. F. Aliferis and G. F. Cooper. A structurally and temporally extended bayesian Belief network model: Definitions, properties, and modeling techniques. Proceedings of the 12th Conference on Uncertainty in Artificial Intelligence, pages 28-39, 1996. Morgan Kaufmann.

[2] Remi Badonnel and Mark Burgess: Fault Detection in Autonomic Networks Using the Concept of Promised Cooperation. DSOM 2007, pages 62-73, 2007.

[3] Y. Breitbart, M. Garofalakis, C. Martin, R. Rastogi, S. Seshadri and A. Silberschatz. Topology Discovery in Heterogeneous IP Networks. Proceedings of IEEE INFOCOM, pp. 265-274, 2000.

[4] C. S. Chao, D. L. Yang and A. C. Liu. An automated fault diagnosis system using hierarchical reasoning and alarm correlation. Journal of Network and Systems Management, 9(2):183-202, 2001.

[5] Jianguo Ding, Ningkang Jiang, Xiaoyong Li, Bernd Krämer, Franco Davoli and Yingcai Bai. Construction of Simulation for Probabilistic Inference in Uncertain and Dynamic Networks Based on Bayesian Networks. Proceedings of the 6th International Conference on ITS Telecommunications, pages 983-986, IEEE Communication Society Press, June 2006.

[6] Jianguo Ding, Bernd J. Krämer, Yingcai Bai and Hansheng Chen. Probabilistic Inference for Network Management. M.M. Freire, P. Chemouil, P. Lorenz (Eds.) Universal Multiservice Networks: Third European Conference, ECUMN 2004, Lecture Notes in Computer Science, Volume 3262, pages 498-507, Springer-Verlag Heidelberg, 2004.

[7] Jianguo Ding, Bernd Krämer, Yingcai Bai and Hansheng Chen. Backward Inference in Bayesian Networks for Distributed Systems Management. Journal of Network and Systems Management, 13(4), pages 409-427, Dec 2005.

[8] Jianguo Ding, Bernd Krämer, Shihao Xu, Hansheng Chen and Yingcai Bai. Predictive Fault Management in the Dynamic Environment of IP Networks. Proceedings of 2004 IEEE International Workshop on IP Operations & Management, pages 233-239, Oct. 2004.

[9] Jianguo Ding, Xiaoyong Li, Ningkang Jiang, Bernd J. Krämer and Franco Davoli. Prediction Strategies for Proactive Management in Dynamic Distributed Systems. Proceedings of the International Conference on Digital Telecommunications, IEEE Computer Science Society Press, p.74(6 pages), August 2006.

[10] A.S. Franceschi, L.F. Kormann and C.B. Westphall. Performance Evaluation for Proactive Network Management. Proceedings of the ICC'96 International Conference on Communications, vol. I, pages 22-26, 1996.

[11] N. Friedman. Learning the Structure of Dynamic Probabilistic Networks. Proceedings of the 14th Annual Conference on Uncertainty in AI, pages 139-147, 1998.

[12] J. Gao, G. Kar and P. Kermani. Approaches to Building Self Healing Systems using Dependency Analysis. Proceedings IEEE/IFIP Network Operations and Management Symposium (NOMS'04), Vol. 1, pages 119-132, 2004.

[13] M. Gupta, A. Neogi, M. K. Agarwal and G. Kar. Discovering Dynamic Dependencies in Enterprise Environments for Problem Determination. Proceedings of 14th IEEE/IFIP International Workshop on Distributed Systems Operations and Management (DSOM 2003), LNCS 2867, pages 221-233, 2003.

[14] Peer Hasselmeyer. Managing Dynamic Service Dependencies. 12th International Workshop on Distributed Systems: Operations & Management (DSOM 2001), Nancy, France, ISBN 2-7261-1190-4, pages 141-150, 2001.

[15] K. Houck, S. Calo and A. Finkel. Towards a practical alarm correlation system. A.S. Sethi, F. Faure-Vincent and Y. Raynaud (Eds.), Integrated Network Management IV, Chapman and Hall, London pages 226-237, 1995.

[16] C. Hill. High-availability systems boost network uptime: Part 1. http://www.eetasia.com/ARTICLES/2001JUL/2001JUL01_NTEK_ST _QA_TA.PDF. Motorola Telecom Business Unit, 2001.

[17] C. Hood and C. Ji. Proactive Network Fault Detection. IEEE Transactions on Reliability, Vol. 46, No.3, pages 333-341, Sept. 1997.

[18] D. Heckerman and M. P. Wellman. Bayesian networks. Communications of the ACM, 38(3):27-30, Mar. 1995.

[19] ITU-T. Recommendation X. 700: Management Framework for Open Systems Interconnection (OSI) for CCITT applications, September 1992.

[20] A. Keller, U. Blumenthal and G. Kar. Classification and Computation of Dependencies for Distributed Management. Proceedings of 5th IEEE Symposium on Computers and Communications. Antibes-Juan-les-Pins, France, July 2000.

[21] S. Kätker and K. Geihs. A Generic Model for Fault Isolation in Integrated Management System. Journal of Network and Systems Management, Special Issue on Fault Management in Communication Networks, Vol. 5, No. 2, June 1997.

[22] T. Koch, B. Kramer and G. Rohde. On a Rule Based Management Architecture. Proceedings of the 2nd International Workshop on Services in Distributed and Networked Environment, pages 68-75, June 1995.

[23] Thomas Koch. Automated management of distributed systems. Ph. D thesis, FernUniversität in Hagen. Shaker Verlag, ISBN: 3-8265-2594-9, 1997.

[24] Jin H. Kim and Judea Pearl. A computational model for combined causal and diagnostic reasoning in inference systems. Proceedings of the Eighth International Joint Conference on Artificial Intelligence (IJCAI-83), pages 190-193, 1983. Morgan Kaufmann.

[25] I. Katzela and M. Schwarz. Schemes for Fault Identification in Communication Networks. IEEE/ACM Transactions on Networking, vol. 3, pages 753-764, 1995.

[26] C. Krugel, T. Toth and C. Kerer. Decentralized event correlation for intrusion detection. Proceedings of Information Security and Cryptology (ICISC), pages 114-131, December, 2001.

[27] F. L. Koch and C. B. Westphall. Decentralized Network Management Using Distributed Artificial Intelligence. Journal of Network and Systems Management, Vol. 9, No. 4, December 2001.

[28] J. Lunze. Diagnosis of quantized systems based on a timed discrete-event model. Systems, Man and Cybernetics Part A, IEEE Transactions 30 (3)(2000) 322-335.

[29] S. Kliger, S. Yemini, Y. Yemini, D. Oshie and S. Stolfo. A Coding Approach to Event Correlation. Proceedings of the Fourth IEEE/IFIP International Symposium on Integrated Network Management, pages 266-277, Chapman and Hall, London, UK, May 1995.

[30] C. Lo, S. H. Chen and B. Lin. Coding-based schemes for fault identification in communication networks. Journal of Network and Systems Management, 10(3), pages 157-164, 2000.

[31] L. Lewis. A case-based reasoning approach to the resolution of faults in communication networks. Integrated Network Management, III, pages 671-682. Elsevier Science Publishers B.V., Amsterdam, 1993.

[32] Lundy Lewis. Managing Business and Service Networks. Kluwer Press, ISBN 0306465590, 2001.

[33] S. L. Lauritzen and D. J. Spiegelhalter. Local Computations with Probabilities on Graphical Structures and Their Application to Expert Systems. Journal of the Royal Statistical Society, Series B 50:157-224, 1988.

[34] Gianfranco Lamperti and Marina Zanella. Diagnosis of discrete-event systems from uncertain temporal observations. Artif. Intell. 137(1-2): 91-163, 2002.

[35] R. E. Miller and K. A. Arisha. FaultManagement Using Passive Testing for Mobile IPv6 Networks. Proceedings of 2001 IEEE Global Telecommunications Conference. Vol. 3, pages 1923-1927, 2001.

[36] M. Mansouri-Samani. Monitoring of Distributed Systems, Ph.D. thesis, Imperial College, UK, 1995.

[37] R. Maxion and F. Feather. A Case Study of Ethernet Anomalies in a Distributed Computing Environment. IEEE Transactions on Reliability. Vol. 39, No. 4, pages 433-443, October, 1990.

[38] R. M. Neal. Probabilistic Inference Using Markov Chain Monte Carlo Methods, Tech. Rep. CRG-TR93-1, University of Toronto, Department of Computer Science, 1993.

[39] Y. A. Nygate. Event Correlation Using Rule and Object Based Techniques. Proceedings of IFIP/IEEE International Symposium on Integrated Network Management, pages 278-289, May 1995.

[40] A. Osmani and F. Krief. Model-Based Diagnosis for Fault Management in ATM Networks. Proceedings of International Conference on ATM ICATM 99, pages 91-99, 1999.

[41] J. Pearl. A constraint-propagation approach to probabilistic reasoning, Uncertainty in Artificial Intelligence. North-Holland, Amsterdam, pages 357-369, 1986.

[42] J. Pearl. Probabilistic Reasoning in Intelligent Systems: Networks of Plausible Inference. Morgan Kaufmann, San Mateo, CA, 1988.

[43] J. Pearl. Causality: Models, Reasoning, and Inference. Cambridge, England: Cambridge University Press. New York, NY, 2000.

[44] Wei Peng, Tao Li and Sheng Ma. Mining log files for data-driven system management. Natural language processing and text mining. ACM SIGKDD Explorations Newsletter archive, Volume 7, Issue 1, June 2005.

[45] G. Pemido, J. Nogueira, and C. Machado. An Automatic Fault Diagnosis and Correction System for Telecommunications Management. Proceedings of 6th IFIP/IEEE International Symposium on Integrated Network Management, pages 777-791, 1999.

[46] Shunshan Piao, Jeongmin Park and Eunseok Lee. Problem Localization for Automated System Management in Ubiquitous Computing. LNCS 4809, pages 158-168, 2007.

[47] http://www.rfc-editor.org

[48] I. Rouvellou and G. W. Hart. Automatic Alarm Correlation for Fault Identification. Proceeding of IEEE INFOCOM95, pages 553-561, 1995.

[49] R. Sterritt and D. W. Bustard. Fusing hard and soft computing for fault management in telecommunications systems. IEEE Transactions on Systems, Man, and Cybernetics, Part C, Vol. 32, No. 2, pages 92-98, 2002.

[50] R. Sterritt and D. W. Bustard. Autonomic Computing - a Means of Achieving Dependability? Proc. IEEE Int. Conf. Engineering of Computer Based Systems, pages 247-251, 2003.

[51] P.H. Schow. The Alarm Information Base: A Repository for Enterprise Management. Proceedings of Second IEEE International Workshop on Systems Management, pages 142-147, 1996.

[52] M. Steinder and A. S. Sethi. Non-deterministic Diagnosis of End-to-end Service Failures in a Multi-layer Communication System. Proceedings of ICCCN, Scottsdale, pages 374-379, 2001.

[53] William Stallings. SNMP, SNMPv2, SNMPv3 and RMON 1 and 2, 3rd Edition, Addison-Wesley, ISBN: 0-201-48534-6, 1998.

[54] J. Stearley. Towards informatic analysis of syslogs. Proceedings of 2004 IEEE International Conference on Cluster Computing, pages 309-318, 2004.

[55] S. A. Yemini, S. Kliger, E. Mozes, Y. Yemini and D. Ohsie. High speed and robust event correlation. IEEE Communications, 34(5):82-90, May 1996.

[56] J. D. Young and J. E. Santos. Introduction to Temporal Bayesian Networks. Online Proceedings of the 1996 Midwest Artificial Intelligence and Cognitive Science Conference. http://www.cs.indiana.edu/event/maics96/Proceedings/Young/maics-96.ps.

[57] J. Zupan and D. Medhi. An Alarm Management Approach in the Management of Multi-Layered Networks. 3rd IEEE International Workshop on IP Operations & Management, pages 77-84. 2003.

Index

Locators for figures are in italics, and locators for tables are in bold.

T - #0346 - 071024 - C10 - 234/156/18 - PB - 9780367385842 - Gloss Lamination